NEUROSCIENCE RESEARCH PROGRESS

ENCYCLOPEDIA OF NEUROSCIENCE RESEARCH

VOLUME 2

NEUROSCIENCE RESEARCH PROGRESS

Additional books in this series can be found on Nova's website
under the Series tab.

Additional e-books in this series can be found on Nova's website
under the e-books tab.

NEUROSCIENCE RESEARCH PROGRESS

ENCYCLOPEDIA OF NEUROSCIENCE RESEARCH

VOLUME 2

EILEEN J. SAMPSON
AND
DONALD R. GLEVINS
EDITORS

Nova Science Publishers, Inc.
New York

Copyright © 2012 by Nova Science Publishers, Inc.

All rights reserved. No part of this book may be reproduced, stored in a retrieval system or transmitted in any form or by any means: electronic, electrostatic, magnetic, tape, mechanical photocopying, recording or otherwise without the written permission of the Publisher.

For permission to use material from this book please contact us:
Telephone 631-231-7269; Fax 631-231-8175
Web Site: http://www.novapublishers.com

NOTICE TO THE READER

The Publisher has taken reasonable care in the preparation of this book, but makes no expressed or implied warranty of any kind and assumes no responsibility for any errors or omissions. No liability is assumed for incidental or consequential damages in connection with or arising out of information contained in this book. The Publisher shall not be liable for any special, consequential, or exemplary damages resulting, in whole or in part, from the readers' use of, or reliance upon, this material. Any parts of this book based on government reports are so indicated and copyright is claimed for those parts to the extent applicable to compilations of such works.

Independent verification should be sought for any data, advice or recommendations contained in this book. In addition, no responsibility is assumed by the publisher for any injury and/or damage to persons or property arising from any methods, products, instructions, ideas or otherwise contained in this publication.

This publication is designed to provide accurate and authoritative information with regard to the subject matter covered herein. It is sold with the clear understanding that the Publisher is not engaged in rendering legal or any other professional services. If legal or any other expert assistance is required, the services of a competent person should be sought. FROM A DECLARATION OF PARTICIPANTS JOINTLY ADOPTED BY A COMMITTEE OF THE AMERICAN BAR ASSOCIATION AND A COMMITTEE OF PUBLISHERS.

Additional color graphics may be available in the e-book version of this book.

Library of Congress Cataloging-in-Publication Data

Encyclopedia of neuroscience research / editors, Eileen J. Sampson and Donald R. Glevins.
p. cm.
Includes index.
ISBN 978-1-61324-861-4 (hardcover)
1. Neurosciences--Research. I. Sampson, Eileen J. II. Glevins, Donald R.
RC337.E53 2011
616.80072--dc23
2011017613

Published by Nova Science Publishers, Inc. † New York

Contents

VOLUME 1

Preface		ix
Chapter I	Prefrontal Morphology, Neurobiology and Clinical Manifestations of Schizophrenia *Tomáš Kašpárek*	1
Chapter II	Prefrontal Cholinergic Receptors: Their Role in the Pathology of Schizophrenia *M. Udawela, E. Scarr and B. Dean*	47
Chapter III	Participation of the Prefrontal Cortex in the Processing of Sexual and Maternal Incentives *Marisela Hernández González and Miguel Angel Guevara*	73
Chapter IV	From Conflict to Problem Solution: The Role of the Medial Prefrontal Cortex in the Learning of Memory-Guided and Context-Adequate Behavioral Strategies for Problem-Solving in Gerbils *Holger Stark*	115
Chapter V	The Orbitofrontal Cortex and Emotional Decision-Making: The Neglected Role of Anxiety *Sabine Windmann and Martina Kirsch*	141
Chapter VI	PEEF: Premotor Ear-Eye Field. A New Vista of Area 8B *Bon Leopoldo, Marco Lanzilotto and Cristina Lucchetti*	153
Chapter VII	Prefrontal Cortex: Its Roles in Cognitive Impairment in Parkinson's Disease Revealed by PET *Qing Wangab, Kelly A. Newella, Peter T. H. Wongd and Ying Luc*	171

Chapter VIII	Noradrenergic Actions in Prefrontal Cortex: Relevance to AD/HD Amy F. T. Arnsten	177
Chapter IX	Prefrontal Cortex: Brodmann and Cajal Revisited Guy N. Elston and Laurence J. Garey	199
Chapter X	Developmental Characteristics in Category Generation Reflects Differential Prefrontal Cortex Maturation Julio Cesar Flores Lázaro and Feggy Ostrosky-Solís1	215
Chapter XI	Common Questions and Answers to Deep Brain Stimulation Surgery Fernando Seijo, Marco Alvarez-Vega1, Beatriz Lozano, Fernando Fernández-González, Elena Santamarta and Antonio Saíz	227
Chapter XII	Deep Brain Stimulation and Cortical Stimulation Methods: A Commentary on Established Applications and Expected Developments Damianos E. Sakas and Ioannis G. Panourias	255
Chapter XIII	Cortical Stimulation versus Deep Brain Stimulation in Neurological and Psychiatric Disorders: Current State and Future Prospects Damianos E. Sakas and Ioannis G. Panourias	271
Chapter XIV	Invasive Cortical Stimulation for Parkinson's Disease and Movement Disorders B. Cioni, A. R. Bentivoglio, C. De Simone, A. Fasano, C. Piano, D. Policicchio, V. Perotti and M. Meglio	303
Chapter XV	Deep Brain Stimulation in Epilepsy: Experimental and Clinical Data M. Langlois, S. Saillet, B. Feddersen, L. Minotti, L. Vercueil, S Chabardès, O. David, A. Depaulis P. Kahane and C. Deransart	319
Chapter XVI	Psychosurgery of Obsessive-Compulsive Disorder: A New Indication for Deep Brain Stimulation? Dominique Guehl, Abdelhamid Benazzouz, Bernard Bioulac and Pierre Burbaud,Emmanuel Cuny and Alain Rougier, Jean Tignol and Bruno Aouizerate	339
Chapter XVII	Deep Brain Stimulation in Adult and Pediatric Dystonia Laura Cif, Simone Hemm, Nathalie Vayssiere and Philippe Coubes	353

VOLUME 2

Chapter XVIII	Deep Brain Stimulation of the Subthalamus: Neuropsychological Effects *Rita Moretti, Paola Torre, Rodolfo M. Antonello and Antonio Baval*	**365**
Chapter XIX	Subthalamic High-Frequency Deep Brain Stimulation Evaluated by Positron Emission Tomography in a Porcine Parkinson Model *Mette S. Nielsen, Flemming Andersen, Paul Cumming, Arne Møller, Albert Gjedde, Jens. C. Sørensen and Carsten R. Bjarkam1*	**381**
Chapter XX	Current and Future Perspectives on Vagus Nerve Stimulation in Treatment-Resistant Depression *Bernardo Dell'Osso, Giulia Camuri, Lucio Oldani and A. Carlo Altamura*	**395**
Chapter XXI	Cognitive Aspects in Idiopathic Epilepsy *Sherifa A. Hamed*	**407**
Chapter XXII	Cognitive Impairment in Children with ADHD: Developing a Novel Standardised Single Case Design Approach to Assessing Stimulant Medication Response *Catherine Mollica, Paul Maruff and Alasdair Vance*	**443**
Chapter XXIII	Novel Therapies for Alzheimer's Disease: Potentially Disease Modifying Drugs *Daniela Galimberti, Chiara Fenoglio and Elio Scarpini*	**475**
Chapter XXIV	Cognitive Interventions to Improve Prefrontal Functions *Yoshiyuki Tachibana, Yuko Akitsuki and Ryuta Kawashima*	**499**
Chapter XXV	Insights from Proteomics into Mild Cognitive Impairment, Likely the Earliest Stage of Alzheimer's Disease *Renã A. Sowell and D. Allan Butterfield*	**521**
Chapter XXVI	Animal Models for Cerebrovascular Impairment and its Relevance in Vascular Dementia *Veronica Lifshitz and Dan Frenkel*	**541**
Chapter XXVII	The Critical Role of Cognitive Function in the Effective Self-administration of Inhaler Therapy *S. C. Allen*	**561**
Chapter XXVIII	Foetal Alcohol Spectrum Disorders: The 21st Century Intellectual Disability *Teresa Whitehurst*	**573**

Chapter XXIX	Where There are no Tests: A Systematic Approach to Test Adaptation *Penny Holding, Amina Abubakar and Patricia Kitsao Wekulo*	587
Chapter XXX	Paid Personal Assistance for Older Adults with Cognitive Impairment Living at Home: Current Concerns and Challenges for the Future *Claudio Bilotta, Luigi Bergamaschini, Paola Nicolini and Carlo Vergani*	599
Chapter XXXI	Neurotoxicity, Autism, and Cognitive Impairment *Rebecca Cicha, Brett Holfeld and F. R. Ferraro*	609

VOLUME 3

Chapter XXXII	Molecular Mechanisms Involved in the Pathogenesis of Huntington Disease *Claudia Perandones, Martín Radrizzani and Federico Eduardo Micheli*	617
Chapter XXXIII	Huntingtin Interacting Proteins: Involvement in Diverse Molecular Functions, Biological Processes and Pathways *Nitai P. Bhattacharyya, Moumita Datta, Manisha Banerjee, Srijit Das and Saikat Mukhopadhyay*	655
Chapter XXXIV	DNA Repair and Huntington's Disease *Fabio Coppedè*	677
Chapter XXXV	Putting Together Evidence and Experience: Best Care in Huntington's Disease *Zinzi Paola1, Jacopini Gioia1, Frontali Marina and Anna Rita Bentivoglio*	691
Chapter XXXVI	Oral Health Care for the Individual with Huntington's Disease *Robert Rada*	705
Chapter XXXVII	Suicidal Ideation and Behaviour in Huntington's Disease *Tarja-Brita Robins Wahlin*	717
Chapter XXXVIII	Making Reproductive Decisions in the Face of a Late-Onset Genetic Disorder, Huntington's Disease: An Evaluation of Naturalistic Decision-Making Initiatives *Claudia Downing*	745
Chapter XXXIX	The Control of Adult Neurogenesis by the Microenvironment and How This May be Altered in Huntington's Disease *Wendy Phillips and Roger A. Barker*	799
Index		857

In: Encyclopedia of Neuroscience Research
Editors: Eileen J. Sampson and Donald R. Glevins
ISBN 978-1-61324-861-4
© 2012 Nova Science Publishers, Inc.

Chapter XVIII

Deep Brain Stimulation of the Subthalamus: Neuropsychological Effects[*]

Rita Moretti, Paola Torre, Rodolfo M. Antonello and Antonio Bava1

[1] Dip. Fisiologia e Patologia, Università di Trieste, Italy

The basal ganglia were considered as a site for integrating diverse inputs from the entire cerebral cortex and funneling these influences via the ventrolateral thalamus to the motor cortex (Alexander, 1994). The subthalamic nucleus (STN) has generally been considered as a rely station within frontal-subcortical motor control circuitry. The function of the STN in the current model of the basal ganglia organization was developed from animal model and neurodegenerative diseases with hypo- and hyperkinetic movement disorders (Alexander et al, 1986): in this model, the STN represents a relay in the so-called "indirect" pathway of the parallel basal ganglia-thalamocortical circuits. STN excitatory efferents reinforce the inhibition of basal ganglia output nuclei, the internal globus pallidus (GPi) and the reticular substantia nigra (SNr) on thalamo-frontal neurons. A "direct" striato-pallidal pathway opposes the STN by normally inhibiting basal ganglia outflow, facilitating thalamo-frontal drive (Schroeder et al., 2002).

Clinicians maintain firm conviction on the fundamental role of the basal ganglia, and of STN, in motor syndromes such as Parkinson Disease. Lesions at the level of STN lead to hemiballism and involuntary movements of the contralateral extremities (Guridi and Obeso, 2001).

The limits of drug therapy in severe forms of Parkinson's Disease have led to a renewed interest in functional surgery of the basal ganglia and the thalamus. Deep brain stimulation of these structures was developed with the aim of reducing the morbidity of surgery and of

---

[*] A version of this chapter was also published in *Basal Ganglia and Thalamus: Their Role in Congition and Behavior*, edited by Rita Moretti published by Nova Science Publishers, Inc. It was submitted for appropriate modification in an effort to encourage wider dissemination of research.

offering an adaptive treatment. This type of stimulation was first applied to the thalamus in patients with severe tremor and has recently been applied to the subthalamic nucleus and the internal pallidum. The physiological expectation for this approach is based on the nigro-striatal dopaminergic defect common of Parkinson Disease: i.e., that it induces over-activity of the excitatory glutaminergic subthalamic nucleus-globus pallidus pathway and thus, when the activity is decreased by stimulation, it should lead to an improvement of levodopa-responsive symptoms (akinesia, rigidity, rest tremor, and gait impairment) (Pollak, 2001).

Deep Brain Stimulation (DBS) has become widely used to treat medically refractory tremor, Parkinson's Disease (PD) and primary dystonia. DBS of the thalamus for tremor and of the subthalamic nucleus (STN) and the internal segment of the Globus Pallidus (GPi) for PD is now FDA approved. As the number of DBS procedures for movement disorders, epilepsy and pain increase, more neurologists will need to acquire the knowledge of DBS programming.

DBS comprises chronic high frequency electrical stimulation of the target nucleus, a brain pacemaker, which might work by entraining neuronal firing patterns and pathological synchronization of neuronal activity. The DBS electrode has 4 contacts, which can be used singly in a monopolar mode of stimulation or in combinations of contacts in a bipolar mode. The adjustable parameters of stimulation are the rate or frequency of stimulation, the pulse width, and the voltage. Thus the number of stimulation parameters possible for programming are huge and DBS programming can be a time consuming and daunting task unless a clear protocol is followed. An advantage of DBS compared to ablative procedures is that the stimulation parameters can be adjusted over time, to optimize the clinical benefit while minimizing side effects (Bronte-Stewart, 2003).

As general accordance, the bilateral stimulation of subthalamic nucleus has as a direct consequence a significant bilateral reduction in Parkinson' disability (Kumar et al., 2000). High frequency deep brain stimulation of the STN improves motor function in Parkinson's Disease patients and has therefore become a strategy to treat medically intractable PD patients (Deep Brain Stimulation for Parkinson's Disease Study Group, 2001). Nevertheless, the physiologic mechanism underlying the beneficial effect of deep brain stimulation is not completely understood, but one can argue that it has reversible suppressive effects on the nuclei, and it leads to the hypothesis that high frequency electrical stimulation is neuro-inhibitive, due to the stimulation of GABAergic fibers. In particular, focusing our interest on subthalamic stimulation, there is evidence that, in epileptic animals, modifications in subthalamic nucleus activity can block the ectopical discharges of epileptic foci (Vercueil et al., 1998).

Using functional imaging, it has previously been shown that STN stimulation enhances movement associated activation of supplementary motor area (SMA) (Ceballos-Baumann et al., 1999). Overactivation of the lateral premotor cortex has been shown in an event-related fMRI study investigating joystick movements as well as in block design fMRI studies on complex finger movements (Haslinger et al., 2001; Sabatini et al., 2000). However, facilitation of movement related frontal activity is not necessarily coupled with improvement in frontal executive function as suggested by neuropsychological studies in Parkinson's disease with DBS. Still controversial and debated are the possible consequences on cognition due to DBS-STN, considering that STN role in cognition has not yet been investigated in detail. Clinical studies suggest that the STN participates in non-motor functions which can now be probed in Parkinson's disease patients with deep brain stimulation of the STN, allowing selective and reversible modulation of this nucleus. Bilateral DBS-STN for PD can cause subtle impairments in various aspects of frontal executive functioning, particularly in older patients (Saint Cyr et al., 2000; Jahanshahi et al.; 2000 Pillon, 2002), but results are quite controversial.

In these aspects the striatum may play a more central role, providing important information for subsequent cortical processing.

Though, linguistic aspects have never been strenuously investigated in the subjects with subthalamic nucleus stimulation. There have been attempts to follow some patients with subthalamic nucleus deep brain stimulation (Zanini, et al., 2000; Moretti et al., 2001; Moretti et al., 2002), during which different patients underwent linguistic assessment, before and after surgery. These Authors have found a small, but significant amelioration in linguistic performance during electrical stimulation.

In order to better define the eventual modification of the cognitive profile of patients who underwent to subthalamic nucleus-deep brain stimulation, we have followed nine Parkinson disease patients before and after surgery.

Our study (Moretti et al., 2003- A) included 9 patients (6 men and 3 women) with idiopathic Parkinson's disease (UK Parkinson's Disease Society Brain Bank Clinical diagnostic criteria) (Hughes et al., 1992). All the patients could be fully studied (mean age 68.7 ± 7.89 years, range= 60-76 years; average age at onset= 48.34 ± 7.89 years, range= 41-56 years). All the patients suffered for 8.91 ± 2.34 years from Parkinson disease and had been treated for several years with levodopa preparations, then developed motor fluctuations and dyskinesias difficult to control by adapting their drug regimen. They were selected as candidates for deep brain stimulation of the subthalamic nucleus because of severe motor fluctuations. In the off medication state, the mean Hohen and Yahr (Hohen and Yahr, 1967) **score was 4.01 and the mean Unified Parkinson's Disease Rating Scale (UPDRS II)** (Fahn, Elton, and the members of the UPDRS Development Committee, 1987) score was 32.1(activity of daily living) and UPDRS III score was 61.72 (motor examination). The mean preoperative levodopa equivalent dosage was 1432 mg/day. Before surgery, three and eight months later they underwent a neuropsychological evaluation.

All the subjects were right-handed following the criteria of Briggs and Nebes (+ 22.34 ± 1.32, for the Briggs and Nebes test, 1975). Their average educational levels, represented by school years, is of 11.34 ± 5.67 years.

The stereotaxic implant was carried out bilaterally in the subthalamic nucleus, with platinum-iridium electrodes, model Medtronic ® 3389, connected to a subcutaneous impulse generator. Stimuli were provided continuously at 140-180 Hz (square waves of 0.2 ms, voltage less than 3.6 V, intensity 40 µA). **Patients responded to neurosurgery**, in particular decreasing motor fluctuations and increasing daily living independency.

The stereotaxic implant was carried out bilaterally in the subthalamic nucleus, with platinum-iridium electrodes, Model Medtronic® 3389. The target visualization was determined using a preoperative stereotactic T1 and T2 weighted MRI. The STN was best identified on the coronal T2 weighted sequences.

The AC-PC line and the midline of the third ventricle were used to define the target coordinates corresponding to the STN in a brain atlas (3 mm posterior to the midpoint of the biscommissural line, 4 mm inferior to AC-PC, 12.5 mm lateral to midline). The electrodes were implanted stereotactically in a single session in accordance (left: x=-13.80; y=8.01; z= 1.75; right: x= 9.87; y=7.71; z= 1.75) with preoperative MRI and intraoperative microrecordings and connected to a subcutaneous impulse generator stimulation. The implanted quadripolar electrodes were positioned as closely as possible to the location where motor benefit was induced by the lowest electrical intensity and adverse effects by the highest electrical intensity using monopolar stimulation. Electrical parameters (pulse width, frequency, and voltage) were progressively adjusted by telemetry, using a console programmer, until an optimal effect was reached, in both the "on" and "off" drug conditions. The stimulation was unipolar, using one contact of the quadripolar electrode. At the time of the study, the mean voltage of stimulation was 2.3 ± 1.02 V; the mean pulse width was of 68 ± 11.34 µs; the mean frequency was 130 ± 23.45 Hz. MRI of the brain was performed the day after electrode implantation to verify the final electrode placement and the lack of significant bleeding. No haemorrhage, infection or ischaemic stroke directly related to the surgery was noted (Moretti et al., 2003- A).

Patients were examined before surgery, after three, six and twelve months after surgery (Moretti et al., 2003-A).

The patients underwent to a Zung self-evaluation for depression (Zung, 1965).

The general cognitive profile was tested by a battery of tests, which comprises: Stroop Test (Stroop, 1935; Perret, 1974), Raven Standard Progressive Matrices (Raven, 1960), digit span backwards and forwards (Wechsler, 1981), retrieval of a story (Wechsler 1945), Syntactic comprehension test and Morphological test (from the Bilingual aphasia test, part B –italian language- Paradis and Canzanella, 1990), alternating verbal semantic fluency (Wechsler, 1981), alternating verbal phonological fluency (Wechsler, 1981), Benton Line Orientation Test (Benton et al, 1983). Moreover, the patients underwent to an alternating syllabic-phonological test, which we have prepared.

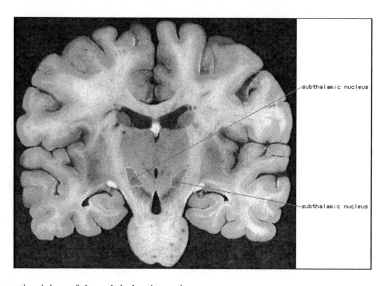

Figure 1. A schematic vision of the subthalamic nucleus.

Subjects were strained to produce in 1 minute time the highest number of words beginning with a given syllable ("per", "tra", and "di"): we test the total number of words produced and the number of intrusion mistakes (e.g.: if syllabic given target is "tra", we count the number of different nouns beginning with "ter", or "ta", or "tr") eventually produced (Moretti et al., 2002). The order of test' presentation was constant in all the sessions.

The scores obtained by the patients were compared with those obtained by a group of 9 patients (6 men and 3 women) with idiopathic Parkinson's disease (UK Parkinson's Disease Society Brain Bank Clinical diagnostic criteria) (Hughes et al., 1992) not submitted to DBS, matched for age, educational level, handedness, and disease degree, who previously agreed to participate and gave their informed consent for testing. All the patients could be fully studied (mean age 69.45 ± 7.12 years, range= 60-76 years; average age at onset= 47.64 ± 7.12 years, range= 41-56 years). All the patients suffered for 8.71 ± 1.46 years from Parkinson disease and had been treated for several years with levodopa preparations.

Statistical analyses were performed using the Statistical Package for the Social Sciences (SPSS, version 10.0). Changes from baseline between Groups with DBS and control group were compared using the two-sample Wilcoxon test. This was done for each efficacy variable. Results are presented as means with standard deviations and *p*-values are presented where appropriate (Table 1).

Table 1. Statistical differences obtained on the three trials: a comparison has been made with the results obtained before surgery during the 2nd and 3rd session. Differences from the control population have been made during the 1st session

Tests	1st session	2nd session	3rd session	controls
Stroop	67.45 seconds ($p<0.05$; χ^2 test); 1 mistakes	71.24 sec. 1 mistakes (n.s)	78 sec. 2 mistakes (n.s)	58.2 seconds; 1 mistake
Raven Standard Progressive Matrices (correct)	36/60 (ns)	36/60 (n.s)	36/60 (n.s)	39/60
Digit span forward (WAIS-R)	4 ($p<0.05$; χ^2 test)	6 ($p<0.05$)	6 ($p<0.05$)	7.5
Digit span backwards (Wais-r)	4 (ns)	3 (n.s)	3 (n.s)	4.3
Retrieval of a story (Wais-memory)	13/22 (ns)	14/22 (n.s)	14/22 (n.s)	17.8/22
Synctactic comprehension (Bilingual aphasia test-b: errors)	17 ($p<0.05$; χ^2 test)	14 (n.s)	14 (n.s)	11
Morphological tasks (Bilingual aphasia test-b: errors)	8 (ns)	8 (n.s)	8 (n.s)	8
Alternating verbal semantic fluency (WAIS-R)	46	38 ($p<0.05$)	37 ($p<0.05$)	49.32
Alternating verbal phonological fluency	38	26 (ns)	24 ($p<0.05$)	42.34
Alternating syllabic verbal fluency	26 (3 interference mistakes) ($p<0.01$ χ^2 test)	21 (3 interference mistakes) ($p<0.05$)	23 (0 interference mistakes) ($p<0.01$)	36.9
Benton line orientation test (errors)	1 (ns)	1 (n.s)	1 (n.s)	0.2

Testing before surgery was conducted when the patients were on medication. At baseline, there were no statistical differences between the two groups on any of the scales, a part from an evident increase in time of Stroop execution (72.34 ± 1.42 seconds) in DBS group, when compared to control group (p<0.05, paired samples, t-test). No other differences were evident, indicating that the two groups were matched for cognitive performances.

At one-month, patients submitted to DBS showed significant increase in the total time to execute Stroop test, of 5.55 points over baseline (p<0.01) and -10.89 over control Group (p<0.001), and in number of mistakes in Stroop test, of 3.44 over baseline (p<0.01) and –3.34 over control Group (p<0.01); there was a significant decrease of correct recall, of –2.3 over baseline (p0.05), of total words produced in semantic fluency tasks, of –7.19 over baseline (p<0.001), and of –5.89 over control (p<0.001), in phonological fluency tasks, of –3.9 over baseline (p<0.01), and of –3.27 over control (p<0.01), in syllabic fluency, of –8.3 over baseline (p<0.001), and of –9.6 over control (p<0.001); on the contrary, there was a significant increase in number of intrusion mistakes, of 5.78 over baseline (p<0.001), and of -5.89 over control group (p<0.001).

At six-month, patients with DBS showed significant increase in the total time to execute Stroop test, of 12.1 points over baseline (p<0.001) and -16.11 over control Group (p<0.001), and in number of mistakes in Stroop test, of 2.33 over baseline (p<0.01) and –2 over control Group (p<0.01); there was a significant decrease of correct recall, of –1.6 over baseline (p<0.05), of total words produced in semantic fluency tasks, of –10.97 over baseline (p<0.001), and of –9.23 over control (p<0.001), in phonological fluency tasks, of –6.2 over baseline (p<0.01), and of –6.67 over control (p<0.01), in syllabic fluency, of –10.52 over baseline (p<0.001), and of –9 over control (p<0.001); on the contrary, there was a significant increase in number of intrusion mistakes, of 4.34 over baseline (p<0.001), and of –4.56 over control group (p<0.001).

At twelve-month, patients with DBS showed significant increase in the total time to execute Stroop test, of 16.33 points over baseline (p<0.001) and -20.11 over control Group (p<0.001), and in number of mistakes in Stroop test, of 1.88 over baseline (p<0.01) and –1.44 over control Group (p<0.01); there was a significant decrease of total mistakes in syntactic comprehension tasks, of –3.12 over baseline (p<0.01), and of -3.11 over control (p<0.01), of total words produced in semantic fluency tasks, of –10.63 over baseline (p<0.001), and of –9 over control (p<0.001), in phonological fluency tasks, of –5.87 over baseline (p<0.01), and of –6.34 over control (p<0.01), in syllabic fluency, of –10.41 over baseline (p<0.001), and of – 8.89 over control (p<0.001); on the contrary, there was a significant increase in number of intrusion mistakes, of 1.34 over baseline (p<0.01), and of –1.56 over control group (p<0.05).

Subthalamic stimulation has been shown to present an advantage as it induces a higher antiakinetic effect and has positive effects on all parkinsonian symptoms (Dujardin et al., 2001; Iansek et al., 2002).

The subthalamus has generally been considered as a rely station within frontal-subcortical motor control circuitry. It has connections to the substantia nigra, pars reticulata, output pathways from the basal ganglia (Jahanshahi, et al., 2000): the dorsolateral prefrontal cortex is the final target of the thalamic projection from the substantia nigra pars reticulata. A PET study (Limousin et al. 1997) showed that there was a greater activation of dorsolateral prefrontal cortex during effective stimulation of the subthalamus nucleus. Direct cortico-STN projections and indirect connections, via the nucleus accumbens and ventral pallidum from prelimbic and medial orbital cortex, have an excitatory influence on the STN (Maurice et al., 1998).

On the other hand, subthalamic nucleus may be connected with the frontal-caudate-pulvinar circuits, and therefore be involved in verifying the semantic acceptability of the linguistic production and in the so-called language planning-loop (Wallesch and Papagno, 1988; Cummings, 1995; Fabbro et al., 1997). Information processing in the circuits is not strictly sequential, and it seems to control the preparatory linguistic activity (Cummings, 1995).

Having accepted that bilateral stimulation of the subthalamic nucleus has as a direct consequence a significant bilateral reduction in Parkinson's disability, the most interesting aspect of deep brain stimulation of the subthalamic nucleus is the possible consequence on cognitive process, extending basal ganglia's **role from motor to cognitive control**. Different authors sustained that there was no significant change in memory or executive functions in a sample of 62 patients treated by bilateral subthalamic nucleus or globus pallidus stimulation (Ardouin et al., 1999) when compared to results obtained before surgery, however other authors obtained previously opposite results (Troster et al., 1997) on the same points. Two recent studies have attempted to study cognition in deep brain stimulation. The first (Pillon, **2002) showed that in "on" phases, deep brain stimulation of the subthalamic nucleus patients** had no cognitive deficit at 12 months, except for a decrease in lexical fluency (not explicated by the authors), and no differential effect associated with subthalamic nucleus or globus pallidus stimulation. The second study (Jahanshahi et al., 2000) assessed the effects of bilateral deep brain stimulation on subthalamic nucleus and globus pallidus on executive function, assessed with Reitan Trail Making test or on Wisconsin Card Sorting Test (data and literature in: Jahanshahi, et al., 2000; Schubert et al, 2002). The results showed a significant better performances obtained in subthalamic nucleus stimulation, when compared to globus pallidus stimulation, in particular of executive function. The conditional associative learning, and in general change in performance became more erroneous too. Thus, the significant deterioration in learning of arbitrary conditional associations lends some support to the proposal of Marsden and Obeso (Marsden and Obeso, 1994) that following subcortical **surgery, patients with Parkinson's disease** should exhibit some deficits in situations that require changing behaviour in novel context. In these aspects the striatum may play a more central role, providing important information for subsequent cortical processing.

As demonstrated by Schroeder et al. (2002), DBS in the STN decreased cerebral blood flow responses in the anterior cingulate cortex during the Stroop task, while impairing task performance at the same time. Decreased activation was also located in the ventral striatum, the structure receiving input from the anterior cingulate cortex in the current model of basal ganglia organization. The angular gyrus is associated with encoding of words (Menard et al., 1996; Peterson et al., 1999). The increased activation during stimulation therefore most likely corresponds to difficulties in suppressing the habitual response, i.e. word processing, when patients had actually to name the font colour of the colour word. Alternatively, during off stimulation, anterior cingulate cortex might effectively modulate language processing in the angular gyrus, resulting in decreased activation compared with the on state. This is in good correspondence to a factorial analysis of cognitive processes in the Stroop task, revealing suppression of language processing to facilitate colour naming (Peterson et al., 1999; Schroeder et al, 2002). Schroeder et al. (2002) results are consistent with the emerging identification of the role of the STN as an important basal ganglia input and output structure not only in the motor, but also in the cognitive/limbic domain.

According to a classical model, STN stimulation should result in facilitation of projections to cortical areas, on the contrary, Schroeder et al. (2002) found the reverse effect, namely impaired cognitive performance coupled to reduced activation in the anterior cingulate cortex circuit. The Stroop task requires a high level of conflict monitoring and response inhibition. The differential impact of STN stimulation on cognitive and motor circuitry supports the two opposite predictions of the so-called paradox of stereotactic surgery in Parkinson's disease (Marsden and Obeso, 1994). The motor system devoid of the important basal ganglia input structure STN could continue to function in routine, predictable and automatic movement. "Releasing the brake" on frontal function with STN stimulation improves aspects of motor functions (Limousin et al., 1995). However, in new and unexpected situations, requiring non-automated behaviour, disruption of basal ganglia input at the level of the TN may lead to inflexibility of mental and motor responses, as shown in the Stroop task.

Thus, we decided to study these aspects of cognitive process, in particular relating them with linguistic process. Therefore, our patients underwent to three different alternating test of verbal fluency, which could give us information on our patients capabilities to adapt themselves in linguistic situations that require changing behaviour and to implement effective strategies to select and produce different words with a given root (in our case a syllable). Attention, working memory, and speediness are indispensable to a good performance on this test. In addition to these aspects, some of the other processes that are necessary for performance of this task are: intrinsic generation of responses partly through suppression of **habitual responses, monitoring of one's output and switching** attention between different stimulus in a set. These are all processes that are strongly dependent on the integrity of the dorsolateral prefrontal cortex.

Our previous work (Moretti, et al., 2001; Moretti et al., 2002) confirmed a general stability of cognitive performances, with a decrease in fluency tasks (confirming the findings of other studies, such those of Troster, et al., 1997 and those of Ardouin, et al., 1999, in the traditional semantic or phonological tasks) associated with a significant reduction of interference errors in syllabic task. This data seem in accordance to what has been found in literature: in addition to being a potential site of storage of learned habits or procedures (Packard and Knowlton, 2002), the basal ganglia also appears to be able to switch between various tasks depending on the demands present. This process is also known as the ability to shift cognitive set in response to the environment. Patients with damage to basal ganglia circuitry exhibit performance costs that are greater than controls when switching from one task to another (Lawrence et al., 2000). In addition, neuroimaging evidence suggests that frontal corticostriatal loops are activated when subjects must select between many possible responses (Desmond et al., 1998). Observations of these stimulated patients supports the prior research; verbal production is decreased in semantic and syllabic tasks, along with an increase in time required to execute the Stroop test, but the quality of production is better preserved, as showed by a significant reduction of interference mistakes in syllabic tasks, and a concomitant reduction of errors in Stroop execution. Mutatis mutandis, syllabic tasks might be considered as a suppression language process task, tightly bound to attention, selection and verbal production. Defects in focal attention has been described as a consequence of the disruption of the fronto-striatal circuit, which can cause a significant impairment in visual search (Chari, et al., 1996). Therefore, better execution on an attention task (Stroop) may be in part dependent on the "releasing the brake", as Jahanshahi said, over frontal cortical function during subthalamus stimulation (Jahanshahi et al., 2000). This situation may be caused by a possible disinhibition of thalamic nuclei, or may simply be related to a facilitation of prefrontal cortical activation. Thus, it appears there is a reduced linguistic output corresponding to a more selective and precise performance.

Chronic stimulation may help restore of the control property on language exerted by deep brain circuits or may implement prefrontal cortex. The result may be consistent with a slowness of cognitive activity, with a reduction of quantitative production, but with an increase of control on the production, which may be more precise and definite. Long term effects on cognition and on language, in particular, need to be studied, and possible different sequel could emerge. This is an issue that requires longitudinal follow up of patients with chronic deep brain stimulation, in order to assess and verify the previous hypothesis.

Few patients were involved in the study, and considering the variability among Parkinson patients, the potential for generalization is seriously limited: though, the results obtained confirm the follow-up previously published (Moretti et al., 2002).

Many different reports seem to evidence the potential side-effect of DBS-STN on cognition, behaviour and social conduct of patients who undergo to this surgery.

It has been accepted that deep brain stimulation (DBS) may help cases of PD where pharmacological therapy could not manage the symptoms any more. Cognitive decline may follow deep brain stimulation (DBS) of the subthalamic nucleus (STN)Changes in cognition between stimulation on and off have also been described(Saint-Cyr et al.,2000; Dujardin et al., 2001; Witt et al., 2004).

We have described a 68-year-old right-handed man, with a 14 – year history of PD, who had severe rigidity, severe akinesia, freezing of gait, and moderate tremor while resting, despite 1000 mg of L-dopa, and 2.1 mg of pramipexole daily, and disabling dyskinesias, underwent DBS-STN (Moretti et al., 2003-B).

His UPDRS III (Fahn et al., 1987) in off was 49, and in on is 25. Brain MRI was normal. The stereotaxic implant was carried out bilaterally in the STN, with platinum-iridium electrodes (Medtronic ® 3389). The electrodes were implanted stereotactically in a single session in accordance with preoperative MRI and intraoperative microrecordings. The stimulation was unipolar, using one contact of the quadripolar electrode (2.0 V, 90 μs, 130 Hz). MRI of the brain was performed the day after electrode implantation to verify the final electrode placement and the lack of significant bleeding.

After surgery the patient interrupted all the oral therapy: his UPDRS-III in off (DBS −) was 35; in off, (DBS +) was 14. Stimulation status was significantly associated with a better motor score in tremor, rigidity, bradykinesia, gait and postural stability. The general cognitive profile was tested by a battery of tests, as seen in table 1. One month and after five months, we resubmitted the patient to further evaluation. The scores obtained were compared with those of a group of 15 volunteers (data have been presented before). In this period, language was fluid, without semantically deviant sentences. Auditory and reading comprehension was maintained, and repetition skills were preserved. One month after surgery, a CT of the head was normal. The patient reported a difficulty of initiate his speech. Initial stuttering is very limiting for the patient, who, began a target word three or four time, repeating the first phoneme, or the initial syllable, but not being keen of proceeding. The hesitation was not correlated with "freezing of speech". He knew what he had to say (since he could point to, or indicate in a multiple choice, either visually either verbally presented, the defined word and his verbal working memory was well preserved). Automatic speech was conserved (both when the patient cursed to his misfortune, or when he pronounced automatic expression). Spared were his writing and reading capabilities. The situation remains stable five months later (Table 2). A single study reported one case of worsening of dysartria and gait freezing which had existed before the implantation (Thobois et al., 2002). To our knowledge, speech start-hesitation has not been previously described as a consequence of DBS-STN (Moretti et al., 2003-B).

Table 2. Neurolinguistic evaluation of 5 minutes spontaneous speech
A comparison with healthy control

	1st session	2nd session	3rd session	Control
Number of utterances	29	28	26	32
Total word's number	345	231	267	478
Words/minute	78	56	61	92
Types	123	109	76	137
Neologism	0	0	0	0
Semantic paraphasia	0	0	0	0
Verbal paraphasia	0	0	0	0
Substitution of free grammatical mormorphemes	0	0	0	1
Omission of Free grammatical morphemes	0	0	0	0
Circumlocutions	2	4	5	1
Perseverations of words	1	9	11	0
Semantically deviant sentences	0	5	2	0

Trying to make some speculation, start-hesitation (by definition that of gait) is a typical initial symptom of the freezing phenomenon (Fahn and Quinn, 2000). Our patient, who had freezing of gait before surgery, manifested speech start hesitation, without other motor symptoms, after DBS-STN. One might say that the mechanism of action of DBS remains to be defined (Ashby, 2000). By whatever mechanism, stimulation mirrors the effects of a destructive lesion.

Subthalamus may be related to speech control by its connections to the substantia nigra, pars reticulata, output pathways from the basal ganglia (Ojemann, 1982): the dorsolateral prefrontal cortex is the final target of the thalamic projection from the substantia nigra pars reticulata. A PET study (Ojemann, 1982) showed that there was a greater activation of dorsolateral prefrontal cortex during effective stimulation of the subthalamus nucleus. With the cortico-striato-thalamo-cortical loop the cortical structures of all cerebral lobes can control the cognitive – motor activity of the frontal lobe structures. The latter are involved in the generation and coordination of voluntary movements (Ojemann, 1982), and maybe in language control. Echolalia, i.e., additional characteristic disorders presented in left basal ganglia lesions, results from the inability of the speaker to inhibit the motor internalisation of verbal expressions produced by the patient (Ojemann, 1982). Chronic stimulation may help to restore the control property on language exerted by deep brain circuits or may implement prefrontal cortex. The result may be consistent with a slowness of cognitive activity, with a reduction of quantitative production . May we say that speech start-hesitation is an excess of control? If so, why has this control been exercised in speech output and not in motor act? (Moretti et al. 2003)

Smeding et al (2007) describe a patient with severe decline after STN DBS and postoperative electrode displacement. The patient is a 43-year-old woman diagnosed with Parkinson disease 8 years prior to DBS STN for severe response fluctuations and dyskinesias. Neuropsychological evaluation was normal. She used levodopa/carbidopa slow release 400/100 mg daily. Surgery was performed using MRI, four-tract microrecordings, and macrostimulation for target localization (12 mm lateral, 2 mm posterior, and 4 mm inferior to the midcommissural point [MCP]). The left electrode (model 3389; Medtronic) was implanted in the central trajectory, with the deepest contact 7 mm below MCP. There was considerable CSF leakage from the burr holes. On the right side, no typical STN activity could be recorded. Because test stimulation along the medial trajectory induced a large reduction in rigidity and bradykinesia, the DBS electrode was implanted with the deepest contact 4 mm below MCP (Smeding et al., 2007). Postoperatively, a stereotactic CT scan showed intracranial air bifrontally with posterior and caudal brain shift. The electrode contacts were at the intended position. Confusion resolved within 1 day. At discharge, the stimulation settings were monopolar, contact 0, pulse width 60 microseconds, frequency 130 Hz, and on the right side amplitude 1.8 V and left side 1.3 V. The levodopa/carbidopa slow-release dosage was reduced to 200/50 mg; 2 mg of pergolide was added. Six months after surgery, motor functioning was satisfactory. Neuropsychological testing showed decline of verbal memory, selective and divided attention, and verbal fluency Her relatives confirmed increased forgetfulness, word-finding difficulties, slowed comprehension, and increased irritability. One year postoperatively, she had progressive memory decline. Neurologic examination showed no changes in motor functioning. Compared with 6 months after surgery, there was a progressive decline in verbal fluency but not in memory. Subsequently, both neurostimulators were turned off. Neuropsychological examination 1 week later showed complete recovery of her cognitive functioning (Smeding et al., 2007). The neurostimulator settings were changed to monopolar at contact 2. One month later, she had no cognitive complaints and motor functioning was satisfactory. Formal testing demonstrated a decline in verbal fluency, but other scores were comparable with the preoperative state. A new stereotactic CT scan was made and co-registered with the preoperative MRI. On the left side, contact 0 was located 12.5 mm lateral, 2 mm anterior, and 3 mm inferior to MCP, in the internal capsule lateral to the anterodorsal STN. Contact 2 was located at the border of the internal capsule and the zona incerta adjacent to the globus pallidus internus (13 mm lateral, 4 mm anterior, 0.5 mm superior relative to MCP). On the right side, contact 0 was in the dorsomedial globus pallidus externus (17 mm lateral, 8.5 mm anterior, 3.5 mm superior to MCP). Contact 2 was in the dorsolateral globus pallidus externus close to the putamen (17 mm lateral, 10.5 mm anterior, 7.5 mm superior to MCP). Cognitive decline resolved after switching off DBS. Adjustment of stimulation settings to the dorsal contacts led only to a minor setback in cognition (Smeding et al., 2007). The electrodes were probably displaced as a result of brain shift caused by CSF leakage during surgery and subsequent unfolding of the brain in the postoperative period. Since this incident, we close the burr holes during surgery with Tissuecol to reduce CSF leakage. Cognitive decline was worst with the ventral contacts activated, which were located outside the STN, on the left side in the internal capsule close to the anterodorsal lateral STN, and on the right side in the dorsomedial globus pallidus externus. Adjustment of stimulation settings did not lead to an improvement in off-phase parkinsonism measured with the Unified Parkinson's Disease Rating Scale, though there was an evident decrease of dyskinesias and less on–off fluctuations possibly related to stimulation of the globus pallidus (Smeding et al., 2007).

Concerns have been raised about the possibility that subthalamic deep brain stimulation (STN-DBS) could also produce cognitive changes and mood and behavioural alterations. Valldeoriola et al. (2006) report the clinical and neuropathological features of a patient with advanced PD who developed behavioural changes and dementia while on STN-DBS. A 74 year old man suffering from PD for 11 years presented troublesome dyskinesias and unpredictable motor fluctuations that did not respond to multiple changes in medication. The Hoehn and Yahr stage was IV while "off" and III while "on" medication. The Schwab and England scale score was 40% in the "off" and 80% in the "on" condition. The UPDRS-III score was 56 while "off" and 21 while "on". No clinically evident signs of cognitive impairment were present. The Mini-Mental State Examination (MMSE) score was 28/30. Neuropsychological assessment was considered to be normal except for the presence of mild cognitive processing slowness and free recall impairment.

Dyskinesias and motor fluctuations had disappeared 3 months after STN-DBS. The Hoehn and Yahr stage was III and the Schwab and England score was 70%. Dopaminergic medication was initially reduced to 20% but the patient later developed restless legs syndrome (attributed to the reduction in dopaminergic medication) and levodopa was reintroduced (400 mg/day). Motor performance remained stable during the following months. There were no significant changes in neuropsychological performance 6 months after STN-DBS. However, shortly afterwards the patient developed mood changes consisting of apathy, anhedonia without sadness, and diurnal hypersomnolence. Levodopa was then increased and antidepressants were started. Changing stimulation electrical parameters and stimulation poles **did not change the patient's** mood. The patient developed fluctuating confusion, visual hallucinations, and paranoid ideations 1.5 years after surgery. The MMSE score was 22/30. He became violent with his relatives. Clozapine (100 mg/day) reduced aggression but confusion persisted. The possible role of STN-DBS on these cognitive and behavioural changes was assessed by switching off the stimulators for 1 week. Mental status remained unchanged but parkinsonism worsened until the stimulators were again switched on. Later in the course of the disease, the patient received galantamine which improved psychiatric symptoms and temporal-spatial orientation. The patient died from bronchopneumonia 3.5 years after surgery (Valldeoriola et al., 2006). Macroscopic examination of the brain disclosed important loss of pigmentation in the substantia nigra pars compacta and locus coeruleus. The electrode tips were placed within the boundaries of the subthalamus on both sides. On microscopic examination moderate inflammatory infiltrates of T lymphocytes and mild astrocytic gliosis were observed surrounding the leads. Lewy bodies (LB) and Lewy neurites were found in the substantia nigra pars compacta, locus coeruleus, raphe nuclei, the dorsal nucleus of the vagus, and the hypoglossal nerve. Dystrophic neurites and cytoplasmatic inclusions were found in the nucleus of Meynert and the subthalamus (Valldeoriola et al., 2006). Cortical type LB were observed in the gyrus cinguli, transentorhinal cortex, the amygdala, and the parietal and temporal cortex. The density of LB was maximal in the subcortical nuclei and the limbic areas. A few neurons showing neurofibrilar degeneration were found in the transentorhinal and entorhinal cortex, amygdalar complex, locus coeruleus, and raphe nuclei. Neocortical senile plaques were scarce and signs of amyloid angiopathy were absent. The pathological diagnosis was diffuse Lewy body disease, transitional type (Valldeoriola et al., 2006). This report (Valldeoriola et al., 2006) is the second clinicopathological description of a patient with PD, dementia, and STN-DBS. The first report by Jarraya et al (2003) described a parkinsonian patient who had some cognitive impairment before STN-DBS and whose mental condition worsened after surgery. Post-mortem examination confirmed the diagnosis of PD. The authors concluded that the bad outcome after surgery was due to the inadequate selection of a candidate who had cognitive deterioration before surgery, but the cause of the worsening mental status after STN-DBS remained unclear. In the case reported here, post-mortem examination disclosed changes typical of diffuse Lewy body disease. In addition, the cognitive and behavioural alterations were not thought to be related to STN-DBS since mental problems appeared late after STN-DBS and changes in stimulation parameters or even disconnection did not result in any mental or behavioural change. The clinical manifestations, the course of the disease, the presence of hallucinations, and the good response to cholinesterase inhibitors were also typical of so called Parkinson's disease-dementia in which the main pathological substrate is the presence of LB in limbic and neo-cortical areas (Cummings, 2005). The case report (Valldeoriola et al., 2006) supports the view that mental problems observed after STN-DBS in some patients are not related to STN-DBS and suggests alternative explanations. A better understanding of the predictive factors for the development of dementia in PD will help to improve candidate selection for STN-DBS (Valldeoriola et al., 2006).

References

Alexander G. A. Basal ganglia thalamocortical circuits: their role in control of movement. *Journal of Clinical Neurophysiolgy* 1994; 11: 420-431.

Alexander G.E., DeLong M.R, Strick P.L Parallel organization of functionally segregated circuits linking basal ganglia and cortex. *Annu. Rev. Neurosci.*1986; 9, 357-381.

Ardouin C., Pillon B., Peiffer E., Bejjani P., Limousin P., Damier P., Arnulf I., Benabid, A.L., Agid Y., Pollak P. Bilateral subthalamic or pallidal stimulation for Parkinson's Disease affects neither memory nor executive functions: a consecutive series of 62 patients. *Annals of Neurology* 1999; 46(2): 217-223.

Ashby P. What does stimulation in the brain actually do? *Prog. Neurol. Surg.*, 2000, 15, 236-245.

Benton AL., Hamshe K., Varney N., Spreen O. Contributions to neuropsychological assessment. New York: Oxford University Press, 1983.

Briggs G., Nebes R. Patterns of hand preference in a student population. *Cortex* 1975; 11: 230-238.

Bronte-Stewart H Deep brain stimulation for movement disorders- general principles. American Academy of Neurology, Minneapolis, 2003. (Ac.003 pdf. 1-23.

Ceballos-Baumann AO, Boecker H, Bartenstein P, von Falkenhayn I, Riescher H, Conrad B et al. A positron emission tomographic study of subthalamic nucleus stimulation in parkinson disease: enhanced movement-related activity of motor association cortex and decreased motor cortex resting activity. *Arch. Neurol.* 1999; 56: 997-1003.

Chari G., Shaw PJ., Sahgal A. Non-verbal visual attention but not recognition memory of learning processes are impaired in motor neuron disease. *Neuropsychologia* 1996; 34(5): 377-385.

Cummings J.L. Anatomic and behavioural aspects of frontal-subcortical circuits. In: Grafman J., Holyoak KJ, Boller F (Eds) Structure and function of the human prefrontal cortex. *Ann. NY Acad. Sci.* 1995; 769: 1-14.

Cummings JL. Cholinesterase inhibitors for treatment of dementia associated with Parkinson's disease. *J. Neurol. Neurosurg. Psychiatry* 2005; 76: 903–4.

DEEP BRAIN STIMULATION FOR PARKINSON'S DISEASE STUDY GROUP. Deep Brain Stimulation of the Subthalamic nucleus or the pars interna of the globus pallidus in Parkinson's Disease. *New Engl. J. Med.* 2001; 345: 956-963.

Desmond JE, Gabrieli JDE, Glover GH. Dissociation of frontal and cerebellar activity in a cognitive task: evidence for a distinction between selection and search. *Neuroimage* 1998; 7: 368-378.

Dujardin K., Defebvre L., Krystkowiak P., Blond S., Destée, A. Influence of chronic bilateral stimulation of the subthalamic nucleus on cognitive function in Parkinson's disease. *Journal of Neurology* 2001; 248: 603-611.

Fabbro F. Subcortical aphasia. *Journal of Neurolinguistics*, 1997; 10: 251-367.

Fahn S, Quinn N. Unusual movement disorders. American Academy of Neurology, San Diego, 2000; 3AG.003-1-24.

Fahn S., Elton R.L., the members of the UPDRS Development Committee The Unified Parkinson's disease rating scale. In: S. Fahn, C. D. Marsden, D. B. Calne, M. Goldstein (Eds) Recent developments in Parkinson's disease. Florham Park, NJ: *Macmillan Healthcare*. 1987. Vol. 2. Pp. 153-163.

Guridi J, Obeso JA. The subthalamic nucleus, hemiballismus and Parkinson's disease: reappraisal of a neurosurgical dogma. *Brain* 2001; 124: 5-19.

Haslinger B, Erhard P., Kampfe N, Boecker H, Rummeny E, Schwaiger M, et al. Event-related functional magnetic resonance imaging in Parkinson's Disease before and after levo-dopa. *Brain* 2001; 124: 558-570.

Hohen, M.M, Yahr MD. Parkinsonism: onset, progression and mortality. *Neurology.* 1967; 17: 427-442.

Hughes AJ, Daniel SE, Kilford L, Lees AJ. Accuracy of clinical diagnosis of idiopathic Parkinson's disease: a clinico-pathological study of 100 cases. *J. Neurol. Neurosurg. Psychiatry* 1992; 55: 181-184.

Iansek R, Rosenfeld JV, Huxham FE. Deep brain stimulation of the subthalamic nucleus in Parkinson's disease. *MJA,* 20002: 177 (3): 142-146

Jahanshahi M, Ardouin CMA., Brown RG., Rothwell JC., Obeso J., Albanese A., Rodriguez-Oroz .MC., Moro E, Benabid AL., Pollak P, Limousin-Dowsey P. The impact of deep brain stimulation on executive function in parkinson's disease. *Brain* 2000; 123, 1142-1154.

Jarraya B, Bonnet A-M, Duyckaerts C, et al. Parkinson's disease, subthalamic stimulation and selection of candidates: a pathologic study. *Mov. Disord.* 2003; 18: 1517–20.

Kumar R., Lozano A. M., Montgomery E., Lang A. E. Pallidotomy and deep brain stimulation of the pallidum and subthalamic nucleus in advanced Parkinson's disease. *Movement Disorders*, 1998; 13(S1): 73-82.

Lawrence AD, Watkins LHA, Sahakian BJ Hodges JR, Robbins TW. Visual object and visuospatial cognition in Huntington's disease: implications for information processing in corticostriatal circuits. *Brain* 2000 ; 123: 1349-1364

Limousin P, Pollack P, Benazzouz A, Hoffmann D, Le Bas JF, Brousolle E et al. Effect of parkinsonian signs and symptoms of bilateral subthalamic nucleus stimulation. *Lancet* 1995; 345: 91-95

Limousin P., Greene J., Pollak P., Rothwell J., Benabid A.L., Frackowiak R. Changes in cerebral activity pattern due to subthalamic nucleus or internal pallidum stimulation in Parkinson's disease. *Annals of Neurology*, 1997, 42, 283-291.

Marsden C.D., Obeso J.A. The functions of the basal ganglia and the parodox of stereotaxic surgery in Parkinson's disease. *Brain* 1994; 117: 877-897.

Maurice N, Deniau JM, Glowinski J, Thierry AM Relationships between the prefrontal cortex and the basal ganglia in the rat: physiology of the corticosubthalamic circuits. *J. Neurosci.* 1998; 18: 9539-9546.

Menard MT, Kosslyn SM, Thompson WL, Alert NM, Rauch SL. Encoding words and pictures: a positron emission tomography study. *Neuropsychologia* 1996; 34: 185-194.

Moretti R, Torre P, Antonello R M.,Capus L, Gioulis M, Zambito Marsala S,Cazzato G, Bava A. "Speech start-hesitation" following subthalamic nucleus stimulation in Parkinson Disease. *European Neurology*, 2003; 49: 251-253.- B

Moretti R, Torre P., Antonello RM, Capus L, Zambito Marsala S, Cattaruzza T, Cazzato G., Bava A. Neuropsychological changes after subthalamic nucleus stimulation: a 12-month follow-up in nine patients with Parkinson Disease. *Parkinsonism and Related disorders*, 2003; 10: 73-79. A

Moretti R, Torre P., Antonello RM, Cazzato G., Bava A. Subcortical-cortical lesions and two-step aplasia in a bilingual patient. In: S.P.Shohov (Ed). *Advances in psychology research.* Vol. 23. Novascience Publisher, New York, 2003. Pp. 33-44. C

Moretti R, Torre P., Antonello RM., Capus L, Gioulis M, Zambito Marsala S., Cazzato G, Bava A. Cognitive changes following subthalamic nucleus stimulation in two patients with parkinson disease. *Perceptual and motor skills*, 2002, 95: 477-486.

Moretti R., Torre P., Antonello R.M., Capus L., Gioulis M., Zambito Marsala, S., Cazzato G, Bava A. Effects on cognitive abilities following subthalamic nucleus stimulation in Parkinson Disease. *European Journal of Neurology* 2001; 8: 726-727.

Ojemann G.A. Models of the brain organization for higher integrative functions derived by electrical stimulation techniques. *Human Neurobiology*, 1982; 1,:243-249.

Packard MG, Knowlton BJ. Learning and memory functions of the basal ganglia. *Ann. Rev. Neurosci.* 2002; 25: 563-593.

Paradis M., Canzanella M. Test per l'afasia in un bilingue. Versione italiana. Hillsdale, N.J:. 1990. Erlbaum.

Peterson BS, Skudlarski P, Gatenby JC, Zhang H, Anderson AW, Gore JC. An fMRI study of Stroop word-color interference: evidence for cingulated subregions subserving multiple distributed attentional systems. *Biol. Psychiatry* 1999; 45: 1237-1258

Pillon B, Boller F, Levy R, Dubois B Cognitive deficits in Parkinson's disease. In: Boller F, Grafman J (Eds) Handbook of neuropsychology Amsterdam: *Elsevier*. 2002.

Pollak, P. Deep brain stimulation. American Academy of Neurology (AAN), Philadelphia, 2001: 3PC. 005-16-52.

Raven J. Guide to the Standard Progressive Matrices. London: H. K. Lewis 1960.

Sabatini U, Boulanouar K, Fabre N, Martin F, Carel C, Colonnese C et al.. Cortical motor reorganization in akinetic patients with Parkinson's Disease: a functional MRI study. *Brain* 2000; 123: 394-403.

Saint-Cyr JA, Trepanier LL, Kumar R, Lozano AM, Lang AE. Neuropsychological consequences of chronic bilateral stimulation of the subthalamic nucleus in Parkinson's Disease. *Brain* 2000; 123: 2091-2108.

Schroeder U, Kuehler A, Haslinger B, Erhard P, Fogel W, Tronnier VM, Lange KW, Boecker H, Ceballos-Baumann AO. Subthalamic nucleus stimulation affects striato-anterior cingulated cortex circuit in a response conflict task: a PET study. *Brain* 2002; 125 (9): 1995-2004.

Schubert T, Volkmann J, Muller U, Sturm V, Voges J, Freund HJ, von Cramon DY (2002). Effects of pallidal deep brain stimulation and levodopa treatment on reaction-time performance in Parkinson's disease. *Exp. Brain Res.* 144: 8-16.

Smeding H.M.M., van den Munckhof P., Esselink R. A.J., Schmand B., Schuurman P. R., Speelman J. D. Reversible cognitive decline after DBS STN in PD and displacement of electrodes. *Neurology* 2007; 68 (15): 1235-1236.

Stroop JR. Studies of interference in serial verbal reactions. *Journal of Experimental Psychology* 1935; 18: 643-662.

Thobois S, Mertens P, Guenot M, Hermier M, Mollion H, Bouvard M, Chazot G, Broussolle E, Sindou M. Subthalamic nucleus stimulation in parkinson's disease. Clinical evaluation of 18 patients. *Journal of Neurology*, 2002; 249, 529-534.

Troster AI, Fields JA, Wilkinson SB, Pahwa R., Miyawaki E., Lyons K.E., Koller WC. Unilateral pallidal stimulation for Parkinson's disease: neurobehavioural functioning before and 3 months after electrode implanatation. *Neurology* 1997; 49: 1078-1083.

Valldeoriola F, Tolosa E, Alegret M, Rey MJ, Morsi O, Pilleri M, Rumià J. Cognitive changes in Parkinson's disease during subthalamic stimulation: a clinicopathologic study Journal of Neurology, *Neurosurgery, and Psychiatry* 2006;77:565-566

Vercueil L., Benazzouz A., Deransart C., Bressand K., Marescaux C., Depaulis A., Benabid A. L. (1998) High frequency stimulation of the subthalamic nucleus suppresses absence seizures in the rat: comparison with neurotoxic lesions. *Epilepsy Research* 1998; 31(1): 39-46.

Wallesch C. W., Papagno C. Subcortical aphasia. In F. Clifford Rose, R. Whurr and M. Wukye (Eds.), Aphasia. London: Whurr. 1988. Pp. 256-287.

Wechsler D. WAIS- Revised manual. New York: The Psychological Corp 1981.

Wechsler D. A standardized memory scale for clinical use. *Journal of Psychology* 1945; 19: 87-97.

Witt K, Pulkowski U, Herzog J, et al. Deep brain stimulation of the subthalamic nucleus improves cognitive flexibility but impairs response inhibition in Parkinson disease. *Arch. Neurol.* 2004;61:697–700.

Zanini S., Melatini A, Vassallo A., Moretti R., Bava A. Speech and language improvement in Parkinson disease after electrical stimulation of the subthalamic nucleus: a case study. *European Journal of Physiology*, 2000; 440: R35-63.

Zung WWK. A self-rating depression scale. *Archives of General Psychiatry* 1965; 12: 63-70.

In: Encyclopedia of Neuroscience Research
Editors: Eileen J. Sampson and Donald R. Glevins

ISBN 978-1-61324-861-4
© 2012 Nova Science Publishers, Inc.

Chapter XIX

Subthalamic High-Frequency Deep Brain Stimulation Evaluated by Positron Emission Tomography in a Porcine Parkinson Model

Mette S. Nielsen[*1], *Flemming Andersen*[2], *Paul Cumming*[2], *Arne Møller*[2], *Albert Gjedde*[2], *Jens. C. Sørensen*[3] *and Carsten R. Bjarkam*[1]

[1]Department of Neurobiology, Institute of Anatomy,
University of Aarhus, Denmark
[2]PET Center, University Hospital of Aarhus, Denmark
[3]Department of Neurosurgery, University Hospital of Aarhus, Denmark

Abstract

Background: Subthalamic high-frequency deep brain stimulation (STN DBS) has during the last decade been widely used in the treatment of Parkinson's disease (PD) complicated by motor fluctuations and medicine-induced adverse effects. The exact mechanism of STN DBS is, however, still unelucidated. Objective: To evaluate whether STN DBS changes regional cerebral blood flow (rCBF) and oxygen consumption, by positron emission tomography (PET) in a non-primate large animal PD model of STN DBS. Methods: Three MPTP (1-methyl-4-phenyl-1,2,3,6-tetrahydropyridine) intoxicated female Göttingen minipigs (age 8-12 months, weight 16-20 kg) were stereotaxically implanted unilaterally with a DBS electrode (Medtronic, model 3387) connected to a pulse generator (Medtronic, model 7424) placed subcutaneously in the neck region. Four to six weeks later the animals were anesthetized and placed in a PET scanner. Three water ($H_2^{15}O$) and three oxygen ($^{15}O_2$) scans were performed, before stimulation with clinical parametres (continuous unipolar stimulation (electrode negative, case positive),

---

[*] Correspondence to: Mette Slot Nielsen. Department of Neurobiology, Institute of Anatomy, University of Aarhus, DK-8000 Aarhus C, Denmark. Phone: (+45) 8942 3027. Fax: (+45) 8942 3060. E-mail: ml@neuro.au.dk

amplitude 3V, frequency 160 Hz, and pulse-width 60 μs) was initiated and followed by 5 water and oxygen scans 5, 30, 60, 120 and 240 min thereafter. The obtained data (the three baseline scans versus the five poststimulation scans) were analysed by parametric DOT-analysis after semiautomatic coregistration to an average MRI pig brain. Results: rCBF was significantly increased (t-value = 5.47, p-value < 0.05) at the electrode tip after initiation of stimulation, and non-significant increases of oxygen consumption occurred in the ipsilateral- (t-value = 3.67, p-value < 0.1), and contralateral cortex (t-value = 3.34, p-value < 0.1). Conclusion: Our results indicate that STN DBS increases local midbrain rCBF and oxygen consumption in centrally placed cortical areas. The minipig may thus be a well-suited animal model for further studies of the mechanism of STN DBS in PD.

Keywords: animal model, basal ganglia, cerebral oxygen consumption, Göttingen minipig, MPTP, regional cerebral blood flow (rCBF)

Introduction

Subthalamic high-frequency deep brain stimulation (STN DBS) has during the last 15 years proven its value in the treatment of Parkinson's disease (PD) complicated by motor fluctuations and levodopa-induced dyskinesias [14,18,35-37,40,49,53,55]. However, many aspects of subthalamic DBS such as the mechanism of action, potential long-term adverse or neuroprotective effect still remain unelucidated [6,10,26,34,42,45,46,48,54].

Clinical studies may resolve some of these unanswered questions, but are often hampered by the inclusion of a limited number of patients exhibiting a diverse disease and medication pattern. Many experimental procedures are likewise impossible to perform in humans due to ethical concerns. Reliable animal models are therefore necessary for further elucidation of the role of STN DBS in the treatment of PD. Few studies of STN DBS have been conducted in animals and only two animal models of STN DBS in PD have been described. One model is based on primates [7,9,28,31] whereas the other is based on rats [8,10,20,23,28,43,50,57].

The use of rats is economically advantageous compared to use of larger laboratory animals. Stereotaxic targeting of the STN is furthermore eased by the availability of a superb rat brain atlas of stereotaxic coordinates related to external skull structures [44]. However, the small size of the rat brain naturally excludes the use of DBS devices intended for human use, and functional neuroimaging with PET or MRI for such small brains require special equipment. Due to the small size of the rat STN, one would also expect considerable spillover of the stimulation effect to adjacent brain structures.

The use of primates as an experimental model for STN DBS is on the other hand hampered by the high cost of these animals, ethical considerations and the danger of transfer of simian viruses [29]. Thus, many studies using primate models include no more than three to four animals [7,9,28,31]. To address these concerns we have recently developed a new large non-primate animal model for STN DBS in PD employing 1-methyl-4-phenyl-1,2,3,6-tetrahydropyridine (MPTP) intoxicated Göttingen minipigs implantated with clinical DBS devices [17]. The present chapter will accordingly describe the first positron emission tomography (PET) results from this animal model and relate these findings to existing PET results on DBS in humans.

Materials and Methods

Animals: Three female young adult Göttingen minipigs (age 8-12 months, weight 16-20 kg) were used in the study, in accordance with a protocol approved by the Danish Council on Animal Research Ethics (DANCARE). All animals were kept separately in a large stable on a 12-hour light/dark cycle at constant temperature and humidity, with free access to food and water.

MPTP-intoxication: The three animals were initially intoxicated with MPTP injected subcutaneously at a dose of 1 mg/kg bodyweight every other day over a period of 17 days, receiving a total dose of 8 mg/kg bodyweight [24]. All intoxicated animals displayed behavioral signs of parkinsonism by the end of the treatment period, evaluated by scores for motility, rigidity and coordination [41].

Stereotaxic DBS electrode implantation: Six months after MPTP-intoxication the animals were anesthetized with ketamine and midazolam, intubated and artificially ventilated with 60/40% N_2O/O_2 and 1.5% isoflurane. The animals were then stereotaxically implanted unilaterally with a DBS electrode (Medtronic, model 3387) that was connected to a pulse generator (Medtronic, model 7424) placed subcutaneously in the neck region [17]. The electrode was fixated to the skull with dental cement. The position of the electrode was verified by postoperative MRI before the animals were extubated and awakened.

PET: Four to six weeks later the animals were anesthetized as described above. The animals were placed prone in a PET scanner (Siemens ECAT HR+ tomograph). Three water ($H_2^{15}O$) and three oxygen ($^{15}O_2$) scans were then performed before DBS stimulation with clinical parametres (continuous unipolar stimulation (electrode negative, case positive), amplitude 3V, frequency 160 Hz, and pulse-width 60 µs) was initiated by telemetry and followed by 5 water and oxygen scans at 5, 30, 60, 120 and 240 min (Fig. 1). The duration of each scan was 3 min and consisted of a 3 min long dynamic positron emission sequence in 3D acquisition mode preceeded by an intravenous injection of min 300 MBq $H_2^{15}O$ or inhalation by single-breath of $^{15}O_2$ followed by 10 s breath holding [47]. Arterial blood samples were collected from the femoral artery every 5 s for the first min, every 10 s for the next min, and every 20 s for the last min during the scan sequence for measurement of the respective plasma input function for $H_2^{15}O$ or $^{15}O_2$. Attenuation correction was applied on the basis of an initially performed 15 min transmission scan (Fig. 1).

The obtained data (the three baseline scans versus the five poststimulation scans) were analysed by parametric DOT-analysis after semiautomatic coregistration to an average MRI pig brain [4,5,56] (Fig. 2).

Perfusion and postmortem analysis: The animals were euthanized with pentobarbital 6-12 months after the described PET-procedure, followed by a transcardial perfusion with 5 l of phosphate buffered 4% paraformaldehyde (pH 7.4). The brain was removed from the skull and the electrode was retracted. The brain was then embedded and oriented in HistOmer (HistOtech ApS, Denmark) according to an MRI-atlas of the Göttingen minipig [56] followed by sectioning into 6-9 mm thick coronal tissue slabs [15,52] enabling identification of the exact electrode position and related changes in the brain tissue.

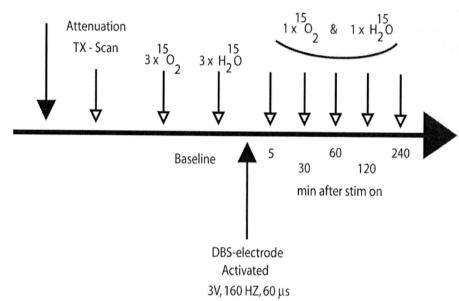

Figure 1. Flow-chart illustrating the time course and interventions of the PET-study. The total procedure would last approximately 6-7 hrs. During this period the animal would be intubated and anesthetized with its ECG, blood pressure and temperature closely monitored. Approximately 1 L of isotonic saline was administered during the procedure, and the body temperature was maintained at 37-38°C with an electrical heating blanket.

Results

Surgical procedure: The three animals survived for more than 6 months with active DBS electrodes. Postoperative infections were not encountered. Unilateral stimulation caused initial rotational behavior in all animals, which dissapeared after a few days. The implanted electrodes were generally tolerated well by the minipigs, and all animals were able to resume normal daily activities shortly after the surgical procedure without any change in consciousness or other non-motor functions.

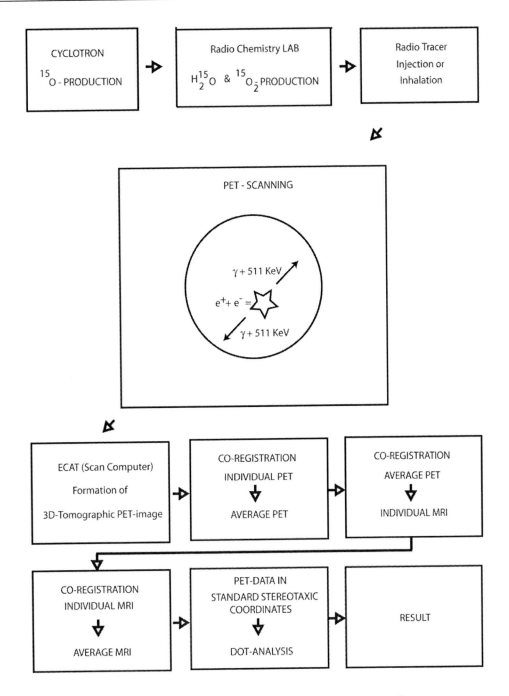

Figure 2. Schematic drawing illustrating the PET procedure from production of ^{15}O in the cyclotron to the formation of the individual 3D-tomographic PET-image and the elaborate process of co-registration [4,5,56] necessary for the DOT-analysis.

Postoperative MRI (Fig. 3): In all three animals, the electrode was located ventrally in the caudal diencephalon (MPTP 2 and 9) or the rostral mesencephalon (MPTP 4) within 5 mm of the subthalamic area depicted with arrow on MPTP 2 and 9.

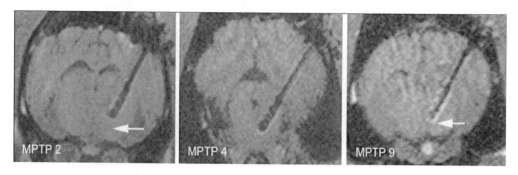

Figure 3. Postoperative MR images illustrating the electrode placement in the three animals MPTP 2, 4 and 9, respectively. The electrode was in all three cases located ventrally in the caudal diencephalon (MPTP 2 and 9) or the rostral mesencephalon (MPTP 4) within 5 mm of the subthalamic area (depicted with arrow on MPTP 2 and 9).

Postmortem examination (Fig. 4). After brain removal 6-12 months postoperatively, it was noticed that the electrodes in MPTP 2 and MPTP 9 had migrated due to insufficient fixation of the electrodes, while the electrode of MPTP 4 had kept its position in the mesencephalon. The meninges and the brain surface around the electrodes were, however, devoid of any apparent signs of inflammation or bleeding.

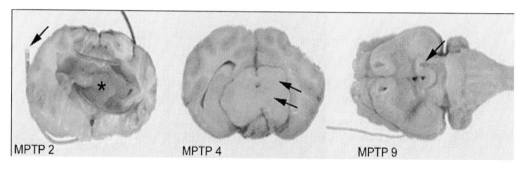

Figure 4. Photographic montage of brain images from MPTP 2, 4 and 9 after postmortem removal. Severe electrode migration was noted in MPTP 2, resulting in the formation of a large cavity in the forebrain (*). Arrow points to electrode tip. Electrode migration was also seen but to a lesser extent in MPTP 9 where the electrode penetrated out through the ventral brain surface at the mesencephalic/diencephalic transition (arrow). It was accordingly only possible to verify the postoperative MRI-visualized electrode placement in MPTP 4 at the postmortem examination (arrows: electrode tract).

PET results (Figs. 5 and 6). Regional cerebral blood flow (rCBF) was significantly increased (t-value = 5.47, p-value between 0.05 and 0.02) after stimulation onset at the mesencephalic/diencephalic transition zone (Fig. 5) corresponding to the MRI-verified postoperative placement of the electrode tip (Fig. 3). No significant increase in rCBF was noted elsewhere in the minipig brain after stimulation onset (Fig. 5).

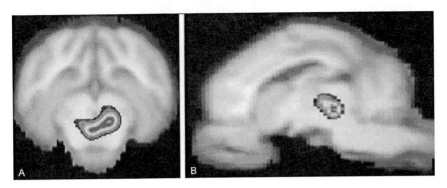

Figure 5. t-statistical coronal (A) and sagittal (B) maps showing change in rCBF after stimulation onset. Note that rCBF is significantly increased (t-value = 5.47, corresponding to a p-value between 0.05 and 0.02) at the mesencephalic/diencephalic transition zone corresponding to the MRI-visualized postoperative placement of the electrode tip.

A non-significant increase in cerebral oxygen consumption after stimulation onset was noted in the ipsilateral centrally placed dorsolateral neocortex (t-value = 3.67, p-value < 0.1) (Fig. 6B, C and D), and in the contralateral dorsomedial neocortex situated close to the central (Fig. 6A, B and D) and posterior part (Fig. 6A and B) of the cingulate sulcus (t-value = 3.34, p-value < 0.1).

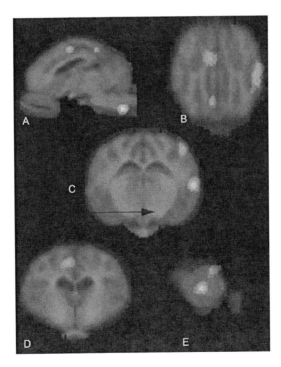

Figure 6. t-statistical mid-sagittal (A), dorsal (B), mid-coronal (C), anterior-coronal (D) and lateral-sagittal (E) maps showing change in cerebral oxygen consumption after stimulation onset. An non-significant increase in cerebral oxygen consumption is seen in the ipsilateral dorsolateral neocortex (t-value = 3.67, corresponding to a p-value between 0.1 and 0.05), and contralateral dorsomedial cortex (t-value = 3.34, corresponding to a p-value between 0.1 and 0.05). The increase noted in the brain stem at the ventral ponto-medular junction is a commonly occurring artifact caused by the anatomy of minipig vascular system. Arrow on (C) indicates the position of the subthalamic nucleus [38].

Discussion

This chapter describes for the first time that immediate initiation of subthalamic DBS may lead to a local increase in rCBF around the electrode tip and increased cerebral oxygen consumption in the centrally placed ipsi- and contralateral neocortex. It is tempting to interpret the results on cerebral oxygen consumption after stimulation onset as support to the hypotheses claiming that the activity of the somatomotor neocortex is suppressed in PD due to decreased thalamocortical activity caused by an increased inhibitory output from the basal ganglia globus palidus pars interna (GPi) originating partly from excessive activity of the subthalamic nucleus in Parkinson disease [1-3,12-14,18,21,25]. Presuming that subthalamic DBS results in decreased subthalamic activity leading to reduced inhibitory output from GPi, causing a release of thalamocortical motor activity, increased neocortical motor activity would be expected after STN DBS, leading to increased neocortical oxygen consumption. The rCBF which normally is closely linked to cerebral oxygen consumption is, however, not increased in the neocortex in this study but only around the electrode tip, which could reflect that the immediate initiation of DBS with full clinical stimulation parameters could cause local vasodilatation on a direct electrical or inflammatory basis, or in fact increases neuronal firing (see below). Caution should be made, however, when interpreting these data due to several factors. Firstly, placement of clinical DBS electrodes is difficult in the relative small subthalamic nucleus of the minipig [17,38]. None of our electrodes were directly situated within the STN but all within 5 mm of the target. The electrodes should, accordingly, be close enough to the STN to influence it, but influence on neighboring structures such as the thalamus and medial lemniscus, which may also lead to increased neocortical oxygen consumption, cannot be excluded. Secondly, two of the electrodes migrated (Fig. 4) during the 6-12 months postoperative follow-up period. The migration occurred between the day of the postoperative MRI visualization of the electrode placement (Fig. 3) and the day of sacrifice, which furthermore makes interpretation of the PET results difficult, if this migration occurred before the PET study was performed. We cannot prove that the electrode migration took place after the PET study was carried out, but the clear increase in rCBF (Fig. 5) at the level corresponding so well with the electrode tip visualized on the postoperative MRI (Fig. 3) gives us some support that the electrode migration had not occurred at the time of the PET study. Thirdly, although the use of MPTP-intoxicated minipigs as an experimental model for PD has been well characterized behaviorally [41], biochemically [5,24,41] and histologically [41], precise mapping of the minipig neocortex has yet to be performed. We are therefore unable to precisely identify the functional importance of the neocortical areas that display a tendency towards increased oxygen consumption after stimulation onset. It is tempting to identify the ipsilateral dorsolateral neocortex as being sensory-motor, but in reality such as statement would require histological staining and neural tracing studies that are ongoing, but have not yet been completed.

Previous PET-studies examining rCBF changes in PD patients undergoing effective STN DBS find, in accordance with our results from the minipig, an increase in rCBF directly in or in close proximity to the stimulated STN, globus pallidus or ventral thalamus [22,32,33]. A SPECT-study by Sestini et al. [51] showed increased rCBF in several cortical areas such as the rostral supplementary motor area (SMA), the anterior cingulate cortex and dorsolateral prefrontal cortex (DLPFC), which is in accordance with the hypotheses on basal ganglia circuitry and mechanism of effect of STN DBS described above. However, several other PET studies of cortical rCBF in PD patients at rest undergoing clinically effective STN DBS describe a decreased rCBF in the SMA, sensorimotor cortex (pre- and postcentral gyrus), and lateral premotor cortex and to a varying extent prefrontal cortex [22,30,32,33,39]. When studying rCBF in STN-stimulated PD patients performing motor tasks, there is general agreement of increased rCBF in the SMA [22,30,39], and some authors additionally find increased rCBF in the DLPFC and cingulated cortex [39], premotor cortex [22] and superior parietal cortex and cerebellum [30]. This suggests a normalization of cortical activity in STN DBS-treated PD patients when performing motor tasks. The decrease in cortical rCBF seen at rest is consistent with the opposite proposal of mechanism of effect of STN DBS described above, suggesting that DBS in fact increases activation of STN output neurons, leading to increased inhibitory pallidal output and a following decrease in cortical activity [33]. Increased activation of STN output neurons may thus serve to normalize pathological neuronal networks as reflected by decreased cortical rCBF [30]. Thus, relief of tremor, rigidity and dystonia in PD in addition to bradykinesia may restore an otherwise attenuated motor cortical activity, resulting in an overall decreased rCBF despite a clearly improved motor performance [39]. In addition it has been suggested that direct cortico-subthalamic projections could very well play an important role in the modulation of subcortical-cortical pathways [32].

The result on rCBF in the minipig corresponds well with the human findings of increased activity in the stimulated midbrain area. Significant results from cortical areas could perhaps have been obtained if it had been possible to include more animals. The observed trend towards increased oxygen consumption in central cortical areas does not correspond with the substantial evidence of reduced rCBF at rest in STN DBS-treated patients, but perhaps the cortical response is different immediately following activation of electrical stimulation rather than in the chronic state examined in patients. In addition, a larger number of animals would be preferred just as caution should be made when comparing anesthetized animals directly with awake though resting patients. Naturally with the minipig we do not have the opportunity to study response to motor tasks with PET, but this animal model still may prove helpful in future studies of STN DBS in PD as the animals can be studied in a variety of different parkinsonian stages without results being hampered by previous medical treatments as seen in patients.

Conclusion

Initiation of STN DBS in a porcine model of PD leads to a local increase in rCBF around the electrode tip and trend towards increased cerebral oxygen consumption in centrally placed neocortical areas. Our study illustrates that MPTP-intoxicated minipigs are suitable for long-term implantation of DBS electrodes and related devices intended for human use, if these are properly placed and fixated [16,19]. The Göttingen minipig may under such circumstances serve as a useful experimental model for further elucidation of the mechanisms underlying the benefits of acute and chronic DBS in PD.

Acknowledgments

The authors acknowledge with gratitude the skillful assistance of Ms. D. Jensen, Mr. A. Meier, and the staff at the Påskehøjgård Animal Care Facility, Clinical Institute, University of Aarhus. This study was supported by the Lundbeck Foundation, the Aarhus University Research Foundation, the Alice Brenaa Foundation, the Danish Medical Research Council, the Danish Parkinson Association, the Foundation for the Advancement of Medical Science, the Lily Benthine Lund Foundation, the Medical Society of Aarhus County, the Novo Nordisk Foundation, the Aase and Ejnar Danielsen Foundation, the Foundation of 2/7 1984 against Parkinson's Disease, the Foundation of Ib Henriksen and the Foundation of King Christian the 10th.

References

[1] Albin, RL; Young, AB; Penney, B. (1989). The functional anatomy of basal ganglia disorders. *Trends Neurosci 12*, 366-375

[2] Alexander, GE; Crutcher, MD. (1990). Functional architecture of basal ganglia circuits: neural substrates of parallel processing. *Trends Neurosci 13*, 266-271.

[3] Alexander, GE. (1994). Basal ganglia-thalamocortical circuits: Their role in control of movements. *J Clin Neurophysiol 11*, 420-431

[4] Andersen, F. (2001). Automation methods for analysis of the porcine brain: A study on the Göttingen minipig brain. Ph.D. Thesis, Faculty of Health Sciences, University of Aarhus, Denmark.

[5] Andersen, F; Watanabe, H; Bjarkam, CR; Danielsen, EH; Cumming, P; The DaNeX Study Group. (2005). Pig brain stereotaxic standard space: Mapping of cerebral blood flow normative values and effect of MPTP-lesioning. *Brain Res Bull 66(1)*, 17-29.

[6] Benabid, AL; Benazzouz, A; Pollak, P. (2002). Mechanisms of deep brain stimulation. *Mov Disorders 17/Suppl. 3*, S73-S74.

[7] Benazzouz, A; Gross, C; Feger, J; Boraud, T; Bioulac, B. (1993). Reversal of rigidity and improvement in motor performance by subthalamic high-frequency stimulation in MPTP-treated monkeys. *Eur J Neurosci 5*, 382-389.

[8] Benazzouz, A; Piallat, B; Pollak, P; Benabid, AL. (1995). Responses of substantia nigra pars reticulata and globus pallidus complex to high frequency stimulation of the subthalamic nucleus in rats: electrophysiological data. *Neurosci Lett 189*, 77-80.

[9] Benazzouz, A; Boraud, T; Feger, J; Burbaud, P; Biolac, B; Gross, C. (1996). Alleviation of experimental hemiparkinsonism by high-frequency stimulation of the subthalamic nucleus in primates: A comparison with L-dopa treatment. *Mov Disord 11/6*, 627-632.

[10] Benazzouz A, Gao DM, Ni ZG; Piallat, B; Bouali-Benazzouz, R; Benabid AL. (2000). Effect of high-frequency stimulation of the subthalamic nucleus on the neuronal activities of the substantia nigra pars reticulata and ventrolateral nucleus of the thalamus in the rat. *Neurosci 99/2*, 289-295.

[11] Benazzouz, A; Hallett, M. (2000). Mechanism of action of deep brain stimulation. *Neurology 55(Suppl 6)*, S13-S16.

[12] Bergman, H; Wichmann, T; DeLong, MR. (1990). Reversal of experimental parkinsonism by lesions of the subthalamic nucleus. *Science 249*, 1436-1438.

[13] Bergman, H; Wichmann, T; Karmon, B; DeLong, MR. (1994). The primate subthalamic nucleus. II. Neuronal Activity in the MPTP model of Parkinsonism. *J Neurophysiol 72/2*, 507-520.

[14] Bjarkam, CR; Sørensen, JC; Sunde, NÅ; Geneser, FA; Østergaard, K. (2001). New strategies for the treatment of Parkinson's disease hold considerable promise for the future management of neurodegenerative disorders. *Biogerontology 2*, 193-207.

[15] Bjarkam, CR; Pedersen, M; Sørensen, JC. (2001). New strategies for embedding, orientation and sectioning of small brain specimens enable direct correlation to MR-images, brain atlases, or use of unbiased stereology. *J Neurosci Methods 108*, 153-159.

[16] Bjarkam, CR; Cancian, G; Larsen, M; Rosendal, F; Ettrup, KS; Zeidler, D; Blankholm, AD; Østergaard, L; Sunde, N; Sørensen JC. (2004). A MRI-compatible stereotaxic localizer box enables high-precision stereotaxic procedures in pigs. *J Neurosci Methods 139(2)*, 293-298.

[17] Bjarkam, CR; Larsen, M; Watanabe, H; Röhl, L; Simonsen, CZ; Pedersen, M; Ringgaard, S; Andersen, F; Cumming, P; Dalmose, AL; Møller, A; Jensen, LW; Danielsen, EH; Dalmau, I; Finsen, B; Öttingen, Gv; Gjedde, A; Sørensen, JC. Title: A porcine model of subthalamic high-frequency deep brain stimulation in Parkinson's disease. Willow MJ, editor. *Title: Parkinson's Disease: New Research.* New York, USA: Nova Science Publishers; 2005; Chapter 5. ISBN-1-59454-352-6.

[18] Bjarkam, CR; Sørensen, JC. Title: Progress and development in Parkinson disease therapy. Rattan S, Kassem M editors. *Title: Prevention and Treatment of Age-related Diseases.* Dordrecht, The Netherlands: Springer Publishers; 2006; Chapter 3; 31-48. ISBN 1-4020-4884-X.

[19] Bjarkam, CR; Jørgensen, RL; Jensen, KN; Sunde, NAa; Sørensen, JC. (2008). Deep brain stimulation electrode anchoring using BioGlue®, a protective electrode covering, and a titanium microplate. *J Neurosci Methods 168*, 151-155.

[20] Bruet, N; Windels, F; Bertrand, A; Feuerstein, C; Poupard, A; Savasta, A. (2001). High frequency stimulation of the subthalamic nucleus increase the extracellular contents of striatal dopamine in normal and partially dopaminergic denervated rats. *J Neuropathol Exp Neurol 60(1)*, 15-24.

[21] Ceballos-Baumann, AO; Obeso, JA; Vitek, JL; DeLong, MR; Bakay, R; Linazasoro, G; Brooks, DJ. (1994). Restoration of thalamocortical activity after posteroventral pallidotomy in Parkinson's disease. *Lancet 344*, 814.

[22] Ceballos-Baumann, AO; Boecker, H; Bartenstein, P; Falkenhayn, Iv; Riescher, H; Conrad, B; Moringlane, JR; Alesch, F. (1999). A positron emission tomographic study of subthalamic nucleus stimulation in Parkinson disease. *Arch Neurol 56*, 997-1003.

[23] Chang, J-Y; Shi, L-H; Luo, F; Woodward, DJ. (2003). High frequency stimulation of the subthalamic nucleus improves treadmill locomotion in unilateral 6-hydroxydopamine lesioned rats. *Brain Res 983*, 174-184.

[24] Danielsen, EH; Cumming, P; Andersen, F; Bender, D; Brevig, T; Falborg, L; Gee, A; Gillings, NM; Hansen, SB; Hermansen, F; Johansen, J; Johansen, TE; Dahl-Jørgensen, A; Jørgensen, HA; Meyer, M; Munk, O; Pedersen, EB; Poulsen, PH; Rodell, AB; Sakoh, M; Simonsen, CZ; Smith, DF; Sørensen, JC; Østergaard, L; Zimmer, J; Gjedde, A; Møller, A. (2000). The DaNeX study of embryonic mesencephalic, dopaminergic

tissue grafted to a minipig model of Parkinson's disease: preliminary findings of effect of MPTP poisoning on striatal dopaminergic markers. *Cell Transplant 9*, 247-259.

[25] DeLong, MR. (1990). Primate models of movement disorders of basal ganglia origin. *Trends Neurosci 13*, 281-285.

[26] Dostrovsky, JO; Lozano, AM. (2002). Mechanisms of Deep Brain Stimulation. *Mov Disorders 17/Suppl.3*, S63-S68.

[27] Gao, D; Benazzouz, A; Bressard, K; Piallat, B; Benabid, AL. (1997). Roles of GABA, glutamate, acetylcholine and STN stimulation on thalamic VM in rats. *Neuroreport 8*, 2601-2605.

[28] Gao, D; Benazzouz, A; Piallat, B; Bressard, K; Ilinsky, IA; Kultas-Ilinsky, K; Benabid AL. (1999). High-frequency stimulation of the subthalamic nucleus suppresses experimental resting tremor in the monkey. *Neurosci 88/1*, 201-212.

[29] Goodman, S; Check, E. (2002). The great primate debate. *Nature 417*, 684-687.

[30] Grafton, ST; Turner, RS; Desmurget, M; Bakay, R; Delong, M; Vitek, J; Crutcher, M. (2006). Normalizing motor-related brain activity. *Neurology 66*, 1192-1199.

[31] Hashimoto, T; Elder, CM; Okun, MS; Patrick, SK; Vitek, JL. (2003). Stimulation of the subthalamic nucleus changes the firring pattern of pallidal neurons. *J Neurosci 23(5)*, 1916-1923.

[32] Haslinger, B; Kalteis, K; Boecker, H; Alesch, F; Ceballos-Baumann, AO. (2005). Frequency-correlated decreases of motor cortex activity associated with subthalamic nucleus stimulation in Parkinson's disease. *Neuroimage 28*, 598-606.

[33] Hershey, T; Revilla, FJ; Wernle, AR; McGee-Minnich, L; Antenor, JV; Videen, TO; Dowling, JL; Mink, JW; Perlmutter JS. (2003). Cortical and subcortical blood flow effects of subthalamic nucleus stimulation in PD. *Neurology 61*, 816-821.

[34] Krack, P; Pollak, P; Limousin, P; Hoffmann, D; Benazzouz, A; Le Bas, JF; Koudsie, A; Benabid, AL. (1998). Opposite motor effect of pallidal stimulation in Parkinson's disease. *Ann Neurol 43*, 180-192.

[35] Krack, P; Pollak, P; Limousin, P; Hoffmann, D; Xie, J; Benazzouz, A; Benabid AL. (1998). Subthalamic nucleus or internal pallidal stimulation in young onset Parkinson's disease. *Brain 121*, 451-457.

[36] Krack, P; Poepping, M; Weinert, D; Schrader, B; Deutschel, G. (2000). Thalamic, pallidal, or subthalamic surgery for Parkinson's disease? *J Neurol 247(Suppl 2)*, II/122-II/134.

[37] Kumar, R; Lozano, AM; Kim, YJ; Hutchison, WD; Sime, E; Halket, E; Lang, AE. (1998) Double-blind evaluation of subthalamic nucleus deep brain stimulation in advanced Parkinson's disease. *Neurology 51*, 850-855.

[38] Larsen, M; Bjarkam, CR; Østergaard, K; West, MJ; Sørensen, JC. (2004). The anatomy of the porcine subthalamic nucleus evaluated with immunohistochemistry and design based stereology. *Anat Embryol 208(3)*, 239-47.

[39] Limousin, P; Greene, J; Pollak, P; Rothwell, J; Benabid, AL; Frackowiak, R. (1997). Change in cerebral activity pattern due to subthalamic or internal pallidum stimulation in Parkinson's disease. *Ann Neurol 42*, 283-291.

[40] Limousin, P; Krack, P; Pollak, P; Benazzouz, A; Ardouin, C; Hoffman, D; Benabid, AL. (1998). Electrical stimulation of subthalamic nucleus deep brain stimulation in advanced Parkinson's disease. *N Engl J Med 339*, 1105-1111.

[41] Mikkelsen, M; Møller, A; Jensen, LH; Pedersen, A; Harajehi, JB; Pakkenberg, H. (1999). MPTP-induced Parkinsonism in minipigs: A behavioral, biochemical, and histological study. *Neurotoxicol Teratol 21*, 169-175.

[42] Montgomery Jr, EB; Baker, KB. (2000). Mechanisms of deep brain stimulation and future technical developments. *Neurol Res 22/3*, 259-266.

[43] Paul, G; Reum, T; Meissner, W; Marburger, A; Sohr, R; Morgenstern, R; Kupsch, A. (2000). High frequency stimulation of the subthalamic nucleus influences striatal dopaminergic metabolism in the naive rat. *Neuroreport 11(3)*, 441-444.

[44] Paxinos, G; Watson, C. The rat brain in stereotaxic coordinates, 4th edition. New York, USA: Academic Press. 1998

[45] Piallat, B; Benazzouz, A; Benabid, AL. (1996). Subthalamic nucleus lesion in rats prevents dopaminergic nigra neuron degeneration after striatal 6-OHDA injection: Behavioral and immunohistochemical studies. *Eur J Neurosci 8*, 1408-1414.

[46] Piallat, B; Benazzouz, A; Benabid, AL. (1999). Neuroprotective effect of chronic inactivation of the subthalamic nucleus in a rat model of Parkinson's disease. *J Neural Transm 55*, 71-77.

[47] Poulsen, PH; Smith, DF; Østergaard, L; Danielsen, EH; Gee, A; Hansen, SB; Astrup, J; Gjedde, A. (1997). In vivo estimation of cerebral blood flow, oxygen consumption and glucose metabolism in the pig by [^{15}O]water injection [^{15}O]oxygen inhalation and dual injections of [^{18}F]fluorodeoxyglucose. *J Neurosci Methods 77*, 199-209.

[48] Rodriguez, MC; Obeso, JA; Olanow, CW. (1998). Subthalamic nucleus-mediated excitotoxicity in Parkinson's disease: A target for neuroprotection. *Ann Neurol 44(3 Suppl 1)*, S175-S188.

[49] Rodriguez-Oroz, MC; Gorospe, A; Guridi, J; Ramos, E; Linazasoro, G; Rodriguez-Palmero, M; Obeso, JA. (2000). Bilateral deep brain stimulation of the subthalamic nucleus in Parkinson's disease. *Neurology 55(Suppl 6)*, S45-S51.

[50] Salin, P; Manrique, C; Forni, C; Kerkerian_Le Goff, L. (2002). High-frequency stimulation of the subthalamic nucleus selectively reverses dopamine denervation-induced cellular defects in the output structures of the basal ganglia in the rat. *J Neurosci 22(12)*, 5137-5148.

[51] Sestini, S; Luzio, ASd; Ammannati, F; Cristofaro, MTRD; Passeri, A; Martini, S; Pupi, A. (2002). Changes in regional cerebral blood flow caused by deep-brain stimulation of the subthalamic nucleus in Parkinson's disease. *J Nucl Med 43*, 725-732.

[52] Sørensen, JC; Bjarkam, CR; Danielsen, EH; Simonsen, CZ; Geneser, FA. (2000). Oriented sectioning of irregular tissue blocks in relation to computerized scanning modalities. Results from the domestic pig brain. *J Neurosci Methods 104*, 93-98.

[53] Tronnier, VM; Krause, M; Heck, A; Kronenbürger, M; Bonsanto, MM; Tronnier, J; Vogel, W. (1999). Deep brain stimulation for the treatment of movement disorders. *Neurol Psychiatr Brain Res 6*, 199-212.

[54] Vitek, JL. (2002). Mechanisms of Deep Brain Stimulation: Excitation or inhibition. *Mov Disord 17/Suppl.3*, S69-S72.

[55] Volkmann, J; Allert, N; Voges, J; Weiss, PH; Freund, HJ; Sturm, V. (2001). Safety and efficacy of pallidal or subthalamic nucleus stimulation in advanced PD. *Neurology 355*, 548-551.

[56] Watanabe, H; Andersen, F; Simonsen, CZ; Evans, SM; Gjedde, A; Cumming, P; The DaNeX Study Group. (2001). MR-based statistical atlas of the Göttingen minipig brain. *Neuroimage 14*, 1089-1096.

[57] Windels, F; Bruet, N; Poupard, A; Urbain, N; Chouvet, G; Feuerstein, C; Savasta, N. (2000). Effects of high frequency stimulation of subthalamic nucleus extracellular glutamate and GABA in substantia nigra and globus pallidus in the normal rat. *Eur J Neurosci 12(11)*, 4141-4146.

In: Encyclopedia of Neuroscience Research
Editors: Eileen J. Sampson and Donald R. Glevins

ISBN 978-1-61324-861-4
© 2012 Nova Science Publishers, Inc.

Chapter XX

Current and Future Perspectives on Vagus Nerve Stimulation in Treatment-Resistant Depression

Bernardo Dell'Osso[*]*, Giulia Camuri, Lucio Oldani and A. Carlo Altamura*
Department of Psychiatry, University of Milan, Milano, Italy

Abstract

Treatment-resistant depression (TRD) is a leading cause of disability and therapeutic strategies used for this prevalent and impairing condition include pharmacological augmentation strategies and brain stimulation techniques. Vagus nerve stimulation (VNS) is a brain stimulation intervention that has shown to decrease seizure frequency in partial-onset seizure patients. Subsequent trials performed on treatment-resistant patients with major depression indicated that VNS may be an effective and safe treatment. VNS involves minor surgery in which an electrode pair is wrapped around the left vagus nerve in the neck and connected to a generator implanted in the chest wall subcutaneously. Several methodological approaches have been employed in order to better understand the mechanism of action of VNS and its effects on neurobiological circuits involved in mood regulation. Despite its invasive nature, the tolerability profile of VNS seems to be favourable given that, differently from other brain stimulation interventions (i.e., ECT, rTMS, tDCS), VNS is a continuous and chronic stimulation. Even though previous positive results from clinical trials have led VNS to be approved by the FDA and the EMEA for the use in TRD, researchers are now focusing their efforts in the identification of possible predictors of response to VNS. Further investigation of the effects of VNS in the long-term treatment (>2 years) of TRD is also required.

---

[*] Dr Bernardo Dell'Osso, Department of Psychiatry, University of Milan, Fondazione IRCCS , Ospedale Maggiore Policlinico, Mangiagalli e Regina Elena, Via Francesco Sforza 35, 20122 Milano, Italy. Phone.: 02-55035994. Fax:02-50320310.email:bernardo.dellosso@policlinico.mi.it.

Keywords: vagus nerve stimulation (VNS), treatment resistant depression (TRD), brain stimulation

Introduction

Partial response, lack of response and residual symptoms following antidepressant treatment are common in patients with Major Depression. In fact, it has been estimated that 30 to 45% of depressed patients show either partial response or no response to standard antidepressant treatments [1, 2]. In addition, depressed patients may develop treatment resistant depression (TRD) [3], defined as the lack of response to two antidepressant trials, given in succession, at adequate doses and for an adequate time, in compliant subjects [4]. In patients with TRD, different strategies of intervention have been proposed including pharmacological and psychotherapic augmentation as well as brain stimulation techniques [5, 6].

Brain stimulation interventions include different techniques which share the common feature to provide a selective electric stimulation of specific brain areas. These include electroconvulsive therapy (ECT), deep brain stimulation (DBS), repetitive transcranial magnetic stimulation (rTMS), transcranial direct current stimulation (tDCS), and vagus nerve stimulation (VNS) which, by means of direct or indirect (magnetic) electric stimulation, can alter neurocircuits in the brain. Among these techniques, only ECT and VNS have FDA-approved indications for Major Depression. VNS, in particular, has been approved as adjunctive treatment in patients with TRD, given that clinical studies with VNS were performed in combination with other pharmacological agents. Differently from other brain stimulation techniques, VNS requires an extracranial surgical implant to provide a continuous electric stimulation that can be maintained for many years. VNS is also a reversible and adjustable procedure.

Historically, VNS has been firstly tested in the late 80's in patients who had treatment-resistant epilepsy and were not candidates to surgical interventions [7]. Subsequent multicenter randomized clinical trials allowed the FDA to approve VNS as adjunctive treatment for adults and adolescents with treatment-resistant partial onset seizures in 1997 [8-10].

First anecdotal indications of the potential efficacy of VNS in depressed patients came from direct clinical observation and subsequent clinical evaluations through standard rating scales of improved mood and cognition in patients with epilepsy treated with VNS [11].

The aim of the present review is to provide a comprehensive and updated overview of the more recent neurobiological and clinical findings about VNS therapy with a detailed description of the procedure and a specific emphasis on tolerability issues. Potential new fields of research with VNS in psychiatry are discussed as well.

The Vagus Nerve Anatomy and Its Implications in Major Depression

It is not surprising that an anticonvulsant treatment, such as VNS, may have antidepressant effect: several anticonvulsant medications (Valproate, Carbamazepine, Lamotrigine, etc), in fact, have shown antidepressant properties as well as the capacity to prevent depressive recurrences in euthymic patients [12]. These indirect observations, have been subsequently supported by neurobiological studies which provided the rationale for the potential efficacy of VNS in Major Depression. The vagus nerve (VN), the longest cranial nerve, is a mixed nerve composed by 80% afferent and 20% efferent fibres and regulates several automatic functions of the body, many of which are important in a variety of emotional tasks [13]. The majority of the fibres in the vagus nerve are C-fibres. VN regulates a lot of functions such as heart rate, blood pressure and peristalsis. The sensory afferent cell bodies are placed in the nodose and the jugular ganglia and project to the nucleus of the solitary tract (NTS). Projections from NTS to the brain areas are widespread and many of these reach brain areas such as the Locus Coeruleus, the Amygdala, the Dorsal Raphe, the Hyppocampus etc, which are in turn connected to the Orbitofrontal Cortex, the Insula and the Anterior Cingulate Gyrus [14, 15]. Many of these regions are supposed to be involved in the pathophysiology of Major Depression [16, 17]. In this perspective, VNS is able to provide a widespread intermittent continuous stimulation of these areas by means of a stimulating electrode which is attached to the left VN.

The Surgical Procedure and the Implanted Device

VNS requires a relatively minor surgical intervention that can be completed in about one hour under local or general anaesthesia. A first small incision is made below the collarbone in the left upper chest area in order to place the VNS pulse generator which is controlled and programmed by physicians through a labtop. A second incision is made in the left side of the neck and a small helicoidal electrode, connected to the pulse generator through a wire placed subcutaneously, is wrapped around the VN. The implanted device delivers chronic, intermittent electrical impulses to the left VN. Once the device is turned on, patients can decide to stop the stimulation through a magnet when needed, by placing it in front of the pulse generator. The left VN is chosen in light of fewer potential heart interactions. Different effects between left and right VN, in fact, are due to their connections to the heart: the right vagus innervates the sinoatrial node whereas the left vagus innervates the atrioventricular node [18].

Once VNS is started, classical stimulation parameters include 30 seconds of stimulation followed by 5 minutes of pause with an initial current of 0.25 mA which is progressively increased by 0.25 mA up to 1.50 mA that is considered the optimal stimulation voltage.

Neurobiology and Neuroimaging Data on VNS Therapy

Several studies have been conducted to assess the effects of VNS therapy on major neurotransmitter systems [19]. The most studied neurotransmitters are Dopamine, Serotonin (5-HT) and Norepinephrine (NE).

With respect to a potential influence of VNS on the dopaminergic system, Carpenter and co-workers found a significantly elevated level of homovanillic acid in cerebrospinal fluid (CSF) samples obtained from 21 patients with TRD after 24 weeks of VNS therapy. The CSF was also analyzed in order to assess eventual changes in the concentration of other substrates (5-hydroxyindoleacetic acid, GABA, etc.), but no relevant changes were observed [20].

Recently, Dorr and Debonnel investigated 5-HT and NE neuronal firing rates in the brainstem nuclei—dorsal raphe nucleus (DRN) and locus coeruleus (LC)—after VNS therapy in rats. After short-term VNS treatment, firing rates were significantly higher for LC, whereas basal firing rates in the DRN and LC were significantly increased after long-term treatments. The authors suggested that VNS may have a novel mechanism of action, enabling its effectiveness in TRD [21].

Imaging studies with SPECT, PET, and functional magnetic resonance (fMRI) have been used to determine which areas of the brain are specifically affected by VNS therapy.

Previous SPECT studies have shown regional cerebral blood flow (rCBF) abnormalities in limbic and cortical structures (such as insula and temporal cortex) of depressed people [22]. In this perspective, a recent SPECT study exploring rCBF changes induced by VNS found decreased rCBF in the amygdala, left hippocampus, left subgenual cingulate cortex, left and right ventral anterior cingulum, right thalamus and brain stem. These effects are consistent with those reported after the administration of SSRIs [23].

PET studies assessing VNS effects have shown mixed results. On one hand, a previous study indicated a decreased metabolism compared to baseline in the substantia nigra, ventral tegmentum, hypothalamus, insula/claustrum and superior temporal gyrus after a 1-year course of VNS therapy [24]. On the other hand, Conway and colleagues recently reported that VNS therapy caused an increased metabolic activity in the orbitofrontal cortex, bilateral anterior cingulate cortex and right superior and medial frontal cortex, and a decreased rCBF in the bilateral temporal cortex and right parietal area [25].

As such, definitive conclusions from neuroimaging studies on VNS effects on specific brain areas cannot be drawn due to the small size of the samples studied and further investigation is required.

Clinical Efficacy Studies

With respect to short-term clinical trials conducted on TRD patients with VNS, a first 10-week open label trial was performed on 30 adult outpatients with nonpsychotic, treatment-resistant major depression who had failed at least two medication trials in the current episode (MDE) while on stable pharmacological regimens. Results showed that 40% of the sample achieved response, while 17% remission on the basis of HDRS total scores. In a further analysis, the authors found that more than half of the total reduction in depressive symptoms occurred after the first 6 weeks of VNS [26].

A subsequent open study conducted with a similar design by Sackeim and colleagues showed that 30.5% of the sample met the criteria for response and 15.3% the criteria for remission after 10 weeks of VNS. The authors concluded that VNS appears to be more effective in patients with low to moderate, but not extreme, antidepressant resistance [27].

A subsequent open study conducted on a very small group ($N = 7$) showed that VNS significantly improved depressive symptoms after 10 weeks of treatment with 4 out of 7 patients classified as HDRS responders and 2 of these remitters (HDRS≤ 10) [28].

In the acute, sham-controlled, double-blind randomized trial published by Rush and co-workers and conducted on 235 patients treated with VNS, the authors reported mixed results. In fact, after 10 weeks of VNS, the mean HDRS response rates of both active VNS and sham groups did not show significant difference [29].

In a more recent study, a british group evaluated 11 patients with TRD who were treated with VNS therapy for 12 weeks [30]. Even though the clinical measures of depression decreased significantly, only one patient was classified as responder. Finally, Neuhaus and colleagues, in an exploratory study investigating the effects of VNS on auditory event-related potentials, reported that almost 40% of patients became responders after 10 weeks of treatment [31].

With respect to long-term effects of VNS, a follow-up study by Marangell and colleagues evaluated the efficacy of the stimulation at one year in the same sample previously treated in the short-term by Rush et al. [26]. The authors found a sustained response rate of 40% to 46% after 9 months. After one year of VNS, the remission rate significantly increased from 17% to 29%. Of note, the 91% of the responders at 3 months maintained their response at 12 months and among the non-responders at 3 months, 18% were considered responders at 12 months [32].

In a subsequent naturalistic follow-up study [33], including 205 patients, a significant improvement of depressive symptoms in the first 4 months of VNS therapy was reported. However, during the 12-month study, 30 patients were hospitalized due to a depressive relapse.

The only 2-year VNS follow-up study performed with a significant sample ($n = 59$) showed positive results [34] reporting the majority of antidepressant effects in the first 3 months of stimulation, even though additional improvements were shown at 12 months and maintained at 24 months.

The efficacy of adjunctive VNS to treatment as usual (VNS+TAU) has been compared to TAU in a 12-month, naturalistic, prospective study conducted on 229 treatment-resistant depressed patients [35]. Both groups, however, were different in terms of number of lifetime MDEs (higher in the TAU group), prior or past ECT exposure (greater in the VNS + TAU group), and pharmacological treatment (the VNS + TAU group showed a greater use of tricyclics, stimulants and neuroleptics in the current MDE). Analysis on the HDRS scores showed that 27% of the VNS + TAU group were responders at 12 months, compared to only 13% of the TAU group. In addition, a higher number of remitters, although not statistically significant, was found in the VNS + TAU group.

A recent open European multicenter study investigated the efficacy of adjunctive VNS therapy in 74 patients with TRD [36]. Treatment remained unchanged for the first 3 months whereas in the subsequent 9 months, medications and VNS parameters were altered as clinically indicated. After 3 months of VNS, response and remission rates reached respectively 37% and 17% of the sample on the basis of HDRS-28 scores. Response rates increased to 53% after 12 months of VNS, and remission rates reached 33% of the sample. Of clinical interest, median time to response was found to be of 9 months.

Side-Effects, Compliance and Tolerability of VNS

VNS related side-effects may rarely occur as a consequence of the neurosurgical intervention. Most frequent side-effects associated with VNS, in fact, are those involving the electrical stimulation of the left VN [37, 38]. When the device is implanted and the stimulation is started, rarely the device can cause cardiac rhythm disturbances. If this happens, the surgeon can readjust the electrodes to prevent this. During the initial phase of stimulation – when the current is increased – if patients experience disturbing side-effects (e.g., cough, voice alteration), the voltage can be readjusted to the previous level of good tolerability for additional time before a new attempt to increase it. Of clinical interest, VNS has not been reported to determine any systemic effect such as weight gain, sexual dysfunction, sedation, and other common side-effects associated with pharmacologic treatments.

During the active stimulation, the most common complications include voice alteration such as hoarseness and related symptoms which may occur in approximately the 50% of patients. In published clinical trials, these symptoms, however, were not serious enough to cause significant dropouts.

In the sham controlled trial, shortness of breath, neck pain, dysphasia, and paresthesias occurred in >10% of patients, and 25% to 35% experienced cough [26, 27]. Most of these symptoms gradually subsided, and the percentage of patients who experienced cough dropped to <10% at 2 years. Of clinical interest, in the acute pivotal trial [26], both sham control and active VNS patients reported similar rates of mild, moderate, or severe adverse effects. These results are consistent with subsequent studies showing a progressive reduction (within one year from the implant) of side-effects in terms of frequency and severity [32]. Taken as a whole, these studies indicate VNS as a safe and well-tolerated treatment.

With respect to the potential risk of a manic switch with VNS, published studies included both unipolar and bipolar patients, and it was reported that only a few patients became manic. On one hand, it is difficult to evaluate if patients spontaneously had a manic episode or if VNS induced it. On the other hand, it is well established that most effective antidepressant treatments have a propensity to increase the switch rate from depression to mania and, therefore, one might expect the same event to occur with VNS as well [39].

Finally, it needs to be taken into account that the device battery has a limited duration and that, as some VNS studies with follow-up at 45 months have shown [40, 41, 42], depressive symptoms may reappear with the end of the battery. Even though batteries with a longer duration are being developed, clinicians should expect that, after some years of stimulation, patients might need a new surgical intervention – which does not involve the electrode wrapped to the VN but only the pulse generator placed in the anterior chest – in order to replace battery and maintain the stimulation in the long-term.

Current and Future Perspectives with VNS

VNS may have some advantages over standard treatments of depression. One of these is represented by the compliance that is insured by a continuous stimulation and does not depend on patient. In addition, VNS has no influence on pregnancy because like other non-pharmacological treatments it does not pass through the placenta [43].

There are other clinical issues that should be further investigated with VNS and that might be of great clinical interest. A recent pivotal trial showed promising results in a small sample of patients with treatment-resistant rapid-cycling bipolar disorder. Over the 12-month study period, VNS was associated with a 38.1% mean improvement in the overall illness as compared to baseline, as well as with significant reductions in symptoms as measured by outcome rating scales [44]. In this perspective, it is noteworthy to highlight that VNS may exert a positive effect not only over depressive symptoms but, also, on the cyclicity that characterizes many forms of mood disorders. This putative effect may also be a *trait d'union* between the anticonvulsant effect of VNS and the clinical use of anticonvulsants as mood stabilizers in affective disorders.

Another clinical field of psychiatric interest for the use of VNS therapy is the one related to eating disorders, obesity in particular. In fact, it is well established that mood affects food craving, in particular the amount and kind of food consumed. On the basis of a relationship between the VN in hunger and satiety and the short-term regulation of food intake [45], Bodenlos and colleagues investigated whether acute VNS could temporarily alter food craving in people with TRD [46]. Thirty-three participants were recruited and divided in three groups: depressed patients treated with VNS, depressed patients who did not receive VNS, and healthy controls. At the endpoint, the authors found that VNS was associated with important changes in food cravings.

Recently, Pardo and colleagues, in a study with 14 depressed patients who received VNS and were followed up for 2 years, observed a significant and progressive weight loss, proportional to the initial BMI, hypothesizing for VNS a potential role in the treatment of obesity [47]. These preliminary positive results should encourage further studies specifically assessing the relationship between VNS and food intake and the potential role of this technique in the treatment of some eating disorders.

Finally, given that published trials investigated the efficacy of VNS in a follow-up period not longer than two years, it is important to evaluate the efficacy of the technique in the long-term. Another important issue is that VNS has been so far mostly investigated as augmentative treatment in TRD, and it would be interesting to assess its effect as monotherapy. In addition, it remains to be elucidated the optimal number of clinical trials failed before considering a patient potentially candidate to VNS. In fact, according to the results of the Sequenced Treatment Alternatives to Relieve Depression (STAR*D) study, chances of achieving remission dramatically drop after two failed adequate antidepressant trials [48, 49], indicating that this level of treatment-resistance has limited options. In this perspective, it should be further evaluated the possibility to administer VNS after a lower number of failed clinical trials.

Conclusions

VNS represents an approved adjunctive brain stimulation intervention for patients with TRD. Potential candidates should be patients with a history of chronic, recurrent depression who failed to respond to multiple adequate antidepressant trials during the index episode. Based on published clinical trials, clinicians may expect significant improvement of depressive symptoms in the first two years of VNS. Nevertheless, treatment studies with VNS have demonstrated different degrees of efficacy which varies depending on the study design (i.e., duration of stimulation, degree of treatment resistance, concomitant pharmacological treatment, etc.) which presented methodological limitations in some cases, making it difficult to compare study results.

On one hand, the high cost of the device, the invasive nature of the intervention and the need to perform further controlled trials to assess the efficacy of the technique in the very long-term and in specific subgroups of patients (i.e., unipolar vs bipolar patients, presence of rapid cycling patients) require some important precautions before prescribing such a treatment. On the other hand, the favourable tolerability profile and the lack of systemic side-effects combined to an optimal compliance encourage clinicians to consider VNS a novel and effective intervention for patients with TRD. Finally, it is mandatory to specify that VNS is not aimed to replace other brain stimulation interventions such as Electroconvulsive Therapy (ECT) which has a very different mechanism of action and has been also combined to VNS in a recent case series [50].

References

[1] Nierenberg Aa, Dececco Lm. Definitions Of Antidepressant Treatment Response, Remission, Nonresponse, Partial Response, And Other Relevant Outcomes: A Focus On Treatment-Resistant Depression. *J Clin Psychiatry.* 2001;62(16):5-9.

[2] Gayetot D, Ansseau M, Triffaux Jm. When Depression Does Not End. Resistant Depression: *Recent Clinical and Therapeutic Aspects. Rev Med Liege.* 2007;62:103-11.

[3] Fava M, Davidson Kg. Definition And Epidemiology Of Treatment-Resistant Depression. *Psychiatr Clin North Am.* 1996;19:179-200.

[4] Sackeim Ha. The Definition And Meaning Of Treatment-Resistant Depression. *J Clin Psychiatry.* 2001;62(S16):10-7.

[5] Preskorn Sh, Burke M. Somatic Therapy For Major Depressive Disorder: Selection Of An Antidepressant. *J Clin Psychiatry.* 1992;53(S):5-18.

[6] American Psychiatric Association. Practice Guidelines For The Treatment Of Patients With Major Depressive Disorder. *Am J Psychiatry.* 2000;157:1-45.

[7] Penry Jk, Dean Jc. Prevention Of Intractable Partial Seizures By Intermittent Vagal Stimulation In Humans: Preliminary Results. *Epilepsia.* 1990;31:(2)40-43.

[8] Rutecki P. Anatomical, Physiological, And Theoretical Basis For The Antiepileptic Effect Of Vagus Nerve Stimulation. *Epilepsia.* 1990;31 (S2):1-6.

[9] Ben-Menachem E, Mañon-Espaillat R, Ristanovic R, Wilder Bj, Stefan H, Mirza W Et Al. Vagus Nerve Stimulation For Treatment Of Partial Seizures. A Controlled Study Of

Effect On Seizures. First International Vagus Nerve Stimulation Study Group. *Epilepsia.* 1994;35:616-626.

[10] The Vagus Nerve Stimulation Study Group. A Randomized Controlled Trial Of Chronic Vagus Nerve Stimulation For Treatment Of Medically Intractable Seizures. *Neurology.* 1995;45:224-230.

[11] Handforth A, Degiorgio Cm, Schachter Sc, Uthman Bm, Naritoku Dk, Tecoma Es Et `Al. Vagus Nerve Stimulation Therapy For Partial-Onset Seizures: A Randomized Active-Control Trial. *Neurology.* 1998;51(1):48-55.

[12] Van Der Loos Ml, Mulder Ph, Hartong Eg, Blom Mbj, Vergouwen Ac, De Keyzer H Et Al. Efficacy And Safety Of Lamotrigine As Add-On To Lithium In The Treatment Of Bipolar Depression: A Multi-Center, Double-Blind, Placebo-Controlled Trial. *Am J Psychiatry.* In Press.

[13] Foley Jo, Dubois F. Quantitative Studies Of The Vagus Nerve In The Cat. I. The Ratio Of Sensory Motor Studies. *J Comp Neurol.* 1937;67:49–67.

[14] Chae Jh, Nahas Z, Lomarev M, Denslow S, Lorberbaum Jp, Bohning De Et Al. A Review Of Functional Neuroimaging Studies Of Vagus Nerve Stimulation (Vns). *J Psychiatr Res.* 2003;37:443–455.

[15] Kraus T, Hosl K, Kiess O. Bold Fmri Deactivation Of Limbic And Temporal Brain Structures And Mood Enhancing Effect By Transcutaneous Vagus Nerve Stimulation. *J Neural Transm.* 2007;114(11):1485-1493.

[16] Drevets Wc. Neuroimaging And Neuropathological Studies Of Depression: Implications For The Cognitive-Emotional Features Of Mood Disorders. *Curr Opin Neurol.* 2001;11(2):240-9.

[17] Mayberg Hs. Positron Emission Tomography Imaging In Depression: A Neural Systems Perspective. *Neuroimaging Clin N Am.* 2003;13(4):805-815.

[18] Ramani R. Vagus Nerve Stimulation Therapy For Seizures. *J Neurosurg Anesthesiol.* 2008;20(1):29-35.

[19] Nemeroff Cb, Mayberg Hs, Krahl Se, Mcnamara J, Frazer A, Henry Tr Et Al. Vns Therapy In Treatment-Resistant Depression: Clinical Evidence And Putative Neurobiological Mechanisms. *Neuropsychopharmacology.* 2006;31(7):1345-55.

[20] Carpenter Ll, Moreno Fa, Kling Ma, Anderson Gm, Regenold Wt, Labiner Dm Et Al. Effect Of Vagus Nerve Stimulation On Cerebrospinal Fluid Monoamine Metabolites, Norepinephrine And A-Aminobutyric Acid Concentrations In Depressed Patients. *Biol Psychiatry.* 2004;56:418-426.

[21] Dorr Ae, Debonnel G. Effect Of Vagus Nerve Stimulation On Serotonergic And Noradrenergic Transmission. *J Pharmacol Exp Ther.* 2006;318(2):890-898.

[22] Devous Md, Husain M, Harris Ts, Rush Aj. *Effects Of Vns On Regional Cerebral Blood Flow In Depressed Subjects.* Poster Presented At The 42nd Annual New Clinical Drug Evaluation Unit Meeting. Boca Raton, Fl, 2002; 10–13.

[23] Zobel A, Joe A, Freymann N, Clusmann H, Schramm J, Reinhardt M Et Al. Changes In Regional Cerebral Blood Flow By Therapeutic Vagus Nerve Stimulation In Depression: An Exploratory Approach. *Psychiatry Res.* 2005;139(3):165-79.

[24] Hagen Mc, Sheikh Js, Adson D, Rittberg B, Abuzzahab Fs, Lee Jt Et Al. *Metabolic Changes In Treatment-Resistant Depression Responsive To Vns Therapy* Poster Presented At The 58th Annual Scientific Convention Of The Society Of Biological Psychiatry, San Francisco, Ca,2003: 5-17.

[25] Conway Cr, Chibnall, Jt, Fletcher Jw, Filla-Taylor J, Grossberg Gt, Li X. Et Al. Three Months Of Vagus Nerve Stimulation Is Associated With Increased Limbic And Paralimbic Activity (Fdg Pet) *In Treatment-Resistant Depressed Subjects*. Poster Presented At The 57th Annual Scientific Convention Of The Society Of Biological Psychiatry. Philadelphia, Pa, 2002:16–18.

[26] Rush Aj, George Ms, Sackeim Ha, Marangell Lb, Husain Mm, Giller C Et Al. Vagus Nerve Stimulation For Treatment Resistant Depression: A Multicenter Study. *Biol Psychiatry*. 2000;47:276-286.

[27] Sackeim Ha, Rush Aj, George Ms, Marangell Lb, Husain Mm, Nahas Z Et Al. Vagus Nerve Stimulation (Vns) For Treatment-Resistant Depression: Efficacy, Side Effects, And Predictors Of Outcome. *Neuropsychopharmacology*. 2001;25:713-728.

[28] Armitage R, Husain M, Hoffmann R, Rush Aj. The Effects Of Vagus Nerve Stimulation On Sleep EEG In Depression: A Preliminary Report. *J Psychosom Res*. 2003;54:475-482.

[29] Rush Aj, Marangell Lb, Sackeim Ha, George Ms, Brannan Sk, Davis Sm Et Al. Vagus Nerve Stimulation For Treatment-Resistant Depression: A Randomized, Controlled Acute Phase Trial. *Biol Psychiatr*. 2005;58:347-354.

[30] O'keane V, Dinan Tg, Scott L, Corcoran C. Changes In Hypothalamic-Pituitary-Adrenal Axis Measures After Vagus Nerve Stimulation Therapy In Chronic Depression. *Biol Psychiatr*. 2005;58:963-968.

[31] Neuhaus Ah, Luborzewski A, Rentzsch J, Brakemeier El, Opgen-Rhein C, Gallinat J Et Al. P300 Is Enhanced In Responders To Vagus Nerve Stimulation For Treatment Of Major Depressive Disorder. *J Affect Disord*. 2007;100:123–128.

[32] Marangell Lb, Rush Aj, George Ms, Sackeim Ha, Johnson Cr, Husain Mm Et Al. Vagus Nerve Stimulation (Vns) For Major Depressive Episodes: One Year Outcomes. *Biol Psychiatry*. 2002;51(4):280-7.

[33] Rush Aj, Sackeim Ha, Marangell Lb, George Ms, Brannan Sk, Davis Sm Et Al. Effects Of 12 Months Of Vagus Nerve Stimulation In Treatment-Resistant Depression: A Naturalistic Study. *Biol Psychiatry*. 2005;58(5),355-363.

[34] Nahas Z, Marangell Lb, Husain Mm, Rush Aj, Sackeim Ha, Lisanby Sh Et Al. Two-Year Outcome Of Vagus Nerve Stimulation (Vns) For Treatment Of Major Depressive Episodes. *J Clin Psychiatry*. 2005;66:1097-1104.

[35] George Ms, Rush Aj, Marangell Lb, Sackeim Ha, Brannan Sk, Davis Sm Et Al. A One-Year Comparison Of Vagus Nerve Stimulation With Treatment As Usual For Treatment-Resistant Depression. *Biol Psychiatr*. 2005;58:364-373.

[36] Schlaepfer Te, Frick C, Zobel A, Maier W, Heuser I, Bajbouj M Et Al. Vagus Nerve Stimulation For Depression: Efficacy And Safety In A European Study. *Psychol Med*. 2008;4:1-11.

[37] Janicak Md. Neuromodulation. *Cns Spectrums* 2008;13(5):370-372.

[38] Daban C, Martinez-Aran A, Cruz N, Vieta E. Safety And Efficacy Of Vagus Nerve Stimulation In Treatment-Resistant Depression. A Systematic Review. *J Affect Disord*. 2008;110(1-2):1-15.

[39] Sackeim H.A. (2004). Vagus Nerve Stimulation. *In: Brain Stimulation In Psychiatric Treatment*. Edited By Lisanby Sh. American Psychiatric Publishing, Inc. Washington, Dc; 99-142.

[40] Martinez Jm, Zboyan Ha. Vagus Nerve Stimulation Therapy In A Patient With Treatment-Resistant Depression: A Case Report Of Long-Term Follow-Up And Battery End-Of-Service. *Cns Spectr.* 2006;11:143-147.

[41] Critchley Hd, Lewis Pa, Orth M, Josephs O, Deichmann R, Trimble Mr Et Al. Vagus Nerve Stimulation For Treatment-Resistant Depression: Behavioral And Neural Effects On Encoding Negative Material. *Psychosom Med.* 2007;69:17-22.

[42] Conway, C.R. *Sustained Antidepressive Benefit After Interruption Of Vagus Nerve Stimulation Therapy.* Poster Presented At The Us Psychiatric And Mental Health Congress November 7–10 Las Vegas, Nevada Usa, 2005.

[43] Husain Mm, Stegman D, Trevino K. Pregnancy And Delivery While Receiving Vagus Nerve Stimulation For The Treatment Of Major Depression: A Case Report. *Ann Gen Psychiatry.* 2005;4:16.

[44] Marangell Lb, Suppes T, Zboyan Ha, Prashad Sj, Fischer G, Snow D Et Al. A 1-Year Pilot Study Of Vagus Nerve Stimulation In Treatment-Resistant Rapid-Cycling Bipolar Disorder. *J Clin Psychiatry.* 2008;69(2):183-9.

[45] Havel Pj. Peripheral Signals Conveying Metabolic Information To The Brain: Short-Term And Long-Term Regulation Of Food Intake And Energy Homeostasis. *Exp Biol Med.* 2001;226:963-977.

[46] Bodenlos Js, Kose S, Borckardt Jj, Nahas Z, Shaw D, O'neil Pm Et Al. Vagus Nerve Stimulation Acutely Alters Food Craving In Adults With Depression. *Appetite.* 2007;48:145-53.

[47] Pardo Jv, Sheikh Sa, Kuskowski Ma, Surerus-Johnson C, Hagen Mc, Lee Jt Et Al. Weight Loss During Chronic, Cervical Vagus Nerve Stimulation In Depressed Patients With Obesity: An Observation. *Int J Obes.* 2007;31(11):1756-1759.

[48] Rush Aj, Trivedi Mh, Wisniewski Sr, Nierenberg Aa, Stewart Jw, Warden D Et Al. Acute And Longer-Term Outcomes In Depressed Outpatients Requiring One Or Several Treatment Steps: A Star*D Report. *Am J Psychiatry.* 2006;163(11):1905-1917.

[49] Trivedi Mh, Fava M, Wisniewski Sr, Thase Me, Quitkin F, Warden D Et Al. Medication Augmentation After The Failure Of Ssris For Depression. *N Engl J Med.* 2006;354(12):1243-52.

[50] Sharma A, Chaturvedi R, Sharma A, Sorrell Jh. Electroconvulsive Therapy In Patients With Vagus Nerve Stimulation. *J Ect.* 2008; In Press

In: Encyclopedia of Neuroscience Research
Editors: Eileen J. Sampson and Donald R. Glevins
ISBN 978-1-61324-861-4
© 2012 Nova Science Publishers, Inc.

Chapter XXI

Cognitive Aspects in Idiopathic Epilepsy

Sherifa A. Hamed [*]

Department of Neurology and Psychiatry, Assiut University Hospital, Assiut, Egypt

Abstract

Epilepsy is a common medical problem. Several studies suggest that idiopathic generalized or focal epilepsies can adversely affect mental development, cognition and behavior. Epileptic patients may experience reduced intelligence, attention, problems in memory, language and frontal executive functions. The exact mechanisms of epilepsy-related cognitive dysfunction are poorly understood. Cognitive deficits with epilepsy may be transient, persistent or progressive. Transient disruption of cognitive encoding processes may occur with paroxysmal focal or generalized epileptic discharges while epileptogenesis-related neuronal plasticity, reorganization, sprouting and impairment of cellular metabolism are fundamental determinants for progressive cognitive deterioration. Also antiepileptic drugs (AEDs) have differential, reversible and sometimes cumulative cognitive adverse consequences. AEDs not only reduce neuronal irritability but also may impair neuronal excitability, neurotransmitter release, enzymes and factors critical for information processing and memory. The present article serves as an overview of recent studies in cognition in adult and children patients with epilepsy. In this review, we will also discuss the known adverse mechanisms of epilepsy and AEDs on cognition.

Keywords: Epilepsy; antiepileptic drugs; cognition; memory

---

[*] Corresponding author: Sherifa Ahmed Hamed, P.O.Box 71516, Telephone: +2 088 2371820, Fax: +2 088 2333327, +2 088 2332278, e-mail: hamed_sherifa@yahoo.com.

Abbreviations

AEDs, antiepileptic drugs; TLE, temporal lobe epilepsy; SE, status epilepticus; KA, kainic acid; MRI, Magnetic Resonance Imaging; EEG, Electroencephalogram; BDNF, brain derived neurotrophic factor; CBZ, carbamazepine; PHT, phenytoin; VPA, valproate; PB, phenobarbital; GBP, gabapentin; OXC, oxcarbazepine; LTG, lamotrigine; TPM, topiramate; ZNS, Zonisamide; TGB, tiagabine; VGB, vigabatrin; LEV, levetiracetam; GABA, γ-amino butyric acid; NMDA, N-methyl-D-aspartate; AMPA, α-amino-3-hydroxy-5-methylisoxazole-4-propionate; LTP, long-term potentiation.

1. Introduction

Epilepsy is one of the most common neurological disorders. It nearly affects 1% of the population with the highest incidence during the first year of life [1]. During the last years, the increasing evidence from experimental pathological studies and in vivo imaging, and neuropsychological studies of patients with long-term epilepsy, have substantially increased our awareness about the possibility that briefer recurring focal or generalized seizures may cause a striking progressive cognitive decline which can be more detrimental to an individual's overall life [2-4]. Determining the frequency of cognitive dysfunction due to epilepsy is difficult to estimate. Community based studies reported that approximately 26.4-30% of children with epilepsy when first diagnosed, have evidence of subnormal global cognitive function or mental retardation with inferior academic achievement [5]. Problems in attention and memory are observed in about 30% of newly diagnosed and untreated epileptic patients with single or several seizures of cryptogenic origin [6].

Recent epilepsy researches have uncovered that impairments of neuronal plasticity and accumulating neuronal damage which evolves over a period of time after the initial seizures may underlie cognitive impairment and behavioral changes associated with epilepsy [3,7]. In addition, the deleterious effect of antiepileptic drugs (AEDs), the only available treatment for epilepsy, on cognitive is well documented in epileptic patients and volunteer studies. In general, AEDs suppress or prevent seizure recurrence and secondary spread of epileptic activity to the surrounding normal brain through their blockage action on voltage dependent ion channels or their modulatory effect on excitatory or inhibitory neurotransmitters. In doing so, AEDs may abort some of the toxic mechanisms leading to neuronal damage. It is expected that AEDs could improve patients' cognitive functions by controlling the number of overt and subtle epileptic activities as well as improving the psychosocial environment which provides additional benefit to cognition. However, a deleterious effect on cognitions has been observed with some AEDs even in the therapeutic ranges of the drug. Even AEDs of little negative impact on cognition in normal subjects may have detrimental cognitive effects in patients with epilepsy [8,9]. Some patient groups may be at particular risk (e.g., fetus, children and elderly) [10-13]. Some parents reported difference in child cognition and behavior after starting therapy as impaired attention, vigilance, slowness in response, talkative less, or difficulty to control. Even in patients who do not report cognitive changes, neuropsychological tests have shown significant impairments [9].

The above information make neurologists and neuroscientist clearly recognize that prevention or suppressing seizures by AEDs is alone insufficient without clear predictions of disease outcome and a challenge is how to keep patients free of seizures without interfering with normal brain function.

2. Epilepsy and Cognition

Animal studies have paralleled human studies in that chronic idiopathic epilepsy can result in progressive cognitive decline and behavioral impairment in developing and mature brains. Problems in various cognitive domains are recorded with epilepsy including poor attention, problems in memory, language, frontal executive functions and reduced scores in intelligence tests [2].

2.1. Evaluation of Cognition: Basic Definitions

The term cognition (obtained from the Latin word: cognoscere, "to know or to recognize") refers to the set of integrated and inter-related mental processes and systems involved in acquiring knowledge and comprehending, storing, retrieving and using this knowledge to perform day-to-day activities. Cognition is considered an abstract property of advanced living organisms and is studied as a direct property of the brain or of an abstract mind. Most neuropsychological assessment models require the independent evaluation of 6 specific areas of cognition (domains): 1) attention, 2) perception, 3) memory and learning, 4) executive functions, 5) verbal and language functions, and 6) spatial/constructural processing abilities. During infancy and early childhood, the more basic elements of attention and perception undergo the most rapid development, while in later childhood and adolescence, the development of higher-order linguistic, spatial and executive elements is primary [14-16].

Attention: a process that enables an individual to focus on the relevant information in the stimulus array while also inhibiting further processing of non-relevant information. Attention is the function of the frontal lobe.

Perception: is the central step in the processing of sensory/attentional information perceived through sensory (olfactive, visual, gustative, tactile and auditory) systems. The perception representation areas of the brain (cortical associative areas) have the function to store shapes and structure of the objects, faces and words, making abstraction of their semantic meaning. The right half of the brain cortex stores the biographic memory, emotional recalls, autobiographic and time-related memories, like the first day of the high school. The left half of the brain cortex stores the cognitive and factual memories (like Cairo is the Egypt's capital, words, numbers and so on). The sensations reach the amygdala (brain's emotional center) to decide the autobiographic and emotional meaning of a sensation within seconds.

Memory and learning: memory is the set of processes that temporarily holds new information while it is being utilized or processed for other purposes, or that more permanently holds learned information or experiences of the past generations and that of the others for future reference and use, while learning is defined as developing habits of thought or action. In general, memory may be A) Declarative (explicit): which include all deliberate, effortful and strategic activities with the main goal is to learn or remember, i.e. all the memories that can be recovered voluntary. Declarative memory is subdivided into: a) *the immediate memory:* which lasts milliseconds and allows one to memorize rapidly things like a phone number or a name, b) *the short term or working memory:* which is a temporary easy accessible storing of important information in consciousness to be useful in a particular activity, like the shopping list or dialing a phone number after having looked in your agenda, c) *the long-term memory*: it is subdivided into: i) *the episodic history memory:* which preserves and turned into conscious all the happenings of a lived personal episode which occurred at specific time and place, i.e. autobiographical information or events like "yesterday I made that thing", and ii) *the semantic culture memory:* which concerned with information poorly defined by tempro-spatial context. In other words, it represents what remains recorded in the brain after forgetting things following a learning process, from exams to experiences and decoding processes. This includes depersonalized information or facts as language, rules and general world knowledge. In the end, culture is what persists after we have forgotten it all. The hippocampus and the associated temporal lobe structures, mammillary bodies and dorsomedial nucleus of the thalamus are important for declarative memory [17,18]. B) Non-declarative (implicit or procedural): which include effortless remembering of habits and automatic behaviors with the goal of the activity is other than to remember. The cerebellum and basal ganglia have been found to be important in non-declarative memory. In other words, the voluntary movements may first performed and controlled by relaying on cerebral feedback from sensory organs but after some practice, the same movement will be performed by feed-forward (subconscious control by the cerebellum and basal ganglia). The movements become more quickly and automatically performed with less conscious effort. Thus the cerebral cortex could be omitted from control process once the sequence (motor act) has been learned. The results of anatomical studies clearly indicate that the cerebellum and basal ganglia participate in multiple circuits or 'loops' with cognitive areas of the cerebral cortex. The cerebellum is largely connected via the thalamus, to many brain areas relevant to cognition and behavior, including the dorsolateral prefrontal cortex, the medial frontal cortex, the parietal and superior temporal areas, the anterior cingulate, and the posterior hypothalamus [19-21]. The basal ganglionic-cortical circuits involved in non-motor functions are connected to the dorsolateral prefrontal cortex (area 46), the lateral orbitofrontal cortex (area 12), and the anterior cingulate/medial orbitofrontal cortices (areas 24 and 13). These frontal regions are known to be involved in planning, working memory, rule-based learning, attention, and other aspects of higher executive function [22]. In functional neuroimaging studies of normal subjects, the cerebellum is activated in tasks involving learning and word generation. The cerebellum stores the memory involving movement, like playing guitar or swimming [23]. In a PET study, Jueptner and colleagues [24,25] asked normal subjects to learn sequences of eight finger movements (key presses). They then compared the brain activity during learning of new sequences with that seen during performance of previously learned sequences. An examination of their data showed that rostrodorsal portions of the globus pallidus as well as portions of areas 9 and 46, the dorsolateral caudate, and the ventroanterior nucleus of the thalamus displayed increased activation preferentially during the learning of new sequences. It is important to summarize that the extensive connections between the prefrontal cortex and other brain regions including hippocampus and temporal lobe explain the importance of frontal lobes in memory functions. Memory from the past has been found to be encoded in lateral and medial temporal systems while future memory is related to frontal-prefrontal cortex [26].

Executive functions: are those functions involved in deliberately pursuing any type of goal in the face of difficulty or stress, especially novel non-routine tasks or complex organizationally demanding tasks. This include the following: self-awareness of strengths and limitations, ability to set adequately reasonable goals, ability to plan and organize goal-directed behavior, ability to self-initiate goal-directed behavior, ability to self-inhibit competing behaviors, ability to self-monitor behavior, ability to self-evaluate behavior in relation to goals, ability to solve problems and think and act strategically in the face of obstacles, ability to flexibly shift focus of attention, strategies, behaviors, and perspectives as required by context and goals and as dictated by feedback from previously unsuccessful behavior and strategies. Prefrontal cortex is responsible for executive functions [27,28].

Verbal and language functions versus Spatial/constructural processing abilities: left cerebral hemisphere is responsible for all activities related to verbal materials such as naming, writing, word lists, stories, etc, whereas the right cerebral hemisphere is responsible for all activities related to non-verbal materials such as geometric, drawings, construction, dressing, facial recognition, etc [29].

Neurological examination, computed tomography (CT), magnetic resonance imaging (MRI), electroencephalography (EEG) and positron emission tomography (PET) scans look at the structural, physical and metabolic condition of the brain while the neuropsychological (cognitive) testing is the only way to formally assess brain function. During the past decade neuropsychological testing has become very sophisticated. Specifically designed cognitive tests are used to measure a psychological function known to be linked to a particular brain structure or pathway. They are categorized according to the aspect or "domain" of cognitive ability that they aim to assess. However, the more common approach today, however, is to use a flexible battery based on hypotheses generated through a clinical interview, observation of the patient and review of medical records [30].

It must be remembered that the term cognition is not a synonym of intelligence quotient (IQ) or intelligence. IQ represents the summary scores on standardized "intelligence tests". The IQ score may include measures of cognition, but can also include and often assess sensory, motor and related abilities that are not typically included under the term of cognition. Another important difference is that the traditional intelligence tests measure what is known as convergent thinking which means that there is just one correct answer to each question as definition of words, questions and arithmetic problems, but do not measure what is known as divergent thinking which means that there is multiple answers to one question. Convergent thinking is the function of the parietal and temporal lobes while divergent thinking is the function of the frontal lobe. This means that lesions in parietal or temporal lobes may produce reliable decrease in IQ but in frontal lobe lesions, IQ is often normal [31]. Thus, intelligence is term that can be narrowly employed to refer to those abilities that are evaluated by intelligence tests.

2.2. Evidences for Cognitive Impairment in Epilepsy

Cognitive comorbidity associated with epilepsy is confirmed in pathological, psychological, physiological and imaging studies [32-37].

2.2.1. Experimental Studies

Kindling involves progressive potentiation of electrographic and behavioral seizure activity resulting from controlled repeated application of initially sub-convulsive electrical or chemical stimulation. In kindling models of TLE, the controlled repeated seizure discharges, trigger a precise sequence of complex activity-dependent neurodegenerative changes with features of excitotoxic active cell death [38]. Using a number of behavioral and cognitive tests as the radial arm maze and Morris water maze which assess hippocampus-dependent spatial learning and memory, it was demonstrated that multiple neonatal seizures in kindled rodents can be associated with reduced development of grey and white matter, lowered seizure threshold, memory impairment and significant long-lasting impairment of spatial learning, the severity of which correlates with the number of seizures experienced [7]. Assessment of functional properties of hippocampal circuitry in adult rats which experienced seizures induced by kainic acid on specific days during early postnatal development revealed presence of long-term loss of hippocampal plasticity manifested as reduced capacity of long-term potentiation, reduced susceptibility to kindling, and impaired spatial learning, which was associated with enhanced paired-pulse inhibition in the dentate gyrus [39]. Kindling-evoked seizure activity may thus prime synapses via calcium-dependent mechanisms, thereby affecting threshold, magnitude, and saturation of long-term plasticity at these synapses and this contributes to the alteration in memory performance and emotional behavior observed in TLE patients [40-42].

2.2.2. Neuropsychological Clinical Studies

Large, comprehensive, long-term and cross-sectional neuropsychological studies in TLE patients spanning observational periods as long as 25-30 years have documented that the cumulative cognitive impairments increase proportionally with the duration and poor control of seizures or its intractability to AEDs [42-44]. Children with severe, frequent and prolonged seizures, children with severe epileptic syndromes and those acquire seizure from a variety of initial precipitating insults early in life, tend to show early rapid, marked and progressive intellectual decline [42]. However, patients with TLE may also develop poor performance on tests of memory function, as well as on measures of intelligence, language, and executive functions suggesting that cognitive dysfunction is not limited to limbic-related tasks [44]. Rzezak et al. [45] found frontal lobe dysfunction in children with TLE. The worst performance in those with mesial TLE was associated with early onset, longer duration of disease, and use of polytherapy. The authors suggested that temporal lobe epileptogenic activity affects the extratemporal regions that mediate attentional and executive functions. Guimarães et al. [46] did a comprehensive neuropsychological assessment to a population of children with TLE including: IQ; forward digit; Trail Making Test for Children B; Wisconsin Card Sorting Test; block design; Boston naming test, verbal fluency and wide range assessment of memory and learning testing including visual learning, verbal memory, visual memory, delayed recall of verbal learning, delayed recall of stories and recognition of stories. The authors found that despite normal IQ, TLE presented with several neuropsychological deficits. The authors concluded that dysfunction of cerebral areas other than temporal lobe, particularly the frontal lobes, might be present in TLE. This is supported by the pronounced topographic organization of neocortical inputs along the septotemporal axis of the hippocampus documented in both anatomical and behavioral studies [47]. Moreover, recent studies signify the importance of cerebellum, basal ganglia and thalamus in cognition

[23,26,48]. The adverse consequences of AEDs, on cognition and behavior must not be ignored when evaluating the results of neuropsychological studies [49].

2.2.3. Neuroimaging Studies

A growing number of multi-parametric MRI follow-up and prospective longitudinal imaging studies in TLE indicate that progressive atrophy after the first SE evolves over a prolonged period of time (weeks, months, or even years) in the hippocampus, amygdala, thalamus and piriform cortex [50]. Some cross-sectional cohort studies reported association of cognitive deficits with smaller hippocampal volume in TLE [51]. Neuroimaging clinical studies also provide evidence that hippocampal volume reduction has been linked with the longer duration of epilepsy and considered a marker as well as a predictor of cognitive decline in patients with epilepsy. In the longitudinal study done by Briellmann et al. [37], 24 patients with mild TLE were studied via quantitative MRI over 3.5 years, the volume of the hippocampus decreased by 10%, the degree of volume loss was correlated with the number of generalized seizures. Over a similar period, Fuerst et al. [52] observed hippocampal volume loss in 12 patients with refractory TLE by volumetric MRI. Liu et al. [53] observed progressive cortical volume loss in patients with recurrent neocortical epilepsy. Recent quantitative MRI volumetric studies confirmed the presence of volumetric abnormalities in both temporal and extratemporal regions consistent with the generalized cognitive compromise associated with early-onset localization-related epilepsy syndromes as TLE. Abnormalities were identified in amygdale, fornix, entorhinal cortex, parahippocampus [54], thalamus and basal ganglia [55], cerebellum [51] and whole brain volumes [56]. Hermann et al. [56] reported reduction in total cerebral white matter volume, increased total CSF and reduced gray matter volume, both ipsilateral and contralateral to the side of temporal seizure onset. Seidenberg et al. [57] reported bilateral thalamic volume reduction in chronic unilateral temporal lobe epilepsy. Thalamic atrophy was significantly correlated with performance in memory and non-memory cognitive domains. In support, generalized reduction of cerebral volume has also been observed in children with mixed seizures, as well as focal temporal and frontal lobe epilepsy which are proportionately associated with delayed neurodevelopment [58]. Results of volumetric quantitative MRI studies are in accordance with generalized reduction in neurospsychological function including intelligence, language, visuoperception, memory and executive function in patients with TLE [56]. The extensive networks and interconnections between cortical regions are considered a contributing factor for the demonstrated widespread and remote cerebral atrophy from the putative epileptic focus. Functional MRI studies revealed that retrieval from working memory is associated with activation of dorsolateral frontal cortex. Other cortical and thalamic brain areas are also activated including the anterior cingulate cortex which is associated with executive function and the posterior parietal cortex which is associated with attention [59].

2.2.4 Neurophysiological Studies

The P300 component of event related potential (ERP) is impaired in patients with epilepsy and correlates with the degree of cognitive impairment encountered in neuropsychological testing [5]. P300 is considered as a "cognitive" neuroelectrical phenomenon because it is generated in psychological tasks when subjects attend and discriminate stimuli that differ from one another on some dimensions. P300 is an objective, non-invasive and clinically relevant method for evaluation of mental processing. P300 latency increases as the dementia symptoms increase, while P300 amplitude is depressed in all levels of dementia [60]. The hippocampus, thalamus and frontal cortex are possible locations of the P300 generators, structures important for learning and memory. Also the changes in electroencepahalography (EEG) peak frequency observed in quantitative occipital EEG are correlated with subjective cognitive complaints [61].

2.3. Factors Associated with Cognitive Deterioration in Patients with Epilepsy

Cognitive issues in epilepsy are associated with number of variables including genetics, basic brain lesion, type of epilepsy, site and side of brain lesion, etiology of epilepsy, age at onset, duration of epilepsy, seizure frequency and severity, ictal as well as interictal transient focal or long-lasting EEG epileptic discharges, adverse effects from antiepileptic medications [9,62-65] and psychosocial variables [66,67].

Genetics: hereditary predisposition to abnormal brain activity has been accounted for 30–50% of phenotypic IQ variance of children born to mothers with epilepsy [68,69]. Age at onset: cognitive and behavioral functioning in patients with epilepsy is an important area in various age groups. Earlier studies reported that seizure onset before the age of 14 years is a risk factor for cognitive decline. In controlled studies, significant neuropsychological impairment has been demonstrated in children and adolescents with chronic epilepsy [10]. However, recent studies indicated that negative effect on cognition, even progressive cognitive deterioration, may occur in older adults with chronic partial or generalized seizure disorders. Type, site and side of epilepsy: some studies found greater cognitive problems in patients with generalized than partial seizures [70], others vice versa [71]. Complaints of memory difficulties are common among patients with TLE where memory-related brain structures are directly involved by seizure activity. TLE is associated with more memory impairments than extratemporal epilepsies and both have more memory impairments than that associated with generalized epilepsy [72,73]. Frontal lobe epilepsy is associated with performance deficits in executive functioning [74]. However, most recent case-control and longitudinal studies revealed that patients with generalized as well as localization-related epilepsies may develop poor performance on tests of memory function, as well as on measures of intelligence, language, and executive functions suggesting that cognitive dysfunction is not limited to limbic-related tasks presented in the hippocampus, amygdala or the piriform cortex but extents to involve diverse brain areas [44,51,75]. The nature and localization of epilepsy is also important determinant of the extent and nature of cognitive deficits. Patients with secondarily generalized seizures showed greater impairment in concentration and mental flexibility than patients with complex partial seizures [76]. Problems of delayed recall of words were observed in newly diagnosed patients with partial seizures prior to medication [60]. TLE is associated with cognitive decline in confrontational naming, visual memory, verbal memory and motor speed [10]. TLE affects declarative memory systems, while non-declarative learning (e.g. procedural learning) appears more or less unaffected. Verbal oriented problems are specifically involved in left-sided epileptogenic foci. Left TLE, especially impairs verbal episodic memory (e.g. word list learning), long-term verbal associations, learning of semantically-related verbal information, speed of learning and delayed memory with deficits in consolidation of verbal information [10]. Visuoconstructive memory dysfunction has been found in patients with right TLE [72,73]. Silva et al. [73] found that epileptic patients with mesial temporal injuries had low cognitive performance in attentional span, memory, speech, daily problems resolution, while patients without injury showed more compensated cognitive performance except mild attentional alterations. Duration of epilepsy: cross-sectional and longitudinal studies of cognitive change in epilepsy suggest that longer duration of epilepsy is associated with decline in many areas of cognition [64]. Etiology of epilepsy: cognitive impairments are more prevalent in symptomatic and cryptogenic compared to idiopathic epilepsy [6]. Psychosocial variables: mood state in epileptic patients may be additional factor that negatively affects cognitive functions. Epileptic patients who are depressed may suffer a double burden of cognitive deficits [77]. Seizures occurring at school or work can result in poor self-perception and reduced social interaction. Stigma resulting from epilepsy and learning problems may lower the parental and teacher expectations. Decreased expectations can negatively affect the academic effort and consequently the performance. Scholastic underachievement, intellectual impairment, lower educational levels and potentially mental retardation are the long-term consequences in children with epilepsy, while low functional status, less educational levels, low rates of employment and poor quality of life are the long-term consequences in adults with epilepsy [66,67].

In some patients with epilepsy, many of the above factors are inter-correlated and independently contributed making it difficult to clearly delineate the relative contribution of any given factor (e.g. cognitive deficits in epilepsy occur regardless patients' age, type and duration of epilepsy or associated diseases).

2.4. The Pathophysiologic Mechanisms Underlying Cognitive Impairment in Epilepsy

The mechanism of cognitive impairment in epilepsy is complex. Negative effects on cognition may occur in presence or absence clinically manifest seizures, convulsive or nonconvulsive SE that occur during awakening or during sleep, and may occur due to focal or generalized EEG epileptic discharges without epileptic symptomatology [63]. Cognitive deficits associated with epilepsy and EEG epileptic discharges may be transient [78,79], persistent [80] or progressive [3,52].

a) The Mechanisms of Transient Epileptic Amnesia Associated with Ictal and Subictal Epileptic Activity

In transient epileptic amnesia, the main manifestation of seizures is recurrent episodes of amnesia. During manifest epilepsy and paroxysmal epileptic activity, transient disruption of cognitive processing is attributed to the following: a) the involvement of a neuronal circuitry in epileptic spiking rendering the same neurons unavailable for normal physiological processes, b) antidromic corticothtalmic backfiring, that would collide and annihilate any incoming information through orthodromic thalamocortical pathways, and 3) prolonged membrane hyperpolarization following paroxysmal depolarization shift mediated by recurrent postsynaptic inhibitory mechanisms which elelctrophysiologically correspond to the after-coming slow wave [81,82]. Presence of slow EEG activity in the same regions showing abundant spike wave has been interpreted as reflecting increased cortical inhibition mediated by hypersynchronous GABAergic inhibitory postsynaptic potentials. This increment of cortical inhibition might temporarily alter normal physiological processing of cognitive disruptions [83]. High seizure frequency disrupts the first encoding state of the memory process and specifically disrupts attention, concentration and working memory. However, in individual cognitive performance, even single seizures can generate long-term attentional slowing in the post-ictal period that exists for at least 24 hours. A single tonic-clonic generalized seizure may have a lasting negative effect on attention for about 30 days [70].

There may be an association between transient cognitive deficits and transient EEG epileptiform (generalized or focal) discharges that are not accompanied by obvious clinical events. This phenomenon is known as a state of transient cognitive impairment (TCI) [78]. It is found in about 50% of patients and regarded as subclinical or interictal [84]. These brief subclinical EEG paroxysms or TCI may cause deficits that usually pass unrecognized by standard memory tests, however, sensitive methods of observation as continuous psychological testing, commonly show brief episodes of impaired cognitive function during such discharges. TCI may adversely affect the patient's psychosocial functioning in daily life as educational skill, learning tasks, attention, behavior, sleep disruption and motor dysfunction [63]. TCI was first demonstrated during 3 cycles/second generalized spike-and-wave discharges [78]. Sirén et al. [85] found that the duration of generalized 3-Hz spike-wave discharges and clinical absence seizures was negatively correlated with performance on the visual memory tasks. TCI was also demonstrated in many cases of benign childhood epilepsy with centrotemporal spikes, a disorder once thought to have no adverse psychological effects [65]. TCI is not simple inattention. The effects of TCI are material and site specific, i.e. lateralized discharges are associated with deficits of functions mediated by the hemisphere in which the discharges occur (e.g. left-sided focal spiking frequently produces errors in verbal tasks, whereas right-sided discharges are often accompanied by impairment in handling nonverbal material). Conversely, specific tasks can activate or suppress focal discharges over the brain regions that mediate the cognitive activity in question. In patients with benign childhood epilepsy with centrotemporal spikes, deficits in IQ were found to be significantly correlated with the frequency of EEG spikes but not with the frequency of seizures [86]. Autistic features observed in come children with epilepsy have been suggested as a consequence of apparently subclinical spikes interfering with specific cerebral processes [87].

b) The Mechanisms of Cognitive Impairment Associated With Continuous Epileptic Activity during Sleep

Recently, the mechanism of cognitive impairment of some specific epileptic syndromes with continuous spikes and waves during sleep (CSWS) has been explored [80]. Landau-Kleffner syndrome (LKS) and the syndrome of continuous spike-and-wave discharges during slow sleep (CSWS) represent a spectrum of epileptic conditions which share many common features including: **1)** onset during childhood, **2)** deterioration of cognitive functions that were normally acquired in the past, **3)** continuous spike-and-wave discharges during slow wave sleep, **4)** pharmacological reactivity, **5)** regression of the neuropsychological symptoms when the EEG abnormalities improves (spontaneously or after drugs as corticosteroids) and **6)** absence of obvious structural lesion detected by CT or MRI scan [88,89]. The cognitive deficits of children with CSWS are long-lasting, present during months or years, and complete recovery is unusual. The pathophysiology of cognitive deficits CSWS and LKF is complex and different from that described with TCI as some patients with CSWS or LKS may have a completely normal awake EEG while cognitive deficits are present in the awake state when interictal epileptiform discharges are rare or absent. Recently, positron emission tomography (PET) studies using [18F]-fluorodeoxyglucose (FDG) during acute and recovery phases of CSWS in a group of children with epilepsy, showed that increased glucose metabolism at the epileptic focus was associated with hypometabolism in distant connected areas and both hypermetabolism and hypometabolism resolved at the recovery phase of CSWS [90,91]. Altered effective connectivity between focal hypermetabolism (centro-parietal regions and right fusiform gyrus) and widespread hypometabolism (prefrontal and orbitofrontal cortices, temporal lobes, left parietal cortex, precuneus and cerebellum) was found at the acute phase of CSWS and markedly regressed at recovery whether spontaneously or with corticosteroids [92]. The parietofrontal altered connectivity observed in patients with hypermetabolism is interpreted as a phenomenon of remote inhibition of the frontal lobes induced by highly epileptogenic and hypermetabolic posterior cortex [90].

c) The Mechanisms of Progressive Cognitive Deterioration with Epilepsy

Many animal and human studies reported persistent and progressive cognitive decline and behavioral impairment in developing and mature brains with epilepsy [52]. In TLE, a specific stereotypical pattern of pathology occurs in the hippocampus, amygdale, entorhinal region, piriform cortex and mesdiodorsal thalamus, areas primarily involved in memory processing. In complex partial and generalized epilepsy, a characteristic pattern of hippocampal sclerosis occur [32] Loss of neural density in the left mesial temporal regions (i.e. CA3 of the hippocampus) and right hippocampal structures can explain the verbal and non-verbal memory impairments in patients with epilepsy [35]. Recent research data indicate that in TLE epilepsy, the loss of hilar cells of the dentate gyrus may be the underlying cause for the lowered seizure threshold [92], whereas damage to hippocampal principal cells is often associated with memory impairment in animals [93] and humans [94]. Electro-physiological studies of the rodent hippocampus show that repeated seizure activity has a profound, deleterious effect on an important form of synaptic plasticity (LTP) which has been suggested to underlie memory formation. Long-term synaptic plasticity (e.g. long-term potentiation or LTP and long-term depression or LTD) is though to be an important cellular mechanism for learning and memory [95]. Long-term potentiation (LTP) is defined as a persistent increase in the efficacy of synaptic connections induced by high frequency stimulation, while in long-term depression, a transient decrease in synaptic strength caused by repetitive stimulation of the presynaptic neuron. Most experimental and human studies of long-term synaptic plasticity have focused on the hippocampus, a structure related to learning and memory [18]. The induction of LTP and LTD requires the appropriate integration of GABAergic inhibitory and glutaminergic excitatory transmission [96]. Human neuro-psychological studies indicates that hippocampal NMDA receptors are necessary for mediating repetition/recognition effects of limbic event-related potentials to continuous word recognition paradigms as well as for intact verbal memory performance [97]. In LTP, a train of postsynaptic potential (PSP) continues to come in, due to the high frequency stimulation, **the NMDA receptor's channel is kept open causing Na^+, K^+ and Ca^{++} flow in through this channel and the Mg^{++} blockade is relieved.** Ca^{++} inside the postsynaptic neuron triggers Ca^{++} dependent kinases (protein kinase C and Ca^{++}/calmodulin kinase II) that induce LTP. This mechanism plays also an important role in memory formation and in epileptogenesis. In LTD, the slow process of building of the synaptic vesicle causes decrease in the neurotransmitter release and decrease in post-synaptic potentiation [98]. Repeated seizure activity has been found to incrementally cause an indiscriminate and widespread induction of LTP, consuming and reducing overall hippocampal plasticity available for information processing [99].

3. Cognitive States with Antiepileptic Medications

3.1. The Differential Effect of AEDs on Cognition

Differential effects on cognition are seen with various AEDs. Carbamazepine (CBZ) [100,101], phenytoin (PHT) [102] and valproate (VPA) [103] can adversely affect cognition to a similar extent and appear to be less than that of barbiturates (PB) and benzodiazepines (BZ) [104,105]. The limited studies done to detect the effect of new AEDs on cognition revealed that topirmate (TPM) reported to have the worst effect on cognition [106] and zonisamide (ZNS) has mild adverse effect on cognition [107] while gabapentin (GBP) [8,107,109], lamotrigine (LTG) [110-112], tiagabine (TGB) [113,114], vigabatrin (VGB) [115,116] and levetiracetam (LEV) [117,118] have no adverse effect on cognition. However, even their modest effects can be clinically significant **and impact the patient's quality** of life. Increased doses of AEDs, rapid initiation and polytherapy entail an increased risk. In general, the cognitive effects of AEDs are less than the sum total of other factors and are usually reversible. Conversion of polytherapy to monotherapy may consequently improve cognitive functioning [119].

Phenobarbital (PB) and Benzodiazepine (BZ)

Animal and human studies confirmed the deleterious effect of PB and BZ on cognition compared to other conventional and new AEDs. The offsprings of pregnant mice treated with PB demonstrated more hyperactivity, less rapid habituation, impaired performance in operant behavior, impaired performance in repeated acquisition task and a conditioned avoidance task compared to control offsprings [120],. Adult rats exposed to PB demonstrated deficits in hippocampal 8-arm maze, spontaneous alternations, and water maze performance [121]. In the largest prospective study done by Shapiro et al. [104] on a large number of children exposed to PB monotherapy in-utero (the number of children exposed to PB with mothers of epilepsy was 35 while 4705 of exposed children had mothers without epilepsy) demonstrated that the latter group did not differ from control children with respect to IQ measured at 4 years of age. The study done by Reinisch et al. [105] on 114 male offspring demonstrated that the effect of PB exposure occurred if maternal treatment lasted for at least 10 days during pregnancy. The authors reported reduced verbal IQ scores (~7 IQ points) in two cohorts of men exposed in-utero. PB and BZ are known to impair cognition in healthy volunteers and patients with epilepsy [9]. Children on PB demonstrated low IQ that was improved with discontinuation of PB [122]. In the study of Farwell et al. [122], the long-term cognitive effects of early postnatal PB exposure was investigated in a randomized, placebo-controlled, blinded study with 217 toddler-aged children having febrile seizures, in which they were randomized to receive either PB (4–5 mg/kg/day) or placebo. Children were examined at age 7, several years after discontinuation of PB, 64% of these children were examined with the Wide Range Achievement Test (WRAT-R) and the Stanford–Binet Intelligence Scale.

Compared with the placebo group, PB exposed children were found to have significantly impaired performance in WRAT-R reading scores, but not in the Stanford–Binet Scale. Sulzbacher et al. [123] demonstrated that the deleterious effect of long-term use of PB remained several years after drug discontinuation when children were tested for cognition 3-5 years later. This suggests that there is persistent complex effect of PB on developmental maturation in addition to interfering with acquired cognitive function.

Phenytoin (PHT)

PHT has shown mild adverse effect on cognition [124,125]. Limited animal studies revealed that PHT resulted in reduced brain weight, impaired startle responses, hyperactivity, alter neuronal membranes in the hippocampus, delay neurodevelopment and impair special memory and motor coordination when given to rats' mothers during pregnancy. The AED-induced dysfunction in rats is related to both the dose and the duration of PHT exposure [104,126]. Shapiro et al. [104] reported reduced IQ results (five points lower than that of control children of mothers without epilepsy) in a cohort on cognition and PHT. To some extent, cerebellar atrophy a known association with chronic exposure or toxicity from PHT may be responsible for cognitive impairment in patients with epilepsy. It has been found that MRI cerebellar volumetric change is proportionately correlated with the chronic exposure or toxicity from PHT [48] Purkinje cells of molecular layer were found to show selective vulnerability to both the excitotoxic and/or toxic PHT models of cerebellar atrophy [32].

Carbamazepine (CBZ) and Oxcarbazepine (OXC)

CBZ utilization may be associated with mild cognitive dysfunction including excessive sedation, compromise in attention, concentration, visual motor coordination and psychomotor slowing [127]. Small risk in learning and memory has been registered especially with high serum concentrations of the drug. Some children are at high risk for developing cognitive side effects due to CBZ [100]. EEG slowing associated with CBZ might be significantly related to the magnitude of cognitive decline on later IQ subset performance [101].

OXC is an analogue of CBZ, with a comparable anticonvulsant efficacy but has better cognitive profile compared to CBZ. It has both neuropsychological impairment and EEG slowing in healthy volunteers but of less magnitude compared to CBZ [128,129].

Valproate (VPA)

Behavioral studies demonstrated that prenatal VPA exposure decreases locomotor activity and increases swimming maze errors in rats tested by the 8-arm radial maze and passive avoidance test [130]. Studies revealed that children exposed to VPA in-utero had learning difficulties, behavioral problems, increased need for special education [103]. In one large prospective study, increased memory deficits, reduced verbal IQ by 8-15 points and excess of additional needs were reported in children exposed to VPA prenatally [131]. Studies in healthy volunteers revealed that VPA may produce a modest but statistically significant cognitive disruption [9].

Topiramate (TPM)

TPM receives greatest concern among the new AEDs due to its documented worst cognitive profile. Its cognitive adverse events are reported in 10-20% [132,133]. The symptoms of cognitive deficits associated with TPM include: concentration/attention difficulty, confusion, abnormal thinking, slow thoughts, dull thinking, mental slowing, blunted mental reactions, word finding difficulties, calculation difficulties and memory impairment. The greatest changes were found in verbal IQ, verbal fluency, verbal learning and digit span [106]. Martin et al. [8] observed that among healthy young adults, the negative effects of TPM on measures of attention, word fluency, verbal memory and psychomotor speed, were greater than those with LTG and GBP when tested 3 hours after large initial doses and its effect persisted for 2- and 4-weeks intervals. Leonard et al. [134] found that motor tasks were affected with TPM as observed by bimanual sequential tapping. This is one of the motor measures that require the most cognitive processing, as patients must tap in a different specific sequential order with the two hands simultaneously. The task involves attention, perception and the capacity to monitor and coordinate out-of-phase movement. Functional MRI and cognitive testing revealed disruption of information processing in prefrontal cortex and more heterogeneous patterns of cortical activation with TPM [135]. Lee et al. [136] observed improvement in both verbal and non-verbal fluency scores by ≥70% after TPM discontinuation. Kockelmann et al. [137] and Huppertz et al. [138] reported significant improvement in performance on tests of verbal fluency, verbal working memory, spatial short-term memory and attentional functions after withdrawal of TPM. Whether TPM side effects are dose-dependent and if they critically depend on the speed of drug titration, are matters of debate [137]. It has been suggested that low starting dose, slow upward drug titration and reduction of polytherapy will control seizures as well as produce tolerance to the drug with minimal cognitive side effects [106,119,136]. In the longitudinal study done by Thompson et al. [106], the authors demonstrated deterioration in verbal IQ, verbal fluency and verbal learning following introduction of higher doses of TPM (150-600mg/d) as adjunctive therapy in patients with epilepsy with improvement in verbal fluency, verbal learning, and digit span occurred when TPM was reduced or withdrawn. Reife et al. [133] observed psychomotor slowing with lower dosage (200mg/day) of TPM and language disturbances in higher doses.

In contrast, recent animal studies have suggested some neuroprotective effects for TPM on cognition. Zhao and colleagues [139] administered TPM or saline during and following series of seizures in 25 neonatal rats. After completion of the TPM treatment, rats treated with TPM performed better in the water maze than rats treated with saline. Koh et al. [140] used a "two-hit" rodent seizure model to study the therapeutic efficacy of a postseizure treatment with TPM in reversing the perinatal hypoxia on later Kainite seizure-induced neuronal damage. The authors observed that repeated administration of TPM given for 48 hours after hypoxia-induced seizures prevented the increased hippocampal neuronal injury induced by Kainite.

Zonisamide (ZNS)

Studies on the effect of ZNS were limited and controversial. Weatherly et al. [141] reported little cognitive decline in some patients with epilepsy on ZNS as add-on therapy. Recently, Park et al. [107] in a prospective randomized and open-labelled study, observed that after one year of starting treatment with ZNS (received as monotherapy in a dose of 100, 200, 300, and 400mg/day), although ZNS decreased seizure frequency and EEG abnormalities, however, mood changes and cognitive deficits were observed in 15% and 47% of patients and were dose related. Cognitive performance was worse on delayed word recall, Trail Making Test Part B and verbal fluency.

3.2. The Known and Hypothesized Mechanisms of Cognitive Impairment with AEDs

In general, the mechanisms of AEDs are to reduce neuronal irritability and increase postsynaptic inhibition or alter synchronization of neural networks to decrease excessive neuronal excitability associated with seizure development and secondary spread of epileptic activity to the surrounding normal brain. AEDs modulate brain activity through their action on voltage dependent ion channels [sodium or Na^+ and low threshold (T-type) calcium or Ca^{++} channels], inhibitory [GABA] and excitatory neurotransmitters and their receptors [142]. However, slowed motor and psychomotor speeds, poor attention and memory processing are common side effects of sodium channel blockade [144,145], increasing GABAergic inhibitory activity and decreasing neuronal excitability [145].

Sodium channel blockade decreases the release of neurotransmitters including excitatory neurotransmitters and contravenes depolarization by interfering with propagation of action potentials i.e. limitation of sustained repetitive firing and stabilizing neuronal membranes [143]. In hippocampal neurons, intracellular increased Na^+ increases the probability to open NMDA receptors and thus might control excitatory synaptic transmission [144]. The cognitive side effects of some AEDs have been attributed to their Na^+ channels blockade activity including: PHT, CBZ and OXC and VPA [143,145].

Increasing GABAergic brain activity by AEDs results in re-establishment of the background level of inhibition and helping the return of the nervous system to its normal balance between excitation and inhibition. Drugs that increase the extracellular levels of brain GABA or mimic GABA transmission are widely used in the treatment of epilepsy in children and adults including PB, BZ, VPA, GBP, TPM, ZNS and VGB [142,145]. In general, the mechanisms of enhancing GABA-mediated inhibition are: 1) Direct modulation of GABAergic neurotransmission induced by allosteric modulation of GABA receptors through direct binding to the receptors and changing its shape or configuration, hence increasing GABAergic inhibitory neurotransmission and inhibitory postsynaptic potential. PB directly binds to $GABA_A$/benzodiazepine receptor complexes (GBRs) while BZ has a unique receptor site on $GABA/Cl^-$ channel complex and are potent GABA agonists, and 2) Indirect modulation of GABAergic neurotransmission through enhancing synthesis and/or decreasing reuptake of GABA.

Drugs that mediate indirect modulation of GABAergic neurotransmission include: VPA, GBP, VGB, TGB, TPM, ZNS. VPA do not directly interact with GABA receptors, but they increase brain levels of GABA, possibly by enhancing glutamate decarboxylase or inhibiting GABA transaminase [146]. GBP is a structural analogue of the inhibitory neurotransmitter GABA. Nuclear Magnetic Resonance (NMR) Spectroscopy indicates that GBP enhances the levels of GABA in the brain and decreases brain glutamate concentration. Its actions include modulation of GABA synthesizing enzyme, glutamic acid decarboxylase (GAD) and glutamate synthesizing enzyme, branched-chain amino acid transaminase [147]. VGB is an AED that elevates brain GABA several-fold by irreversibly inhibiting the GABA-metabolizing enzyme, GABA-transaminase [115]. TGB is selective GABA reuptake inhibitor. It induces its effect by increasing synaptic GABA availability via selective inhibition of the GAT-1 GABA transporter [148]. TPM rapidly raises brain GABA [149]. Mental slowing, memory impairment, inattention and language dysfunction are suggested to be due to increasing GABAergic activity in the prefrontal cortex caused by TPM [135].

Drugs that directly increase GABAergic inhibitory neurotransmission and inhibitory postsynaptic potential as PB and BZ produce significant disruption of short-term memory and attention [5]. Indirect modulation of GABA neurotransmissions have modest (e.g. VPA) or even little effect on cognition than direct GABA modulation, i.e. GBP, VGB, TGB and LEV have good neuropsychological profile or little or negligible effect on cognition.

Decrement in glutamate-mediated excitation as by antagonizing the response mediated by NMDA or AMPA/KA subtype of glutamate receptors are believed to underlie the mechanism of some AEDs as TPM [150]. Slowed motor and psychomotor speeds, poor attention and memory processing are adverse effects of reduced neuronal excitability as well as increasing GABAergic inhibitory activity on the brain.

Some AEDs can cause cognitive deficits through multiple mechanisms. For example: VPA modest effect on cognition has been attributed to 1) indirect modulation of GABA neurotransmissions, 2) An enhancement of $GABA_A$ receptor-mediated hyperpolarizing responses caused by VPA will inhibit the activation of NMDA receptors in a dose-related manner [151-153]. It suppressive effect on synaptic response mediated by NMDA receptors may contribute to the impairment of LTP and LTD caused by VPA [154]. Lee et al. [136] reported that VPA suppresses the expression of LTP in the CA1 region of the hippocampal slices. Gean et al. [154] reported that in rat amygdaloid slices, VPA suppresses response mediated by NMDA receptors in dose related manner, and 3) VPA reduces repetitive neuronal firing via blockade of voltage-dependant sodium (Na^+) channels [146].

Several mechanism(s) has been suggested to explain the adverse effect of TPM on cognition. TPM increases cortical GABA notably in the frontal lobes. GABAergic dysfunction in the prefrontal cortex could lead to mental slowing and subsequent spread of this dysfunction to the dorsolateral areas including Broca's area, could underlie impairments of language production with TPM [132]. Clinical studies demonstrated that TPM adverse effects on cognition including impaired concentration and memory, slowed thinking, and word finding difficulties have been attributed to its inhibitory effect on carbonic anhydrase isoenzymes II and IV, causing increased magnesium-dependent tonic inhibition of NMDA receptors and apoptosis [155]. TPM also has negative modulatory effect on the AMPA/KA subtype of glutamate receptors [149,155]. In addition, TPM blocks Na^+ channels [156].

No studies were done to explore the mechanisms of cognitive deficits observed with Zonisamide (ZNS), however one can speculate that its blockage effect on voltage-sensitive Na^+ channels and modulatory effect on GABA-mediated neurotransmission inhibition may explain the cognitive impairment elicited by ZNS [144].

3.3. Mechanisms of the Beneficial Effect of AEDs on Cognition

Some AEDs are able to improve cognition through mechanisms other than their anticonvulsant effect. In general, the beneficial effects of AEDs on cognition are due to: 1) reduction of seizure activity, 2) modulating effect on neurotransmitters, lowering excitotoxicity associated with a reduction in glutamate release from presynaptic terminals and by preventing anoxic depolarization capacities, 3) inhibition calcium-mediated cellular functions (protein phosphorylation and neurotransmitter release) and calcium-dependent depolarization 4) scavenging of free radicals, and 5) their psychotrophic effect.

Several studies provide information about the beneficial effect of PHT on cognition. In support: 1) PHT was reported to prevent stress and corticosterone-induced reductions in CA3 apical dendritic length and branch point numbers [157], 2) PHT may reverse stress-induced impairment of spatial learning and hippocampal atrophy [158], 3) PHT could keep LTP, an important component of memory, from being inhibited by stress [159], 4) In vivo, PHT decreased the dimension of cerebral infarct in animals with bilateral or unilateral carotid occlusion [160], and 5) Recently, PHT was found to be associated with increased hippocampal volume in patients with post-traumatic stress disorder (PTSD) assessed by MRI and this was associated with improvement in hippocampal-based verbal declarative memory function [161]. PHT has effect on glutamatergic function which affects brain structure. PHT arrests calcium-mediated cellular functions and calcium-dependent depolarization, both associated with neuronal death. It blocks cellular responses to excitatory amino acids. PHT antagonizes glutamate-induced excitation of cerebrocortical neurons and blocks the effect of glutamate at NMDA receptors [159,161,162].

Several studies demonstrated that CBZ does not compromise and even improves the learning performance of non-epileptic animals in different learning and memory tasks [163,164]. Rostock et al. [164] reported that administration of low does of CBZ was able to reverse amnesia induced by electroconvulsive shock as well as to improve learning during an active avoidance test treated with repeated doses with ethanol. CBZ affords significant protection against glutamate neurotoxicity in hippocampal cell cultures and reduces NMDA-mediated brain injury. It inhibits KA-induced Ca^{++} ion elevation [165]. Ambrosio et al. [166] suggested that at concentrations that do not cause toxicity, CBZ has a neuroprotective effect on KA-induced toxicity in hippocampal neurons which is essentially mediated by the activation of AMPA receptors. CBZ increases brain acetylcholine level in the hippocampal structure and simultaneously reduces choline level. The role of cholinergic function in memory and related cognitive processes is well known. Deterioration of cholinergic neurons in the medial septal nucleus that project to the hippocampus, amygdala and cortex (e.g. critical memory areas) was demonstrated in rat models of epilepsy [166]. In support: acetylcholine esterase inhibitors are shown to improve memory functioning in diverse neurological conditions [167]. The hippocampus, a cerebral structure highly involved in learning and memory, is a target for abundant cholinergic innervation and hippocampal nicotinic acetylcholine receptors that modulate synaptic plasticity via mechanisms involved in LTP [168]. The normothymic effect (effect on mood) of CBZ may be related to its impact on the neurotransmitter systems (GABA-ergic, serotoninergic, noradrenergic or adenosine and additionally demonstrate G-protein or inositol phosphate modulating effects [169].

GBP may also promote an improved mood and sense of well-being independent of seizure reduction and hence improvement of cognitive functions [108]. Dimond et al. [170] demonstrated increases in ratings of quality of life and well-being when patients were switched to this drug. Harden et al. [108] demonstrated significant reduction in depressive scores on a dysthymia rating scale in patients receiving GBP independent of seizure reduction.

Although, no studies were done to explore the mechanisms of beneficial effect of LTG on cognition, however, the neuroprotective effect of LTG may be attributed to lowering excitotoxicity mainly in the hilus and the CA3 subfield of the hippocampus as well as the piriform cortex [169,171].

Possible mechanisms underlying mechanisms by which LEV improves cognitive function and quality of life (QOL) remain unknown. LEV is postulated to inhibit seizure activity through a totally different mechanism. LEV seems to partially inhibit N-type high-voltage–activated Ca^{++} currents and reduces the Ca^{++} release from intraneuronal stores. It also reverses inhibition of GABA and glycine gated currents induced by negative allosteric modulators, and effects voltage gated potassium channel conductance. LEV also has a specific stereoselective binding site in the CNS at the synaptic vesicle protein 2A (SV2A) [172]. LEV can reduce neuronal necrosis and maintain LTP in the hippocampus [173], which may also contribute to its effects on cognition. Piracetam and its derivative LEV belong to the pyrrolidine class; drugs in this class can protect against brain insults and have low toxicity [174]. They might enhance the efficacy of higher integration mechanisms in the brain and improve mental function such as learning and memory, while protecting against seizures. Piracetam seems to improve learning, memory, and attention [175] and has been used to treat age-related cognitive disturbances and aphasia [176]. Hence it is reasonable to assume that LEV may also influence the metabolism of some frontal areas leading to improved cognitive function [118]. Zhou et al. [177] evaluated the effect of LEV as an add-on treatment, on cognitive function and QOL in patients with refractory partial seizures. Their study comprised two phases: a) a short-term phase (randomized, double-blind, placebo-controlled design) for 8-week baseline period, 4-week titration interval, and 12-week period at the maximum LEV dose (1500 mg twice daily), and b) a long-term phase (an open-label study) in which the maximum LEV dose was administered for another 24 weeks. After short-term LEV treatment, performance time on the Wisconsin Card Sorting Test (WCST) and Delayed Logic Memory were significantly improved for the patient but not the control group. Subscale scores on the QOLIE-31, including scores on cognitive functioning and social function, were also improved only with LEV group. At the end of the long-term phase, these improvements were maintained, and both groups performed better in more areas, as measured by the Trail Making Test, WCST, Delayed Visual Memory and the QOLIE-31 subscales.

ZNS decreases secretion of excitatory amino acids and reduces post anoxic depolarization as well as scavenges of free radicals including hydroxyl and nitric oxide radicals and these effects are contributed to its neuroprotective effect against cognitive impairment [178].

4. The Vulnerability of Immature Brains to the Cognitive Adverse Consequences of Epilepsy and its Medications

Normally, biological development and organization of the brain in human is very rapid in-utero and start to slow down in the second year of postnatal life [179]. Although gross organization is nearly complete by 2 or 3 years of age, maturation may continue through adolescence and beyond [180]. The period of infancy is characterized by peak hippocampal and cortical regional development, as well as myelinogenesis, dendritogenesis, and synaptogenesis in the brain and changes in these processes underlie deficits in spatial learning and memory processes [181]. Many of the human studies on cognition and behavior have focused on infants, preschool, and school-age children. There is a developmental component to the relation between poor seizure control and mental performance. The presence of epilepsy and its treatment during a period of maximal white-matter growth could affect development of white matter.

4.1. The Vulnerability of Immature Brains to Seizures and its Consequences

Systematic studies indicated that neuronal computation of immature brain is different from mature brain and the vulnerability of the brain to seizures and its consequences is age-specific. Compared to adult brains, experimental studies showed that immature brains are highly prone to develop seizures and SE, but are more resistant to seizure and SE induced damage. Even if epilepsy occurs, it may not be intractable to treatment. In the developing brain, NMDA receptor subunits create populations of receptors that flux more Ca^{++}, open more easily and block less frequently than mature forms, making the immature brains more electrically excitable with increased susceptibility to develop SE compared to mature brains [182,183]. Previous and recent studies indicated that structural, functional, and neurochemical changes occur after brief and prolonged seizures in immature brains. However, not all of them are necessarily detrimental to brain function and the immature brain appears to be resistant to seizure-induced neuronal damage and neurogenesis [184,185]. Little or no neuronal damage in animal models of kindling and SE was observed in animals younger than 3 weeks. Also, synaptic reorganization did not occur until after the third postnatal week. Repeated episodes of pilocarpine-induced SE during post-natal days 7–9 resulted in abnormal distribution of neocortical inter-neurons and reduction of natural apoptosis in rats at post-natal day 35 and there was no hippocampal damage [186]. Occasional hippocampal damage can be detected in rats from the second post-natal week onwards but the pattern was different from that in adult rats. The damage might not be related to the seizure itself but to stress associated with the seizure [187].

Studies examined the effects of electroconvulsive-induced seizures in rats at various developmental stages revealed that in early development, seizures selectively impaired myelin accumulation of proportion to their effect on brain growth [188]. In epileptic rats, examination of cerebroside and proteolipid protein, relatively specific myelin lipids, was found to be reduced by about 11-13% in immature rats [189]. Executive functions, mainly under frontal lobe control, seem to be particularly vulnerable to epileptic EEG activity during the period of maturation; their disruption possibly interferes with the normal development of learning processes [124]. Adults rats experiencing kainic acid-induced seizures on specific days during early postnatal development revealed the presence of a long-term loss of hippocampal plasticity as manifested by reduced capacity in LTP, which has been suggested to underlie memory formation, reduced susceptibility to kindling and impaired special learning [39].

The Proposed mechanisms for the relative resistance of immature hippocampus to repeated seizures and SE-induced damage has been attributed to the fact that in immature brain there are: 1) preservation of GABA synthesis which declines with maturation [190], 2) increased expression of $GABA_A$ receptor-1 subunit in contrast to adults [191], 3) presence of mitochondrial uncoupling protein [192], 4) absence of mitochondrial oxidative stress [193], 5) absence of glia activation and cytokine production [194], 6) absence of GluR2 down-regulation or even up-regulation of GluR2 and down-regulation of GluR3 receptor subunits [195], and 7) presence of higher expression of growth factors and neurotrophins (as BDNF) [196].

4.2. The Vulnerability of Immature Brain to the Cognitive Adverse Consequences of AED

The effects of in-utero exposure to AEDs are increasingly being investigated and differential drug risk is considered for both anatomic and cognitive outcomes. Although information on the role of fetal and postnatal exposure to AEDs is limited in humans, there is a growing body of information from animals suggesting that AEDs may have substantial effects on brain development. The pathophysiologic mechanisms responsible for these deficits remain largely unknown, however, there is evidence that AEDs can adversely affect neuronal proliferation and migration and increase apoptosis [197-201].

Normally, neuroblast migration is influenced by crucially promoting signals (motility, acceleratory and stop signals) from GABA and glutamate neurotransmittors that act on several receptors subtypes (GABA$_A$, GABA$_B$ and NMDA). Neuronal migration may be influenced not only by genetic alterations but also by drug intake. GABA$_A$ agonists are frequently used in mothers with epilepsy or their offspring as sedatives and anticonvulsants. Neuronal migration can adversely be affected by AEDs [201]. Manent and colleagues [201] reported occurrence of hippocampal and cortical dysplasias in rat pups exposed to VGB and VPA in-utero. The authors found that prenatal exposure to VGB (200 mgkg/day), and VPA (100 mg/kg/day) from embryonic days 14 to 19 (in doses which are similar to those used clinically) resulted in neuronal migration defect and neuronal death observed postnatally in the form of hippocampal and cortical dysplasias. These effects were not found with CBZ (20 mg/kg/day).

There may be a relationship between AED-induced apoptosis and cognitive function. Recent discovery of neuronal apoptosis following in-utero AED exposure in animals during a period that corresponds to the third trimester and early infancy in humans raises further concerns. Utilization of PB to rat pups results in significant decreases in brain weight and DNA, RNA, protein, and cholesterol concentrations and reduced neuronal number [120,197]. Chronic exposure of cultured mouse spinal cord neurons to PB leads to reduced cell survival and decreased length and number of dendrite branches. Brain concentrations of dopamine and norepinephrine were reduced and the uptake of dopamine, norepinephrine, serotonin, and GABA into synaptosomal preparations of brain tissue was greater for offspring of pregnant mice treated with PB [198]. Studies demonstrate that PHT, PB, BZ, VGB and VPA cause wide spread apoptotic neurodegeneration in the developing rat brain at plasma concentrations relevant for seizure control in humans. In these studies, the drugs were administered to the fetus or rat pup during a period of intense synaptogenesis [199,200]. Jevtovic-Todorovic and colleagues [202] observed widespread apoptotic neurodegeneration and impaired long-term potentiation in hippocampal slices obtained from 7-day-old infant rats 3 weeks administered midazolam, nitrous oxide, and isoflurane. Persistent deficits in memory and learning were demonstrated when the rats were tested subsequently during the Morris water maze or the radial arm maze. In contrast, similar apoptotic effects were not seen at therapeutic dosages for CBZ, LEV, LTG, or TPM in monotherapy dosages [203-205]. However, preliminary results suggest that CBZ, LTG, and TPM, but not LEV, may potentiate cell death when given in combination with pro-apoptotic agents such as other AEDs.

The apoptotic effect of some AEDs appears to result from reduced neurotrophins and protein kinases, which are important for neuronal survival. Postnatal VPA exposure suppresses the synthesis of the neurotrophins, BDNF and neurotrophin-3 (NT-3) and reduces the levels of survival-promoting proteins in the brain, which reflected an imbalance between neuroprotective and neurodestructive mechanisms in the brain [199]. Lee et al. [206] found that VPA suppressed protein kinase C (PKC) activity in both membrane and cytosol compartments in hippocampal slices. PKC is highly enriched in the brain and plays a major role in regulating both pre- and postsynaptic aspects of neurotransmission, including neuronal excitability, neurotransmitter release, and long-term alterations in gene expression and plasticity. PKC is critical for the induction of LTP and LTD [207]. There is also evidence from lab studies that blockage of NMDA receptors can increase neuronal apoptosis resulting in chronic behavioral, structural and molecular effects. Harris et al. [208] studied the long-term consequences of NK-801 (0.5 mg/kg), NMDA antagonist, in a group rats treated postnatally day 7. The authors observed reduced volume and neuronal number within the hippocampus and altered hippocampal NMDA receptor (NR1 subunit). The same treated adult rats with MK-801 developed prepulse inhibition deficits and increased locomotor activity. The same mechanism for neuronal apoptosis can be applied for AEDs that cause decrement in glutamate-mediated excitation by antagonizing the response mediated by NMDA or AMPA/KA subtype of glutamate receptors.

5. Clinical and Research Implications

a) The Nature, Timing and Course of Cognitive Progression with Epilepsy are of Considerable Concern as Follow

It has been suggested that the first year of life is a critical period for the subsequent development of intellectual abilities. Development of epilepsy in the first year of life is associated with high incidence of intellectual impairment (82.4%) [34]. This highlights early identification and proper evaluation. Neurocognitive deficits may be subtle. Children born to mothers may have subtle deficits that may not be identified for years following delivery. Some learning deficits may not be apparent until the teenage years. Subtle neurocognitive deficits may induce long-term consequences and significantly reduce the child's likelihood of achieving success in school and eventually reduce employment opportunities. Even static cognitive impairment in children and adolescents with epilepsy may have lifespan implications. Research in general population has shown that lower childhood intelligence level at age 11 is associated with the risk of adverse cognitive outcome decades later [209]. Also it is important to note that children born to mothers with epilepsy who appear to be functioning within the normal range may have disorders higher cortical function (as memory, attention, speech and language, abstract thinking, and executive control) that are not manifest themselves until the child reaches grade school. Thus relying solely on IQ is insensitive for assessing cognition. Careful sensitive neuropsychological tests that require appropriate professional input may reveal specific impairments [208]. Hence, comprehensive pre-treatment evaluation and judicious management of all factors that contribute to cognition, behavior and educational problems in epilepsy, are essential for optimal outcome. For old individuals with chronic epilepsy, additional risk factors that are associated with abnormal cognition in which many of these risk factors present as early as midlife. It is important to systematically identify and treat known modifiable risk factors in order to protect and promote cognition in older persons with chronic epilepsy [210].

b) Optimized Therapeutic Modalities

AEDs are the major therapeutic modality for control of seizures. Provided that treatment decision may have lifelong implications, the neurologists should be aware that the lack of cognitive side effects related to an AED should be relevant for their treatment decisions. For example, CBZ and LTG have been widely advocated for the treatment of women with epilepsy during pregnancy. Women on VPA should be discussed for gradual withdrawal of VPA and switch to one of the known safe AEDs during pregnancy. If there is no optimal seizure control with the alternative drugs, then alternative combination of low dose VPA with the additional drug might be acceptable. Also AEDs Levels should be monitored at least each trimester and for 2-3 months following delivery. Pharmacokinetic changes demand monitoring of free levels of highly protein bound AEDs to avoid confusion with increased seizures (or symptoms of toxicity) despite therapeutic total serum drug levels. If dose is increased during pregnancy, medication dose tapering should be anticipated in the postpartum period to avoid medication toxicity.

The argument stated that "the optimal treatment for epilepsy is not only to control seizures but also to reduce the risk or consequences of cognitive impairments and behavioral abnormalities as much as possible", also has to be made for decisions about epilepsy surgery. Preliminary findings indicate that postsurgical training improves memory deficits and encourage further research. Epilepsy surgery is an option for patients intractable for medical treatment with focal seizures that arise from noneloquent brain regions. Because the epileptogenic tissue that is resected is dysfunctional, seizures are reduced and the use of AEDs is reduced, the risk of significant cognitive decline is generally reduced. However, the risks of functional impairment due to tissue ablation need to be weighted carefully against the benefits of surgery on seizure control and overall functional state. Damage of functional tissue, low mental reserve capacity, and poor seizure outcome increase the risk for postsurgical memory impairment whereas functional release due to seizure freedom counteracts negative impact. Patients older than 40 years may be at increased risk of memory impairment postoperatively. Risk for verbal memory decline occurs with left anterior termporal lobectomy (ATL) while visuospatial memory impairment occurs with right ATL [211].

Furthermore, multidisciplinary management strategies in epilepsy have also to address the psychosocial variables: Proper neuropsychological assessment of the child will help the school personnel for planning the academic strategies. Strategies such as social skills training, educational, speech and language interventions, and psychopharmacotherapy are necessary. It is important to indicate epilepsy-specific optimistic orientation and the potential activities for overcoming stigma and increasing education and awareness related to epilepsy in community-based research studies [212,213].

c) Proper Assessment of the Neuropsychological Profile of the Available AEDs

Most neuropsychological drug effects have been incompletely described. A recent American Academy of Neurology (AAN) and Child Neurology Society (CNS) practice guidelines stated that behavioral and cognitive side effects need to be better evaluated especially for new AEDs and individual risks as well as group differences assessed on tests of cognition [214]. Further investigations of mediating factors such as serum concentrations, co-medication, and other potential risk factors are needed to enable appropriate targeting of treatment with the effective AED.

d) The Need to Design Studies that Address the Developmental Issues in Epilepsy Management

Experimental studies indicate that the immature brain responds differently to treatment than does the mature brain. Because AEDs have various adverse effects, which may interfere with normal developmental processes and affect cognitive functions, physicians must weigh up the adverse effects of more aggressive treatments with the benefits of complete seizure control. For example, it is known that glutamate plays key roles in successive steps of brain development. Interfering of glutamate receptors (as by NMDA blockers) may result in deleterious consequences in the developing brain [215], while antagonism of non-NMDA receptors (e.g., AMPA/KA) sites had shown neuroprotective and antiepileptogenic effect in immature as well as mature brains [140,216,217]. Furthermore, some AEDs which seem to be neutral in mature brains have shown neurotoxic effect and can exacerbate neuronal damage in some paradigms in immature brains (as VPA, PB, BZP and TGB) [200]. Large cohort and controlled prospective studies are necessary and should include a sufficient number of women and children exposed to newer AEDs. This demands provision of adequate information and counseling about drug treatment during childbearing years through epilepsy services programs. In addition, the basic mechanisms underlying AED-induced cognitive/behavioral teratogenesis need to be delineated through prospective clinical studies. Demonstrations in animals that AEDs can induce neuronal apoptosis in developing brains raise concern that similar adverse effects may occur in children exposed in-utero or in the neonatal period.

e) The Need for Clear Definitions of the Clinical Paradigm and Adequate Outcome Measures

Assigning cause or effect, or detecting positive therapeutic impact in 'complex systems' of neural circuitry is critical because the aims are not necessarily equivalent. For example, it is difficult to conclude that administration of AEDs during the acute or latency phase would have an effect on the molecular cascades underlying epileptogenesis or predict the risk for later adverse consequences on cognition or behavior. Thus, it is necessary to design specific, sensitive, long-term outcome measures to assess cognition in epilepsy.

References

[1] Hauser, W.A. Incidence and prevalence. In: Engel, J., Pedley, T.A., eds. Epilepsy: a comprehensive textbook. Philadelphia: Lippincott-Raven, 1997, pp 47-57

[2] Elger, C.E., Helmstaedter, C., Kurthen, M. Chronic epilepsy and cognition. *Lancet Neurol.* 2004;3:663-672.

[3] Hamed, S.A. Neuronal Plasticity: Implications in Epilepsy Progression and Management. *Drug Develop. Res.* 2007;68(8):498-511.

[4] Hamed, S.A. The Aspects and Mechanisms of Cognitive Alterations in Epilepsy: The role of antiepileptic medications. *C.N.S. Neurosci. Therap.* 2009: 2009; 15:134-156.

[5] Gokcay, A., Celebisoy, N., Gokcay, F., Atac, C. Cognitive functions evaluated by P300 and visual and auditory number assays in children with childhood epilepsy with occipital paroxysms (CEOP). *Seizure* 2006;15(1):22-27.

[6] Kälviäinen, R., Äikiä, M., Helkala, E.L., Mervaala, E., Riekkinen, P.J. Memory and attention in newly diagnosed epileptic seizure disorder. *Seizure* 1992;1:255-262.

[7] Kotloski, R., Lynch, M., Lauersdorf, S., Sutula, T. Repeated brief seizures induce progressive hippocampal neuron loss and memory deficits. *Prog. Brain. Res.* 2002; 135:95-110.

[8] Martin, R., Kuzniecky, R., Ho, S., Hetherington, H., Pan, J., Sinclair, K., Gilliam, F., Faught, E. Cognitive effects of topiramate, gabapentin, and lamotrigine in healthy young adults. *Neurology* 1999;52:321-327.

[9] Meador, K.J. Cognitive effects of epilepsy and of antiepileptic medications. In: Wyllie E, editor. The treatment of epilepsy: principles and practices. Philadelphia: Lippincott Williams and Wilkins; 2005, pp. 1185–1195.

[10] Hermann, B.P., Seidenberg, M., Bell, B. The neurodevelopmental impact of childhood onset temporal lobe epilepsy on brain structure and function and the risk of progressive cognitive effects. *Prog. Brain Res.* 2002;135:429–438.

[11] Vinten, J., Adab, N., Kini, U., Gorry, J., Gregg, J., Baker, G.A., Liverpool and Manchester Neurodevelopment Study Group. Neuropsychological effects of exposure to anticonvulsant medication in utero. *Neurology* 2005;64:949–954.

[12] Piazzini A, Canevini MP, Turner K, Chifari R, Canger (2006) Elderly people and epilepsy: cognitive function. *Epilepsia* 47 (Suppl 5):82-84.

[13] Titze, K., Koch, S., Helge, H., Lehmkuhl, U., Rauh, H., Steinhausen, H.C. Prenatal and family risks of children born to mothers with epilepsy: effects on cognitive development. *Dev. Med. Child Neurol.* 2008;50(2):117-122.

[14] Guilford, J.P. The nature of human intelligence. New York, McGraw-Hill. 1967.

[15] Sparrow, S.S. Recent advances in assessment of intelligence and cognition. *J. child Psychology and Psychiatry* 2000;41(1):117-131.

[16] Sternberg, R.J. Reasoning, problem slowing and intelligence. In Sternberg, R.J. [ed.] handbook of human intelligence. New Yourk, Cambridge University Press. 1982, pp. 225-307.

[17] Damasio, A.R., Graff-Radford, N.R., Eslinger, P.J., Damasio, H., Kassell, N. Amnesia following basal forebrain *Arch. Neurol.* 1985;42(3):263-271.

[18] Squire, L.R. 1992. Memory and the hippocampus: a synthesis from findings with rats, monkeys and humans. *Psychol. Rev.* 1992;99:195-231.

[19] Daum, I., Ackermann, H. Cerebellar contributions to cognition. *Behav. Brain Res.* 1995;67:201–210.
[20] Middleton, F.A., Strick, P.L. Cerebellar output channels. *Int. Rev. Neurobiol.* 1997; 41:61–82
[21] Dolan, R.J. A cognitive affective role for the cerebellum. *Brain* 1998;121:545–546.
[22] Alexander, G.E., DeLong, M.R., Strick, P.L. Parallel organization of functionally segregated circuits linking basal ganglia and cortex. *Annual Review of Neuroscience* 1986;9:357–381.
[23] Raichle, M.E., Fiez, J.A., Videen, T.O., MacLeod, A.M., Pardo, J.V., Fox, P.T., Petersen, S.E. Practice-related changes in human brain functional anatomy during non-motor learning. *Cereb. Cortex* 1994;4:3–26
[24] Jueptner, M., Frith, C.D., Brooks, D.J., Frackowiak, R.S., Passingham, R.E. Anatomy of motor learning. II. Subcortical structures and learning by trial and error. *Journal of Neurophysiology* 1997;77:1325–1337.
[25] Jueptner, M., Stephan, K.M., Frith, C.D., Brooks, D.J., Frackowiak, R.S., Passingham, R.E. Anatomy of motor learning. I. Frontal cortex and attention to action. *Journal of Neurophysiology* 1997;77:1313–1324.
[26] Friedman, H.R., Goldman-Rakic, P.S. Coactivation of prefrontal cortex *J. Neurosci.* 1994;14(5 Pt 1):2775-2788.
[27] Andrés, P. Frontal cortex *Cortex* 2003;39(4-5):871-95.
[28] Godefroy, O., Jeannerod, M., Allain, P., Le Gall, D. Frontal lobe, executive functions and cognitive control. *Rev. Neurol.* (Paris 2008;164 Suppl 3:S119-127.
[29] Moser, D.C., Fridriksson, J., Healy, E.W. Sentence comprehension *Clin. Linguist. Phon.* 2007;21(2):147-156.
[30] http://www.brainsource.com/nptests.htm
[31] Dennis, M., Francis, D.J., Cirino, P.T., Schachar, R., Barnes, M.A., Fletcher, J.M. Why IQ is not a covariate in cognitive studies of neurodevelopmental disorders. *J. Int. Neuropsychol. Soc.* 2009;15(3):331-343.
[32] Thom, M. Neuropathological findings in epilepsy. *Current Diagnostic Pathology* 2004;10(2):93-105.
[33] Berg, A.T., Langfitt, J.T., Testa, F.M., Levy, S.R., DiMario, F., Westerveld, M., Kulas, J. Global cognitive function in children with epilepsy: A community-based study. *Epilepsia* 2008;49(4):608-614.
[34] Cormack, F., Cross, J.H., Isaacs, E., Harkness, W., Wright, I., Vargha-Khadem, F., Baldeweg, T. The development of intellectual abilities in pediatric temporal lobe epilepsy. *Epilepsia* 2007;48(1):201-204.
[35] Jokeit, H., Ebner, A., Arnold, S., Schuller, M., Antke, C., Huang, Y., Steinmetz, H, Seitz, R.J., Witte, O.W. Bilateral reductions of hippocampal volume, glucose metabolism, and wada hemispheric memory performance are related to the duration of mesial temporal lobe epilepsy. *J. Neurol.* 1999;246:926-933.
[36] Shannon, H.E., Love, P.L. Effects of antiepileptic drugs on learning as assessed by a repeated acquisition of response sequences task in rats. *Epilepsy Behav.* 2007;10:16-25.
[37] Briellmann, R.S., Wellard, R.M., Jackson, G.D. Seizure-associated abnormalities in epilepsy: evidence from MR imaging. *Epilepsia* 2005;46:760-766.

[38] Pollard, H., Charriaut-Marlangue, C., Cantagrel, S., Represa, A., Robain, O., Moreau, J., Ben-Ari, Y. Kainate-induced apoptotic cell death in hippocampal neurons. *Neuroscience* 1994;**63**:7-18.

[39] Lynch, M., Sayin, U., Bownds, J., Janumpalli, S., Sutula, T. Long-term consequences of early postnatal seizures on hippocampal learning and plasticity. *Eur. J. Neurosci.* 2000;12(7):2252-2264.

[40] Schwarcz, R., Witter, M. Is neurodegeneration responsible for memory and behavioral alterations in epilepsy? *Epilepsy Res.* 2002;50(1-2):161-177.

[41] Hecimovic, H., Goldstein, J.D., Sheline, Y.I., Gilliam, F.G. Mechanisms of depression in epilepsy from a clinical perspective. *Epilepsy Behav.* 2003;3:S25-S30.

[42] Helmstaedter, C., Kurthen, M., Lux, S., Reuber, M., Elger, C.E. Chronic epilepsy and cognition: a longitudinal study in temporal lobe epilepsy. *Ann. Neurol.* 2003;54:425-432.

[43] Moser, M.B., Moser, E.I., Forrest, E., Andersen, P., Morris, R.G.M. Spatial learning with minislab in the dorsal hippocampus. *Proc. Natl. Acad. Sci. USA* 1995;92:9697-9701.

[44] Oyegbile, T.O., Dow, C., Jones, J., Bell, B., Rutecki, P., Sheth, R., Seidenberg, M., Hermann, B.P. The nature and course of neuropsychological morbidity in chronic temporal lobe epilepsy. *Neurology* 2004;62:1736-1742.

[45] Rzezak, P., Fuentes, D., Guimarães, C.A., Thome-Souza, S., Kuczynski, E., Li, L.M., Franzon, R.C., Leite, C.C., Guerreiro, M., Valente, K.D. Frontal lobe dysfunction in children with temporal lobe epilepsy. *Pediatr.Neurol.*2007;37(3):176-185.

[46] Guimarães, C.A., Li, L.M., Rzezak, P., Fuentes, D., Franzon, R.C., Augusta Montenegro., Cendes, F., Thomé-Souza, S., Valente, K., Guerreiro, MM. Temporal lobe epilepsy *J. Child Neurol.* 2007;22(7):836-340.

[47] Hampson, R.E., Simeral, J.D., Deadwyler, S.A. Distribution of spatial and nonspatial information in dorsal hippocampus. *Nature* 1999;402(6762):610-614.

[48] De Marco, F.A., Ghizoni, E., Kobayashi, E., Min, L., Cendes, F. Cerebellar volume and long-term use of phenytoin. *Seizure* 2003;12:312-315.

[49] Ortinski, P., Meador, K.J. Cognitive side effects of antiepileptic drugs, *Epilepsy Behav.* 2004;5(S1):S60-S65.

[50] Roch, C., Leroy, C., Nehlig, A., Namer, I.J. Predictive value of cortical injury for the development of temporal lobe epilepsy in 21-day-old rats: an MRI approach using the lithium-pilocarpine model. *Epilepsia* 2002;43:1129-1136.

[51] Lawson, J.A., Vogrin, S., Bleasel, A.F., Cook, M.J., Bye, A.M. Cerebral and cerebellar volume reduction in children with intractable epilepsy. *Epilepsia* 2000;41:1456–1462.

[52] Fuerst, D., Shah, J., Shah, A., Watson, C. Hippocampal sclerosis is a progressive disorder: A longitudinal volumetric MRI study. *Ann. Neurol.* 2003;53(3):413-416.

[53] Liu, R.S., Lemieux, L., Bell, G.S., Hammers, A., Sisodiya, S.M., Bartlett, P.A., Shorvon, S.D., Sander, J.W., Duncan, J.S. Progressive neocortical damage in epilepsy. *Ann. Neurol.* 2003;53:312-324.

[54] Bernasconi, N., Bernasconi, A., Caramanos, Z., Antel, S.B., Andermann, F., Arnold, D.L. Mesial temporal lobe damage in temporal lobe epilepsy: a volumetric study of the hippocampus, amygdale, and parahippocampal region. *Brain* 2003;126:462–469.

[55] Helmstaedter, C. Effects of chronic epilepsy on declarative memory systems. Prog. *Brain. Res.* 2002;135:439–453.

[56] Hermann, B., Seidenberg, M., Bell, B., Rutecki, P., Sheth, R.D., Wendt, G., O'Leary, D, Magnotta, V. Extratemporal quantitative MR volumetrics and neuropsychological status in temporal lobe epilepsy. *J. Int. Neuropsychol. Soc.* 2003;9:353–362.

[57] Seidenberg, M., Hermann, B., Pulsipher, D., Morton, J., Parrish, J., Geary, E., Guidotti, L. Thalamic atrophy and cognition in unilateral temporal lobe epilepsy. *J. Int. Neuropsychol. Soc.* 2008;14(3):384-393.

[58] Lawson, J.A., Cook, M.J., Vogrin, S., Litewka, L., Strong, D., Bleasel, A.F., et al. Clinical, EEG, and quantitative MRI differences in pediatric frontal and temporal lobe epilepsy. *Neurology* 2002;58:723-790.

[59] Manoach, D.S., Greve, D.N., Lindgren, K.A., Dale, A.M. Identifying regional activity associated with temporally separated components of working memory using event-related functional MRI. *Neuroimage* 2003;20:1670–1678.

[60] Polich, J., Ehlers, C.L., Otis, S., Mandell, A.J., Bloom, F.E. P300 latency reflects the degree of cognitive decline in dementing illness. *Electroencephalogr. Clin. Neurophysiol.* 1986;63:138-144.

[61] Tassinari, C.A., Rubboli, G. Cognition and Paroxysmal EEG Activities: From a Single Spike to Electrical Status Epilepticus during Sleep. *Epilepsia* 2006;47(Suppl 2):40–43.

[62] Goode, D.J., Penry, J.K., Dreifuss, F.E. Effects of paroxysmal spike-wave and continuous visual-motor performances. *Epilepsia* 1970;11:241– 254.

[63] Aldenkamp, A.P., Overweg, J., Gutter, T., Beun, A.M., Diepman, L., Mulder, O.G. Effect of epilepsy, seizures and epileptiform EEG discharges on cognitive function. *Acta Neurol. Scand.* 1996;93:253-259.

[64] Dodrill, C. Progressive cognitive decline n adolescents and adults with epilepsy. *Prog. Brain Res.* 2002;135:399-407.

[65] Binnie, C.D. Cognitive Impairment during epileptiform discharges: is it ever justifiable to treat the EEG?. *Lancet Neurol.* 2003;2:725-730.

[66] Sillanpaa, M., Jalava, M., Kaleva, O., Shinnar, S. Long-term prognosis of seizures with onset in children. *N. Eng. J. Med.* 1998;338(24):1715-1722.

[67] Kanner, A.M., Nieto, J.C.R. Depression disorder in epilepsy. *Neurology* 1999;53(suppl 2):S26-S32.

[68] Sattler, J.M. Assessment of Children revd/updated. 3rd ed. San Diego: Jerome M. Sattler; 1992.

[69] Kantola-Sorsa, E., Gaily, E., Isoaho, M., Korkman, M. Neuropsychological outcomes in children of mothers with epilepsy. *J. Int. Neuropsychol. Soc.*2007;13(4):642-652.

[70] Dodrill, C.B. Correlates of generalized tonic-clonic seizures with intellectual, neuropsychychological, emotional, and social function in patients with epilepsy. *Epilepsia* 1986;27:399–411.

[71] Bornstein, R.A., Pakalnis, A., Drake, Jr. M.E., Suga, L.J. Effects of seizure type and waveform abnormality on memory and attention. *Arch. Neurol.* 1988;45:884-887.

[72] Giovagnoli, A.R., Avanini, G. Learning and memory impairment in patients with temporal lobe epilepsy: relation to the presence, type, and location of brain lesion. *Epilepsia* 1999;40:904-911.

[73] Silva, A.N., Andrade, V.M., Oliveira, H.A. Neuropsychological assessment in patients with temporal lobe epilepsy. *Arq. Neuropsiquiatr* .2007;65(2B):492-497.

[74] Helmstaedter, C. Behavioral aspects of frontal lobe epilepsy. *Epilepsy Behav.* 2:384–395.

[75] Natsume, J., Bernasconi, N., Andermann, F., Bernasconi, A. MRI volumetry of the thalamus in temporal, extratemporal, and idiopathic generalized epilepsy. *Neurology* 2003;60:1296–1300.

[76] Prevey, M.L., Delaney, R.C., Cramer, J.A., Mattson, R.H. Complex partial and secondarily generalized seizure patients: cognitive functioning prior to treatment with antiepileptic medication VA epilepsy cooperative study 264 group. *Epilepsy Res.* 1998;30:1-9.

[77] Tracy, J.I., Lippincott, C., Mahmood, T., Waldron, B., Kanauss, K., Glosser, D, Sperling, M.R. Are depression and cognitive performance related in temporal lobe epilepsy? *Epilepsia* 2007;48(12):2327-2335.

[78] Binnie, C.D., Marston, D. Cognitive correlates of interictal discharges. *Epilepsia* 1992; 33 Suppl 6:S11-17.

[79] Butler, C.R., Zeman, A.Z. Recent insights into the impairment of memory in epilepsy: transient epileptic amnesia, accelerated long-term forgetting and remote memory impairment. *Brain* 2008;131(9):2243-2263.

[80] Luat, A.F., Asano, E., Juhász, C., Chandana, S.R., Shah, A., Sood S, Chugani, H.T. Relationship between brain glucose metabolism positron emission tomography (PET) and electroencephalography (EEG) in children with continuous spike-and-wave activity during slow-wave sleep. *J. Child Neurol.* 2005;20(8):682-690.

[81] Haverkamp, F., Hanisch, H, Mayer, H., Noeker, M. Evidence for a specific vulnerability of sequential information processing in children with epilepsy. *J. Child Neurol.* 2001;16:901-905.

[82] Steriade, M. Corticothalamic resonance, states of vigilance and mentation *Neuroscience* 2000;101:243–276.

[83] Massa, R., de Saint-Martin, A., Carcangiu, R., Rudolf, G., Seegmuller, C., Kleitz, C., Metz-Lutz, M.N., Hirsch, E., Marescaux, C. EEG criteria predictive of complicated evolution in idiopathic rolandic epilepsy. *Neurology* 2001;57:1071–1079.

[84] Aarts, J.H., Binnie, C.D., Smith, A.M., Wilkins, A.J. Selective cognitive impairment during focal and generalized epileptiform EEG activity. *Brain* 1984;107:293-308.

[85] Sirén, A., Kylliäinen, A., Tenhunen, M., Hirvonen, K., Riita, T., Koivikko, M. Beneficial effects of antiepileptic medication on absence seizures and cognitive functioning in children. *Epilepsy Behav.* 2007;11(1):85-91.

[86] Croona, C., Kihlgren, M., Lundberg, S., Eeg-Olofsson, O., Eeg-Olofsson, K.E. Neuropsychological findings in children with benign childhood epilepsy with centrotemporal spikes. *Dev. Med. Child Neurol.* 1999;41:813–818.

[87] Clarke, D.F., Roberts, W., Daraksan, M., Dupuis, A., McCabe, J., Wood, H., Snead, O.C.3rd., Weiss, S.K. The prevalence of autistic spectrum *Epilepsia* 2005;46(12):1970–1977.

[88] Roulet Perez, E., Davidoff, V., Despland, P.A., Deonna, T. Mental and behavioral deterioration of children with epilepsy and CSWS: acquired epileptic frontal syndrome. *Dev. Med. Child Neurol.* 1993;35:661-671.

[89] Maquet, P., Hirsch, E., Metz-Lutz, M.N., Motte, J., Dive, D., Marescaux, C., Franck, G. Regional cerebral glucose metabolism in children with deterioration of one or more cognitive functions and continuous spike-and-wave discharges during sleep. *Brain* 1995;118(Pt 6):1497-520.

[90] De Tiège, X., Goldman, S., Laureys, S., Verheulpen, D., Chiron, C., Wetzburger, C., Paquier, P., Chaigne, D., Poznanski, N., Jambaqué, I., Hirsch, E., Dulac, O., Van Bogaert, P. Regional cerebral glucose metabolism in epilepsies with continuous spikes and waves during sleep. *Neurology* 2004;63(5):853-857.

[91] De Tiège, X., Ligot, N., Goldman, S., Poznanski, N., de Saint Martin, A., Van Bogaert, P. Metabolic evidence for remote inhibition in epilepsies with continuous spike-waves during sleep. *Neuroimage* 2008;40(2):802-810.

[92] Sloviter, R.S. Status epilepticus-induced neuronal injury and network reorganization. *Epilepsia* 2000;40:34-41.

[93] Dolorfo, C.L., Amaral, D.G. Entorhinal cortex of the rat: topographic organization of the cells of origin of the perforant path projection to the dentate gyrus. *J. Comp. Neurol.* 1998;398(1):25-48.

[94] Zola-Morgan, S., Squire, L.R., Amaral, D.G. Human amnesia and the medial temporal region: enduring memory impairment following a bilateral lesion limited to field CA1 of the hippocampus. *J. Neurosci.* 1986;6:2950-2967.

[95] Malenka, R.C. Synaptic plasticity in the hippocampus: LTP and LTD. *Cell* 1994; 78:535-538.

[96] Davies, C.H., Collingridge, G.L. The physiological regulation of synaptic inhibition by $GABA_B$ autoreceptors in rat hippocampus. *J. Physiol. (London)* 1993;472:245-265.

[97] Grunwald, T., Beck, H., Lehnertz, K., Blumcke, I., Pezer, N., Kurthen, M., Fernandez, G., Van Roost, D., Heinze, H.J., Kutas, M., Elger, C.E. Evidence relating human verbal memory to hippocampal N-methyl-D-aspartate receptors. *Proc. Natl. Acad. Sci. USA* 1999; 96:12085-12089.

[98] Bliss, T.V., Collingridge, G.L. A synaptic model of memory: Long-term potentiation in the hippocampus. *Nature* 1993;361:31-39.

[99] Reid, I.C., Stewart, C.A. Seizures, memory and synaptic plasticity. *Seizure* 1997;6(5):351-359

[100] Seidel, W.T., Mitchell, W.G. Cognitive and behavioral effects of carbamazepine in children: data from benign rolandic epilepsy. *J. Child. Neurol.* 1999;14:716-723.

[101] Frost, J.D. Jr., Hrachovy, R.A., Glaze, D.G., Rettig, G.M. Alpha rhythm slowing during initiation of carbamazepine therapy: implications for future cognitive performance. *J. Clin. Neurophysiol.* 1995,12(1):57-63.

[102] Hanson, J.W., Smith, D.W. The fetal hydantoin syndrome. *J. Pediatr.* 1975;87:285–90.

[103] Moore, S.J., Turnpenny, P., Qinn, A., Glover, S., Lioyd, D.J., Montgomery, T., Dean, J.C. A Clinical study of 57 children with fetal anticonvulsant syndromes. *J. Med. Genet.* 2000;37:489-497.

[104] Shapiro, S., Hartz, S.C., Siskind, V., Mitchell, A.A., Slone, D., Rosenberg, L., Monson, R.R., Heinonen, O.P. Anticonvulsants and parental epilepsy in the development of birth defects. *Lancet* 1976;1:272–275.

[105] Reinisch, J.M., Sanders, S.A., Mortensen, E.L., Rubin, D.B. In utero exposure to phenobarbital and intelligence deficits in adult men. *JAMA* 1995;274:1518–1525.

[106] Thompson, P.J., Baxendale, S.A., Duncan, J.S., Sander, J.W. Effects of topiramate on cognitive function. *J. Neurol. Neurosurg. Psychiatry* 200;69:636-641.

[107] Park, S.P., Hwang, Y.H., Lee, H.W., Suh, C.K., Kwon, S.H., Lee, B.I. Long-term cognitive and mood effects of zonisamide monotherapy in epilepsy patients. *Epilepsy Behav.* 2008;12(1):102-108.

[108] Harden, C.L., Lazar, L.M., Pick, L.H., Nikolov, B., Goldstein, M.A., Carson, D., Ravdin, L.D., Kocsis, J.H., Labar, D.R. A beneficial effect on mood in partial epilepsy patients treated with gabapentin. *Epilepsia* 1999;40:1129-1134.

[109] Meador, K.J., Loring, D.W., Ray, P.G., Murro, A.M., King, D.W., Nichols, M.E., Deer, E.M., Goff, W.T. Differential cognitive effects of carbamazepine and gabapentin. *Epilepsia* 1999;40(9):1279-1285.

[110] Blum, D., Meador, K., Biton, V., Fakhoury, T., Shneker, B., Chung, S., Mills, K., Hammer, A., Isojärvi, J. Cognitive effects of lamotrigine compared with topiramate in patients with epilepsy. *Neurology* 2006;67(3):378-379.

[111] Pressler, R.M., Binnie, C.D., Coleshill, S.G., Chorley, G.A., Robinson, R.O. Effect of lamotrigine on cognition in children with epilepsy. *Neurology* 2007;68(10):797-798.

[112] Kaye, N.S., Graham, J., Roberts, J., Thompson, T., Nanry, K. Effect of open-label lamotrigine as monotherapy and adjunctive therapy on the self-assessed cognitive function scores of patients with bipolar I disorder. *J. Clin. Psychopharmacol.* 2007;27(4):387-391.

[113] Dodrill, C.B., Arnett, J.L., Deaton, R., Lenz, G.T., Sommerville, K.W. Tiagabine versus phenytoin and carbamazepine as add-on therapies: effects on abilities, adjustment, and mood. *Epilepsy. Res.* 2000;42:123–132.

[114] Äikiä, M., Jutila, L., Salmenpera, T., Mervaala, E., Kälviäinen, R. Comparison of the Cognitive Effects of Tiagabine and Carbamazepine as Monotherapy in Newly Diagnosed Adult Patients with Partial Epilepsy: Pooled Analysis of Two Long-term, Randomized, Follow-up Studies. *Epilepsia* 2006;47(7):1121–1127.

[115] Grant, S.M., Heel, R.C. Vigabatrin A review of its pharmacodynamic and pharmacokinetic properties, and therapeutic potential in epilepsy and disorders of motor control. *Drugs* 1991;41:889-926.

[116] Provinciali, L., Bartolini, M., Mari, F., Del Pesce, M., Ceravolo, M.G. Influence of vigabatrin on cognitive performances and behavior in patients with drug-resistant epilepsy. *Acta Neurol. Scand.* 1996;94(1):12-18.

[117] Piazzini, A., Chifari, R., Canevini, M.P., Turner, K., Fontana, S.P., Canger, R. Levetiracetam: an improvement of attention and of oral fluency in patients with partial epilepsy. *Epilepsy Res.* 2006;68:181–188.

[118] Gomer, B., Wagner, K., Frings., Saar, J., Carius, A., Härle, M., Steinhoff, B.J., Schulze-Bonhage, A. The influence of antiepileptic drugs on cognition: a comparison of levetiracetam with topiramate. *Epilepsy Behav.* 2007;10(3):486-494.

[119] Bootsma, H.P., Coolen, F., Aldenkamp, A.P., Arends, J., Diepman, L., Hulsman, J., Lambrechts, D., Leenen, L., Majoie, M., Schellekens, A., de Krom, M. Topiramate in clinical practice: long-term experience in patients with refractory epilepsy referred to a tertiary epilepsy center. *Epilepsy Behav.* 2004;5:380–387.

[120] Middaugh, L.D., Simpson, LW, Thomas TN, Zemp, J.W. Prenatal maternal phenobarbital increases reactivity and retards habituation of mature offspring to environmental stimuli. *Psychopharmacology (Berl)* 1981;74:349–352.

[121] Yanai, J., Fares, F., Gavish, M., Greenfeld, Z., Katz, Y., Marcovici, G., Pick, C.G., Rogel-Fuchs, Y., Weizman, A. Neural and behavioral alterations after early exposure to phenobarbital. *Neurotoxicology* 1989;10:543–554.

[122] Farwell, J.R., Lee, Y.J., Hirtz, D.G., Sulzbacher, S.I., Ellenberg, J.H., Nelson, K.B. Phenobarbital for febrile seizures-effects on intelligence and on seizure recurrence. N Engl. J. med. 1990;322(6):364-369.

[123] Sulzbacher, S., Farwell, J.R., Temkin, N., Lu, A.S., Hirtz, D.G. Late cognitive effects of early treatment with Phenobarbital. Clin. Pediatr. (Phila) 1999;38(7):387-394.

[124] Metz-Lutz, M.N., de Saint Martin, A., Monpiou, S., Massa, R., Hirsch, E., Marescaux, C. Cognitive development in benign focal epilepsies of childhood. Dev. Neurosci. 1999;21:182-190.

[125] Salinsky, M.C., Spencer, D.C., Oken, B.S., Storzbach, D. Effects of oxcarbazepine and phenytoin on the EEG and cognition in healthy volunteer. Epilepsy Behav. 2004;5(6):894-902.

[126] Schilling, M.A., Inman, S.L., Morford, L.L., Moran, M.S., Vorhees, C.V. Prenatal phenytoin exposure and spatial navigation in offspring: effects on reference and working memory and on discrimination learning. Neurotoxicol. Teratol. 1999;21:567–578.

[127] Rybakowski, J. Leki normotymiczne. Terapia 2004;12:12-17.

[128] Mecarelli, O., Vicenzini, E., Pulitano, P., Vanacore, N., Romolo, F.S., Di Piero, V., Lenzi, G.L., Accornero, N. Clinical, cognitive and neurophysiologic correlates of short-term treatment with carbamazepine, oxcarbazepine and levetiracetam in healthy volunteer. Ann. Pharmacother. 2004;38(11):1816-1822.

[129] Donati, F., Gobbi, G., Campistol, J., Rapatz, G., Daehler, M., Sturm, Y., Aldenkamp, A.P.; Oxcarbazepine, Cognitive Study Group. The cognitive effects of oxcarbazepine versus carbamazepine or valproate in newly diagnosed children with partial seizures. Seizure 2007;16(8):670-679.

[130] Balakrishnan, S., pandhi, P. Effect of nimodipine on the cognitive dysfunction induced by phenytoin and valproate in rats. Methods Find. Exp. Clin. Pharmacol. 1997;19:693-697.

[131] Adab, N., Kini, U., Vinten, J., Ayres, J., Baker, G., Clayton-Smith, J., Coyle, H., Fryer, A., Gorry, J., Gregg, J., Mawer, G., Nicolaides, P., Pickering, L., Tunnicliffe, L., Chadwick, D.W. The longer term outcome of children born to mothers with epilepsy. J. Neurol. Neurosurg. Psychiatry 2004;75:1575–1583.

[132] Tatum, W.O.4th., French, J.A., Faught, E., Morris, G.L.3rd., Liporace, J. Postmarketing experience with topiramate and cognition. Postmarketing experience with topiramte and cognition. Epilepsia 2001;42:1134-1140.

[133] Reife, R., Piedger, G., Wu, S.C. Topiramte as add-on therapy: pooled analysis of randomized controlled trials in adults. Epilepsia 2000;41(supp 1):S66-S71.

[134] Leonard, G., Milner, B., Jones, L. Performance on unimanual and bimanual tapping tasks by patients with lesions of the frontal and temporal lobe. Neuropsychologia 1988;26:79-91.

[135] Aldenkamp, A.P., Bodde, N. Behaviour, cognition and epilepsy. Acta Neurol. Scand. 2005;112(Suppl.182):S19–25.

[136] Lee, S., Sziklas, V., Andermann, F., Farnham, S., Risse, G., Gustafson, M., Gates, J., Penovich, P., Al-Asmi, A., Dubeau, F., Jones-Gotman, M. The effects of adjunctive topiramate on cognitive function in patients with epilepsy. Epilepsia 2003;44(3):339-347.

[137] Kockelmann, E., Elger, C.E., Helmstaedter, C. Cognitive profile of topiramate as compared with lamotrigine in epilepsy patients on antiepileptic drug polytherapy: relationships to blood serum levels and comedication. *Epilepsy Behav.* 2004;5:716–721.
[138] Huppertz, H.J., Quiske, A., Schulze-Bonhage, A. Cognitive impairments due to add-on therapy with topiramate. *Nervenarzt* 2001;72:275–80.
[139] Zhao, Q., Hu, Y., Holmes, G.L. Effect of topiramate on cognitive function and activity level following neonatal seizures. *Epilepsy Behav.* 2005;6:529–536.
[140] Koh, S., Tibayan, F.D., Simpson, J.N., Jensen, F.E. NBQX or topiramate treatment after perinatal hypoxia-induced seizures prevents later increases in seizure-induced neuronal injury. *Epilepsia* 2004;45:569–575.
[141] Weatherly, G., Risse, G.L., Carlson, B.E., Gustafson, M.C., Penovich, P.E. Decline in cognitive functioning associated with zonisamide therapy. *Epilepsia* 2002;43(suppl 7):186.
[142] Temkin, N.R. Antiepileptogenesis and seizure prevention trials with antiepileptic drugs: meta-analysis of controlled trials. *Epilepsia* 2001;42:515–524.
[143] Fisher, J.E., Vorhees, C. Developmental toxicity of antiepileptic drugs: relationship to postnatal dysfunction. *Pharmacol. Res.* 1992;26(3):207–221.
[144] Yu, X.M., Salter, M.W. Gain control of NMDA-receptor currents by intracellular sodium. *Nature* 1998; 396:469-474.
[145] Bradford, H.F. Glutamate, GABA and epilepsy. *Prog. Neurobiol.* 1995;47:477–511.
[146] Löster, W. Basic pharmacology of valproate. *CNS Drugs* 2002;16:669-694.
[147] Satzinger, G. Antiepileptics from gamma-aminobutyric acid. *Arzneim.-Forsch* 1994;**44**:261-266.
[148] Fink-Jensen, A., Suzdak, P.D., Swedberg, M.D., Judge, M.E., Hansen, L., Nielsen. The aminobutyric acid (GABA) uptake inhibitor, tiagabine, increases extracellular brain levels of GABA in awake rats. *Eur. J. Pharmacol.* 1992;220:197–201.
[149] Petroff, O.A., Hyder, F., Rothman, D.L., Mattson, R.H. Topiramate rapidly raises brain GABA in epilepsy patients. *Epilepsia* 2001;42:543–548.
[150] Meldrum, B.S., Craggs, M.D., Dürmüller, C.N., Smith, S.E., Chapman, A.G. The effects of AMPA receptor antagonists on kindled seizures and on reflex epilepsy in rodents and primates. In Engel, Jr., Wasterlain, J., Cavalheiro, C., Heinemann, E.A., Avanzini, U. (ed.), Molecular Neurobiology of epilepsy. Elsevier Science Publishers, Amsterdam. 1992, pp. 307-311.
[151] vanDongen, A.M., VanErp, M.G., Voskuyl, R.A. Valproate reduces excitability by blockage of sodium and potassium conductance. *Epilepsia* 1986;27:177-182.
[152] Zhan, M., Yu, K., Xiao, C., Ruan, D. The influence of developmental periods of sodium valproate exposure on synaptic plasticity in the CA1 region of rat hippocampus. *Neuroscience Letters* 2003;351:165-168.
[153] Mott, D.D., Lewis, D.V. Facilitation of the induction of long-term potentiation by $GABA_B$ receptors. *Science* 1991;252:1718-1720.
[154] Gean, P.W., Huang, C.C., Huang, C.R., Tsai, J.J. Valproic acid suppresses the synaptic response mediated by the NMDA receptors in rat amygdalar slices. *Brain Res. Bull.* 1994;33:333-336.
[155] Shank, R.P., Gardocki, J.F., Streeter, A.J., Maryanoff, B.E. An overview of the preclinical aspects of topiramate: pharmacology, pharmacokinetics, and mechanism of action. *Epilepsia* 2000;41(S1):S3-S9.

[156] Taverna, S., Sancini, G., Mantegazza, M., Franceschetti, S., Avanzini, G. Inhibition of transient and persistent Na+ current fractions by the new anticonvulsant topiramate. *J. Pharmacol. Exp. Ther.* 1999;288:960-968.

[157] Watanabe, Y., Gould, E., Cameron, H.A., Daniels, D.C., McEwen, B.S. Phenytoin prevents stress- and corticosterone-induced atrophy of CA3 pyramidal neurons. *Hippocampus* 1992;2(4):431-435.

[158] Luine, V., Villages, M., Martinex, C., McEwen, B.S. Repreated stress causes reversible impairment of spatial memory performance. *Brain Res.* 1994;639:167-170.

[159] Zhang, H., Yang, Q., Xu, C. Effect of chronic stress and phenytoin on the long-term potentiation (LTP) in rat hippocampal CA1 region. *Acta Biochimica et Biophysica Sinica* 2004;36(5):375-378.

[160] Taft, W.C., Clifton, G.L., Blair, R.E., DeLorenzo, R.J. Phenytoin protects against ischemia-produced neuronal cell death. *Brain Res.* 1989;483:143-148.

[161] Bremner, J.D., Mletzko, T., Welter, S., Quinn, S., Williams, C., Brummer, M., Siddiq, S., Reed, L., Heim, C.M., Nemeroff, C.B. Effects of phenytoin on memory, cognition and brain structure in post-traumatic stress disorder: a pilot study. *J. Psychopharmacology* 2005;19(2):1-7.

[162] Wamil AW, Mclean NJ. Phenytoin blocks N-Methyl-D-Aspartate responses of mouse central neurons. *J. Pharmacol. Exp. Ther.* 1993;267:218-227.

[163] Trimble, M.R. Anticonvulsant drugs and cognitive function: a review of the literature. *Epilepsia* 1987;28(suppl 3):S37-45.

[164] Rostock, A., Hoffmann, W., Siegemund, C., Bartsch, R. Effects of carbamazepine, valproate calcium, clonazepam and piracetam on behavioral test methods for evaluation of memory-enhancing drugs Methods. *Find. Exp. Clin. Pharmacol.* 1989;11:547-553.

[165] Ambrósio, A.F., Silva, A.P., Araujo, I., Malva, J.O., Soares-da-Silva, P., Carvalho, A.P., Carvalho, C.M. Neurotoxic/neuroprotective profile of carbamazepine, oxcarbazepine and two new putative antiepileptic drugs, BIA 2-093 and BIA 2-024. *Eur. J. Pharmacol.* 2000;406:191-201.

[166] Consolo, S., Bianchi, S., Landinski, H. Effect of carbamazepine on cholinergic parameters in rat brain areas. *Neuropsychopharmacology* 1976;15:653-657.

[167] Barkai, E., Hasselmo, M. Acetylcholine and associative memory in the piriform cortex. Molecular *Neurobiology* 1997;15:17–29.

[168] Ji, D., Lape, R., Dani, J.A. Timing and location of nicotinic activity enhances or depresses hippocampal synaptic plasticity. *Neuron* 2001;31:131–141.

[169] Puzynski, S. Anticonvulsants (carbamazepine, valproate, lamotrigine) on bipolar affective disorder. *Psychiatria Polska* 2002;6:52-61.

[170] Dimond KR, Pande AC, Lamoreaux L, Pierce MW (1996) Effect of gabapentin (Neurontin) on mood and well-being in patients with epilepsy. *Prog. Neuropsychopharmacol. Biol. Psychiatry* 1996;20(3):407-417

[171] Xie, X., Hagan, R.M. Cellular and molecular actions of lamotrigine: possible mechanisms of efficacy in bipolar disorder. *Neuropsychobiology* 1998;38:119-130.

[172] Rigo, J.M., Hans, G., Nguyen, L., Rocher, V., Belachew, S., Malgrange, B., Leprince, P., Moonen, G., Selak, I., Matagne, A., Klitgaard, H. The anti-epileptic drug levetiracetam reverses the inhibition by negative allosteric modulators of neuronal GABA- and glycine-gated currents. *Br. J. Pharmacol.* 2002;136(5):659-672.

[173] Klitgaard, H.V., Matagne, A.C., Vanneste-Goemaere, J., Margineanu, D.G. Effects of prolonged administration of levetiracetam on pilocarpine induced epileptogenesis in rat. *Epilepsia* 2001;42(Suppl.7):114–115.

[174] Schindler, U. Pre-clinical evaluation of cognition enhancing drugs. *Prog. Neuropsychopharmacol Biol. Psychiatry* 1989;13:S99–S115.

[175] Genton, P., Van Vleymen, B. Piracetam and levetiracetam: close structural similarities but different pharmacological and clinical profiles. *Epileptic Disord.* 2000;2:99-105.

[176] Flicker, L., Grimley, E.J. Piracetam for dementia or cognitive impairment. Cochrane Database Syst. Rev. 1998;Issue 1:Article No: CD001011.

[177] Zhou, B., Zhang, Q., Tian, L., Xiao, J., Stefan, H., Zhou, D. Effects of levetiracetam as an add-on therapy on cognitive function and quality of life in patients with refractory partial seizures. *Epilepsy Behav.* 2008;12(2):305-310.

[178] Komatsu, M., Hiramatsu, M., Willmore, L.J. Zonisamide reduces the increase in 8-hydroxy-2'-deoxyguanosine levels formed during iron-induced epileptogenesis in the brains of rats. *Epilepsia* 2000;41:1091-1094.

[179] Dobbing, J. Vulnerable periods in developing brain Brain, behavior, and iron in the infant diet London: Springer-Verlag, 1990, 1-25.

[180] Jernigan, T.L., Trauner, D.A., Hesselink, J.R., Talla, P.A. Maturation of the human cerebellum observed in vivo during adolescence. *Brain* 1991;11:2037-2049.

[181] Jorgenson, L.A., Wobken, J.D., Georgieff, M.K. Perinatal iron deficiency alters apical dendritic growth in hippocampal CA1 pyramidal neurons. *Dev. Neurosci.* 2003;**25**:412–420.

[182] Stafstrom, C.E., Moshé, S.L., Swann, J.W., Nehlig, A., Jacobs, M.P., Schwartzkroin, P.A. Models of pediatric *Epilepsia* 2006;47(8):1407-1414.

[183] Sperber, E.F., Haas, K.Z., Romero, M.T., Stanton, P.K., Flurothyl status epilepticus in developing rats: behavioral, electrographic histological and electrophysiological studies. *Brain Res. Dev. Brain Res.* 1999;116:59–68.

[184] Haas, K.Z., Sperber, E.F., Opanashuk, L.A., Stanton, P.K., Moshé, S.L. Resistance of immature hippocampus to morphologic and physiologic alterations following status epilepticus or kindling. *Hippocampus* 2001;11:615-625.

[185] Sperber, E.F., Haas, K.Z., Stanton, P.K., Moshé, S.L. Resistance of the immature hippocampus *Brain Res. Dev. Brain Res.* 1991;60(1):88-93.

[186] da Silva, A.V., Regondi, M.C., Cavalheiro, E.A., Spreafico, R. Disruption of cortical development as a consequence of repetitive pilocarpine-induced status epilepticus in rats. *Epilepsia* 2005;46(Suppl 5):22-30.

[187] Ribak, C.E., Baram, T.Z. Selective death of hippocampal CA3 pyramidal cells with mossy fiber afferents after CRH-induced status epilepticus in infant rats. *Brain Res. Dev. Brain Res.* 1996;91:245-251.

[188] Dwyer, V.E., Wasterlain, C.G. Electroconvulsive seizures in the immature rat adversely affect myelin accumulation. *Exp. Neurol.* 1982;78:616-628.

[189] Jørgensen, O.S., Dwyer, B., Wasterlain, C.G. Synaptic proteins after electroconvulsive seizures in immature rats. *J. Neurochem.* 1980;35:1235-1237.

[190] Sankar, R., Shin, D.H., Wasterlain, C.G. GABA metabolism during status epilepticus in the developing rat brain. *Dev. Brain Res.* 1997;98:60–64.

[191] Zhang, G., Raol, Y., Hsu, F-C., Coulter, D., Brooks-Kayal, A. Effects of status epilepticus on hippocampal GABAA receptors are age-dependent. *Neuroscience* 2004; 125:299-303.

[192] Sullivan, P.G., Dube, C., Dorenbos, K., Steward, O., Baram, T.Z. Mitochondrial uncoupling protein-2 protects the immature brain from excitotoxic neuronal death. *Ann. Neurol.* 2003;53:711-717.

[193] Patel, M., Li, Q.Y. Age dependence of seizure-induced oxidative stress. Neuroscience 2003;118: 431-437.

[194] Rizzi, M., Perego, C., Aliprandi, M., Richichi, C., Ravizza, T., Colella, D., Veliskŏvá, J., Moshé, S.L., De Simoni, M.G., Vezzani, A. Glia activation and cytokine increase in rat hippocampus by kainic acid-induced status epilepticus during postnatal development. *Neurobiol. Dis.* 2003;14:494–503.

[195] Friedman, L.K., Sperber, E.F., Moshé, S.L., Bennett, M.V., Zukin, R.S., Developmental regulation of glutamate and GABA(A) receptor gene expression in rat hippocampus following kainate-induced status epilepticus. *Dev. Neurosci.* 1997;19:529-542.

[196] Tandon, P., Yang, Y., Das, K., Holmes, G.L., Stafstrom, C.E., 1999. Neuroprotective effects of brain derived neurotrophic factor in seizures during development. *Neuroscience* 1999;91:293–303.

[197] Diaz J, Schain, R.J., Bailey, B.G. Phenobarbital-induced brain growth retardation in artificially reared rat pups. *Biol. Neonate* 1977;32:77–82.

[198] Vorhees, C.V. (1983) Fetal anticonvulsant syndrome in rats: dose– and period–response relationships of prenatal diphenylhydantoin, trimethadione and phenobarbital exposure on the structural and functional development of the offspring. *J. Pharmacol. Exp. Ther.* 1983;227:274–287.

[199] Olney, J.W., Wozniak, D.F., Jevtovic-Todorovic, V., Farber, N.B., Bittigau, P., Ikonomidou, C. Drug-induced apoptotic neurodegeneration in the developing brain. *Brain Pathol.* 2002;12:488–98.

[200] Bittigau, P., Sifringer, M., Ikonomidou, C. Antiepileptic drugs and apoptosis in the developing brain. *Ann. N. Y. Acad. Sci.* 2003;993:103-114.

[201] Manent, J.B., Jorquera, I., Mazzucchelli, I., Depaulis, A., Perucca, E., Ben-Ari, Y., Represa, A. Fetal exposure to GABA-acting antiepileptic drugs generates hippocampal and cortical dysplasias. *Epilepsia* 2007;48:684–693.

[202] Jevtovic-Todorovic, V., Hartman, R.E., Izumi, Y., Benshoff, N.D., Dikranian, K., Zorumski, C.F., Olney, J.W., Wozniak, D.F. Early exposure to common anesthetic agents causes widespread neurodegeneration in the developing rat brain and persistent learning deficits. *J. Neurosci.* 2003;23:876–882.

[203] Kaindl, A.M., Asimiadou, S., Manthey, D., Hagen, M.V., Turski, L., Ikonomidou, C. Antiepileptic drugs and the developing brain. *Cell Mol. Life Sci.* 2006;63:399–413.

[204] Glier, C., Dzietko, M., Bittigau, P., Jarosz, B., Korobowicz, E., Ikonomidou, C. Therapeutic doses of topiramate are not toxic to the developing rat brain. *Exp. Neurol.* 2004;187:403–409.

[205] Katz, I., Kim, J., Gale, K., Kondratyev, A. Effects of lamotrigine alone and in combination with MK-801, phenobarbital, or phenytoin on cell death in the neonatal rat brain. *J. Pharmacol. Exp. Ther.* 2007;322(2):494-500.

[206] Lee, G.Y., Brown, L.M., Teyler, T.J. The effects of anticonvulsant drugs on long-term potentiation (LTP) in the rat hippocampus. *Brain Res. Bull.* 1996; 39:39-42.

[207] Ben-Ari, Y., Aniksztejn, L., Bregestovski, P. Protein Kinase C modulation of NMDA currents: an important link for LTP induction. *Trends Neurosci. 1992;*15:333-339.

[208] Harris, L.W., Sharp, T., Gartlon, J., Jones, D.N., Harrison, P.J. Long-term behavioural, molecular and morphological effects of neonatal NMDA receptor antagonism. *Eur. J. Neurosci.* 2003;18(6):1706-1710.

[209] Deary, I.J., Whiteman, M.C., Starr, J.M., Whalley, L.J., Fox, H.C. (2004) The impact of childhood intelligence on later life: following up the Scottish mental surveys of 1932 and 1947. *J. Pers. Soc. Psychol.* 2004;86:130–147.

[210] Hermann, B., Seidenberg, M., Jones, J. The neurobehavioural comorbidities of epilepsy: can a natural history be developed? *Lancet Neurol.* 2008;7(2):151-160.

[211] Seidenberg, M., Hermann, B., Wyler, A.R., Davies, K., Dohan, F.C. Jr., Leveroni, C. Neuropsychological outcome following anterior temporal lobectomy in patients with and without the syndrome of mesial temporal epilepsy. *Neuropsychology* 1998;12:303-305.

[212] Pais-Ribeiro, J., da Silva, A.M., Meneses, R.F., Falco, C. Relationship between optimism, disease variables, and health perception and quality of life in individuals with epilepsy. *Epilepsy Behav.* 2007;11(1):33-38.

[213] Thornton, N., Hamiwka, L., Sherman, E., Tse, E., Blackman, M., Wirrell, E. Family function in cognitively normal children with epilepsy: impact on competence and problem behaviors. *Epilepsy Beha.* 2008;12(1):90-95.

[214] Hirtz, D., Berg, A., Bettis, D., Camfield, C., Camfield, P., Crumrine, P., Gaillard, W.D., Schneider S, Shinnar S. Quality Standards Subcommittee of the American Academy of Neurology and the Practice Committee of the Child Neurology Society. *Neurology* 2003; 60(2):166-175.

[215] Ikonomidou, C., Bosch, C., Miksa, M., Bittigau, P., Vöckler, J., Dikranian, K., Tenkova, T.I., Stefovska, V., Turski, L., Olney, J.W. Blockade of NMDA receptors and apoptotic neurodegeneration in the developing brain. *Science* 1999;283:70–74.

[216] Rigoulot, M.A., Koning, E., Ferrandon, A., Nehlig A., 2004. Neuroprotective properties of topiramate in the lithium–pilocarpine model of epilepsy. *J. Pharmacol. Exp. Ther.* 2004;308:787–795.

[217] Suchomelova, L., Baldwin, R.A., Kubova, H., Thompson, K.W., Sankar, R., Wasterlain, C.G., 2006. Treatment of experimental status epilepticus in immature rats: dissociation between anticonvulsant and antiepileptogenic effects. *Pediatr. Res.* 2006;59(2):237-243.

Former address:
Research Center For Genetic Medicine, Children's National Medical Center, Washington DC (111 Michigan Avenue, NW, Washington, DC 20010 Main: 202-884-5000)

In: Encyclopedia of Neuroscience Research
Editors: Eileen J. Sampson and Donald R. Glevins

ISBN 978-1-61324-861-4
© 2012 Nova Science Publishers, Inc.

Chapter XXII

Cognitive Impairment in Children with ADHD: Developing a Novel Standardised Single Case Design Approach to Assessing Stimulant Medication Response

Catherine Mollica, Paul Maruff and Alasdair Vance *

Department of Paediatrics, University of Melbourne,
Murdoch Childrens Research Institute,
Royal Children's Hospital, Australia

Abstract

Cognitive difficulties are now recognized as a major driver of functional impairment in children with ADHD. However, to date, clinicians remain less aware of feasible and appropriate statistical approaches to measure cognitive and behavioural change in their patients. This chapter presents a practical statistical decision making rule and then outlines a clinician-friendly study design for ascertaining stimulant medication response in children with ADHD.

Introduction

There is limited understanding of the problems associated with repeated neuropsychological assessment in children, including the statistics used to guide decisions about cognitive change. Further, clinicians rarely consider change in cognitive function when evaluating treatment response in individual children with ADHD. This is most likely due to a lack of suitable assessment tasks as well as clinicians' limited awareness of the appropriate statistical techniques for measuring cognitive change in individuals. This chapter outlines a

---

* Parkville, Melbourne VIC 3052, Phone: 61 3 9345-4666, Fax: 61 3 9345-6002. e-mail: avance@unimelb.edu.au.

study investigating the application of a statistically principled decision rule to the cognitive and behavioural measures of individual children with ADHD in order to classify a significant, positive response to medication. The data demonstrate an evidence-based approach to clinical decision-making that can be used to evaluate cognitive and behavioural improvement in individual children with ADHD following treatment with stimulant medication. Then, a study **investigating a novel "intensive design" method for assessing stimulant medication**-related improvement in cognitive function in children with ADHD is presented. The results demonstrate the effectiveness of the medication in improving behavioural symptoms of ADHD, as well as certain features of cognitive function (psychomotor, visual attention and learning). Overall, **these findings support the use of a novel "intensive" within**-subjects design to examine the short-term effects of stimulant medication on cognitive and behavioural functions in children with ADHD. Further, this design is readily utilised in routine clinical practice.

Study 1

Attention-deficit/Hyperactivity disorder (ADHD) is characterised by developmentally inappropriate levels of inattention, and/or impulsiveness/overactivity (American Psychiatric Association, 1994). Stimulant medication is currently the most common and effective treatment for this disorder (Conners et al., 2001; Hoagwood et al., 2000). Successful response to treatment with stimulant medication is typically inferred from a reduction in behavioural symptoms, as determined by clinical judgment and/or parent and teacher behavioural ratings (Aman and Turbott, 1991). Although medication-responsiveness can vary according to many factors such as the severity of ADHD symptoms, medication dose, or the presence of comorbid disorders (Buitelaar et al., 1995; Denney and Rapport, 1999; DuPaul et al., 1994), the incidence of actual non-response is reported to be rare when thorough medication trials are conducted (Elia et al., 1991). In addition to improving behavioural symptoms, there is strong evidence to suggest that performance on tests of psychomotor, attention and memory also improves following treatment with stimulant medication (e.g., Bedard et al., 2003; de Sonneville et al., 1994; Douglas et al., 1995; Douglas et al., 1986; Mollica et al., submitted; for a review, see Rapport and Kelly, 1993). These findings suggest that improvement in cognitive function may be another useful marker of treatment response in individual children with ADHD (Rapport and Denney, 2000); therefore, it is important to improve current methods of assessing and analysing cognitive change in individuals with ADHD in order to accurately measure treatment outcomes.

There are at least three reasons for the limited application of cognitive tests to assist clinical decisions about treatment response in ADHD. First, tests of cognitive function that are used by clinicians must support reliable inferences about the presence or absence of change in cognitive function in individual patients. This is in contrast to the use of neuropsychological tests in research studies, where inferences about change are based on the performance of *groups* of children, and are aimed primarily at developing brain-behaviour models of the disorder. In fact, it has been argued that the optimal characteristics of tests designed to detect cognitive change differ substantially from those of most conventional neuropsychological tests (see Collie et al., 2003; Collie et al., 2003; Silbert et al., 2004).

Collie and colleagues suggest that tests that are administered repeatedly should contain equivalent alternate forms, incorporate simple response requirements, and include a brief administration time to minimise the effects of practice. In addition, it is necessary that the performance metrics of such tests incorporate features that facilitate the detection of subtle cognitive change. These include a large range of scores to allow for performance variation, and scores that yield a normal data distribution. Because decisions about individuals are generally made without reference to a matched control group, it is necessary that the effects of repeated administration on a cognitive test be minimal, or if present be understood thoroughly.

A second reason for the limited application of cognitive tests to assist clinical decisions about treatment response in ADHD concerns the analysis of data from individuals. The statistical methods required to guide decisions about cognitive improvement in individuals are not as well developed as the methods used to define cognitive impairment (typically a comparison to some normative range) or those used to compare the cognitive performance of groups of children before and after medication. However, an increasing specialisation in the identification of cognitive change, such as that which can result from concussion, has led to improved methods for guiding decisions about cognitive change in individuals (for a review, see Collie et al., 2002). These methods include Reliable Change Indices (RCIs) and related techniques (the Modified Reliable Change Indices (MRCIs) and Reliability-Stability Index), as well as simple and multiple regression techniques. However, most of these methods cannot be applied within clinical settings because they require data from some healthy comparison group tested over equivalent intervals to the patient of interest, and such data are rarely available for conventional neuropsychological tests (McCaffrey and Westervelt, 1995).

The third issue for the limited application of cognitive tests to determining treatment response relates to the aspects of cognitive function assessed. At present, simple stimulus-response tests are often used to measure change in the cognitive function of individual patients after medication (e.g., the Test of Variables of Attention (TOVA); Greenberg and Waldman, 1993; the VIGIL; Rodgers et al., 2003). Simple tests are attractive because they are generally brief (i.e., they require less than 30 minutes), deliver data that have good metric properties, and are not associated with substantial practice effects. However, as children with ADHD show impairment across a range of cognitive functions (e.g., attention, executive function, and memory; Barnett et al., 2001; Barnett et al., submitted), the measurement of single aspects of cognition may cause true medication-related improvement to be overlooked. Importantly, assessing medication effects with many different cognitive tests increases the time required for testing.

Even if the time required to assess multiple cognitive functions can be minimised there remains an important statistical issue. When the criterion for classifying change is constant (e.g., improvement >1 standard deviation), the probability of detecting improvement or decline by chance increases with the number of tests given. For example, the probability of detecting change of one standard deviation unit in cognitive function using 10 tests is much greater than the probability of detecting change using a single test. To reduce the probability of Type I error as the number of tests in a battery increases, the criterion used to define change should be made more conservative. Ingraham and Aitken (1996) have shown that the experiment-wise error rate can be controlled as the number of cognitive tests increase by increasing both the cut-off score required for a classification of change and the number of tests for which performance must exceed the cut-off score. According to their simulations, with 10 performance measures, a change rule requiring improvement greater than 1.65 SDs ($p < .05$ one-tailed) on two or more tests retains Type I error at less than .05. Another approach to controlling the experiment-wise error rate in multi-test cognitive batteries is to compute a composite score from all performance measures. The justification for this approach is that a criterion for change, such as improvement greater than 1.65 SD on at least two tests, will classify individuals as responders only if they show marked improvement in specific aspects of cognition, but not if they demonstrate subtle improvement across all or most domains. Small but positive changes across all tests will sum and increase the value of the composite score so that it can be differentiated reliably from zero (Rasmussen et al., 2001). In contrast, for individuals in whom no true cognitive change has occurred, change scores on individual cognitive tests would be both positive and negative, making their sum (i.e., the composite score) close to zero. However conclusions based on composite scores are limited to interpretation **defined broadly by all tests in a battery**, and don't allow consideration of performance changes on the specific measures of cognitive function from which the composite score is derived.

Apart from absolute difference scores and percentage change scores, statistical approaches to defining change (for single tests or composite scores) require some estimate of normal variability in performance over time. This normal variability is used to determine whether any change observed in the performance of the individual patient is meaningful (for a review, see Collie et al., 2002). Different change statistics derive this estimate of normal variability from different data sources. For example, some RCIs use the SD of difference scores as their denominator (Rasmussen et al., 2001). The RCI is considered the superior statistic for guiding decisions about change (Collie et al., 2002), although the clinician is faced with the problem of deciding upon the most appropriate control group. In most circumstances, this problem can be overcome by estimating the stability of performance from a matched control group, assessed on the same cognitive tests at the same test-retest intervals as the patient of interest. For ADHD however, the selection of a control group may be problematic because researchers contend that, in addition to subtle impairments in cognitive function, children with ADHD also show greater variability in performance than healthy, age-matched children without ADHD (Douglas, 1999). In fact, some argue that this increased variability is the cognitive hallmark of ADHD (Castellanos and Tannock, 2002), although the magnitude of this variability has not been quantified. If increased within-individual variability is part of the ADHD cognitive syndrome then the best estimate of normal variability in performance (in the absence of medication) would come from a group of children with ADHD who are assessed repeatedly while unmedicated. Thus, it is important to test the hypothesis that children with ADHD show greater variability in cognitive performance than that of age-matched children without ADHD. If the hypothesis is refuted, the stability data obtained from the group of healthy children would also provide an appropriate denominator to guide decisions about the significance of change in cognitive function in children with ADHD following medication.

The aim of the current study was to apply statistical decision rules to the cognitive test data of individual children with ADHD in order to classify whether each showed a cognitive response to medication. The ability of this technique to identify true treatment response was determined by the extent to which classifications of cognitive response were commensurate with classifications of behavioural response in the same children. Although there is some speculation that cognitive and behavioural responses to medication may be independent, previous published findings indicated that both cognitive and behavioural improvement occurred in the same ADHD group (Mehta et al., 2004; Pietrzak et al., 2006). Furthermore, behavioural response is the current standard by which medication response is determined, both clinically and in medication trials (Conners, 2002; Swanson et al., 1993). Therefore, behavioural response was used as a gold standard to determine the utility of the cognitive test and statistical rules in identifying treatment response. Before this analysis, we sought to determine whether the variability in cognitive performance in medication naïve children with ADHD was greater than that in age-, gender- and IQ-matched control children by computing estimates of test stability in the two groups and comparing them directly. To provide statistical power adequate to detect small differences in variability, we also compared the stability of performance in children with ADHD to that of a larger group of healthy children.

Method

Participants

The data were obtained from children who participated in two separate studies (Mollica et al., 2005; Mollica et al., submitted; refer to these papers for more detailed information about participants, materials and procedures). The first participant group included 87 healthy children aged between 8 and 12 years (42 males; $M = 120.10$ months, ±14.47). The remaining groups included 14 children with ADHD aged between 7 and 12 years (12 males, $M = 110.71$ months ±22.38) and 14 healthy children, each of whom was matched to a child with ADHD for gender and age (12 males, $M = 111.50$ months ±20.73; $t(26) = 0.10$, $p = .92$). The FSIQ of these two groups was also equivalent (healthy mean = 104.43, ±5.67; ADHD mean = 100.07, ±8.44; $t(26) = 1.60$, $p = .12$). Prior to enrolment in the study, all children with ADHD had met the DSM-IV symptom cut-off and impairment criteria for ADHD (combined type), as diagnosed by a child psychiatrist. The exclusion criteria for participants in the three groups have been described previously.

Measures

Behavioural measures: The *Child Behavior Checklist*- parent form (CBCL; Achenbach, 1991) was completed for each participant prior to testing. The *Rutter and Graham Interview Schedule* was completed for each child during each testing session (Rutter and Graham, 1968), from which a composite Hyperactivity (RGIS-H) rating was calculated. In addition, the *Abbreviated Conners Rating Scale* (ACRS; Conners, 1985) and the *Children's Impulsiveness Scale* (CIS; Vance and Barnett, 2002) were completed within each testing session for the ADHD group and their matched controls. Total scores for the ACRS and CIS were used.

Table 1. Schedule of testing for the participants with ADHD and their matched control group

	Session A	Two hour break	Session B
Day One	Assessment 1	2.5mg dexamphetamine	Assessment 2
Day Two	Assessment 3	7.5 mg dexamphetamine	Assessment 4
Day Three	Assessment 5		

Cognitive measures: The participants were repeatedly assessed using the CogState battery. This battery comprises seven tasks that assess psychomotor function (detection task), visual attention (identification, matching and monitoring tasks), executive function (working memory and sorting tasks) and memory (learning task). The dependent variable for the tests of psychomotor function and visual attention was mean reaction time (RT). Accuracy (i.e., the percentage of correct responses) was recorded for the tests of executive function and memory. These performance measures were used because they are the most appropriate for measuring cognitive change with children: they cause minimal practice effects, yield normal distributions, and allow enough variation in performance to detect decline and improvement in performance (Mollica et al., 2005). A description of the battery's administration and the seven cognitive tasks has been detailed previously.

Procedure

The large group of healthy children was tested in groups of ten at their primary school. These children completed four administrations of the cognitive battery within a two-hour testing session, with 10-15 minute rest breaks in between administrations. In a separate study, the ADHD and matched control groups completed five administrations of the cognitive battery over three consecutive weekdays: two on day one, two on day two, and one on day three (see Table 1). Each testing session was completed within 30 minutes and participants had a two-hour break between sessions that were conducted on the same day. Immediately after completion of the first and third test sessions, the children with ADHD were administered a dose of stimulant medication. They received 2.5mg of dexamphetamine after session one and 7.5mg of dexamphetamine following session three (see Table 1; these doses are within the range of standard medication doses prescribed by the treating child psychiatrist). None of the matched control children were administered any medication throughout testing. Five children with ADHD were regularly taking stimulant medication prior to their involvement in the study: these children ceased taking their stimulant medication at least 24 hours prior to the first testing session and did not resume until the study was complete. Following each testing session, the assessor completed the RGIS for each child. In addition, the assessor completed the ACRS and CIS for each child in the ADHD group and their matched controls following each testing session.

Statistical Analysis

Treatment of Data from Cognitive Tasks and Behavioural Rating Scales

Each participant's CBCL ratings were converted to standardised T scores. The healthy children's T scores were inspected to ensure that none exceeded 65 (Achenbach, 1991). The RTs for each correct trial were identified and transformed using a logarithmic base 10 (log10) transformation to ensure that the data were suitable for parametric statistical analyses. The mean transformed RT for each individual was computed for the detection, identification, matching and monitoring tasks. For the working memory, sorting and learning tasks, the number of correct responses were recorded and expressed as a percentage of the total number of trials. Arcsine transformations were then applied to these percentage scores in order to normalise the data distributions (Winer, 1971). Once complete, data analysis proceeded in two stages. The first compared variability in behaviour and cognitive function between the different groups to determine the most appropriate denominator for change equations. The second investigated the cognitive and behavioural change scores for each test within the ADHD group and the matched control group.

Stability of Cognitive Performance and Behavioural Ratings in ADHD

For each measure, performance on the initial baseline was excluded because a practice effects was identified previously in this data set, which reflected familiarisation from the first to second assessment (Mollica et al., 2005). In order to keep the experimental design balanced, behavioural data from the first assessment were also excluded for the ADHD participants and their matched controls. Data from the multiple baseline conditions was then submitted to a series of one-way Analysis of Variance (ANOVA) in order to determine the mean square residual (MSr) on each cognitive task (i.e., assessments two, three and four for the large healthy group, and assessments three and five for the ADHD and matched control groups). The square root of the MSr was then calculated to determine the within-subjects standard deviation (WSD; Bland and Altman, 1996) for each group on all of the cognitive measures. The WSD was also calculated using the behavioural ratings of the ADHD group and their matched controls. The F_{max} test statistic (Winer, 1971) was used to compare the magnitude of the estimates of variability for the behavioural measures between the ADHD group and their matched controls. For the WSD of the cognitive measures, two series of F_{max} tests were conducted: the ADHD group was compared to each of the groups of healthy children in separate analyses.

Defining Improvement on Cognitive and Behavioural Measures

The number of individuals in the control and ADHD groups who showed improvement under the low or high dose condition was computed for each behavioural and cognitive measure. For each individual, data from the two baseline assessments (i.e., assessments three and five) were collapsed to form a single, average baseline measure. Exploratory analysis indicated no difference between the baseline assessments for any cognitive or behavioural measure. Performance for the average baseline condition was then subtracted from performance for the low and high dose conditions, respectively. This difference was expressed as a ratio of the WSD to determine the treatment response ratio. For the cognitive and behavioural measures, ratios greater than 1.65 ($p < .05$ one-tailed) were classified as a significant improvement. The number of individuals in the ADHD group showing improvement indicated the sensitivity of the test to stimulant medication. The false positive rate of the cut-off score was determined empirically by computing the number of individuals in the (untreated) control group who showed improvement on each behavioural and cognitive measure.

A response to treatment with stimulant medication might manifest as subtle improvement (i.e., less than the cut-off score) across a range of cognitive measures rather than as large improvement on two or more specific measures. Therefore, cognitive response was also classified using a *composite* score derived from change in performance on the seven cognitive performance measures (e.g., Rasmussen et al., 2001). This z-composite score was computed by summing the seven treatment response ratios for each individual in the ADHD and control groups for the low and high dose conditions. The *SD* of the *z*-composite score was computed from the control group and used as a denominator in the equation:

z-composite score = $\dfrac{\text{individual's sum of treatment response ratios}}{\text{SD of composite scores from the matched control group}}$

Cognitive *z*-composite scores exceeding 1.65 ($p < .05$ one-tailed) were used to classify general cognitive improvement. This method was also used to compute a *z*-composite score for the behavioural measures.

Definition of Cognitive and Behavioural Treatment Response in Individuals

The binomial probability tables of Ingraham and Aitken (1996) were used to determine the number of individuals that demonstrated significant improvement in cognitive or behavioural function following treatment with stimulant medication. For each individual, a cognitive treatment response was defined as a treatment response ratio of 1.65 or greater ($p < .05$, one-tailed) on two or more of the seven cognitive measures. Given the smaller number of behavioural measures, the less conservative treatment response ratio of 1.50 or greater ($p < .05$, one-tailed) on two or more measures was used to classify a behavioural treatment response. To check the model developed by Ingraham and Aitkin (1996) the same rules were applied to the behavioural ratings and cognitive performance of the non-ADHD, non-medicated control group. Any classification of treatment response in this group would be a false positive classification.

The possibility of an individual experiencing more subtle, but general improvement in function was investigated using the z-composite scores. For the cognitive *and* behavioural *z*-composite scores, treatment response was classified if the value exceeded 1.65 ($p < .05$ one-tailed). Therefore, to be classified as a cognitive responder an individual required a treatment response ratio of 1.65 or greater on two or more cognitive performance measures, a cognitive *z*-composite score of 1.65 or greater, or both. To be classified as a behavioural responder an individual required a treatment response ratio of 1.50 or greater on two or more behavioural rating scales, a behavioural *z*-composite score of 1.65 or greater, or both.

Validity of Cognitive Responder Classification

According to the criteria outlined above, each child could be classified into one of four response categories. These categories were: cognitive responder/behavioural responder (true positive classification), cognitive responder/behavioural non-responder (false positive classification), cognitive non-responder/behavioural responder (false negative classification) and cognitive non-responder/behavioural non-responder (true negative classification). The number of individuals within each of these classification categories was identified for both the low- and high-dose conditions separately. The association between the classification of cognitive and behavioural response was determined using the Chi-square test. The strength of the association between the magnitude of behavioural and cognitive response was then determined by calculating the Pearson's correlation coefficient (r) between the behavioural and cognitive *z*-composite scores at both the low- and high-dose conditions. Finally, the presence and strength of the association between the magnitude of cognitive response at the low- and high-dose conditions was examined by computing the Pearson's correlation coefficient between the cognitive z-composite score for both medication conditions. For all analyses, the alpha level used to indicate statistical significance was maintained at $p < .05$. This criterion was selected because this is a novel approach to investigating cognitive change in individual children, hence it was considered important to generate hypotheses for future investigation.

Results

Stability of Cognitive Performance and Behavioural Ratings in ADHD

The WSD for each cognitive measure in the two control groups and the ADHD group is shown in Table 2. The magnitude of the WSD was similar between the ADHD group and both control groups. In fact, no significant differences were found between the ADHD and control groups for the stability of performance on any cognitive measure. In contrast, behavioural measures were significantly more variable in the ADHD group than in the matched controls. Therefore, the groups' respective WSDs were used as the denominator for computation of standardised response magnitude scores for the cognitive and behavioural measures.

Table 2. A comparison of the within-subjects standard deviation (WSD) between a large sample of healthy control children (Control 1), a group of children with ADHD, and their matched control group (Control 2)

Measure		WSD Control 1 (n = 87)	WSD Control 2 (n = 14)	WSD ADHD (n = 14)	F_{max} statistic Control 1 vs. ADHD	F_{max} statistic Control 2 vs. ADHD
Behavioural						
	ACRS	-	0.57	1.94	-	11.58**
	CIS	-	0.73	1.6	-	4.80*
	RGIS-H	-	0.85	1.7	-	4.00*
Cognitive						
(RT)	Detection	0.05	0.05	0.06	1.44	1.44
(RT)	Identification	0.07	0.07	0.05	0.51	1.96
(RT)	Matching	0.07	0.06	0.05	1.96	1.44
(RT)	Monitoring	0.08	0.08	0.08	1.00	1.00
Working memory (Acc)		0.12	0.14	0.10	0.69	1.96
(Acc)	Sorting	0.16	0.13	0.11	2.12	1.40
(Acc)	Learning	0.11	0.13	0.11	1.00	1.40

Note. ACRS = Abbreviated Conners Rating Scale; CIS = Children's Impulsivity Scale; RGIS-H = Hyperactivity Composite score of the Rutter and Graham Interview Schedule; RT = reaction time; Acc = accuracy; * = $p < .05$; ** = $p < .01$.

Defining Treatment Response for Cognitive and Behavioural Measures

The number of individuals classified as showing improvement on each behavioural and cognitive measure in each group is shown in Table 3. Very few controls met the criteria for improvement on any behavioural measure. For the ADHD group, significantly more improvement was identified for the ACRS at the low- and high-dose and the CIS at the high-dose. However, the behavioural z-composite score showed the greatest amount of improvement in the ADHD group as well as yielding no false positive classifications.

For the cognitive performance measures, the number of individuals from the control group classified as showing improvement was uniformly low. The cognitive performance measures that were most sensitive to treatment response in children with ADHD were the detection task at both the low- and high-dose, and the identification and sorting tasks at the high-dose. Once again, the z-composite score was most sensitive to cognitive improvement following stimulant medication, as well as yielding no false positive classifications.

Defining Cognitive and Behavioural Treatment Response in Individuals

The number of individuals in each group whose improvement in behaviour ratings or cognitive performance was sufficient to be classified as a behavioural or cognitive treatment response is shown in Table 4. None of the control children was classified as either a behavioural or cognitive responder. In comparison, a significant proportion of the ADHD group met the criteria for classification as both behavioural (Chi-square = 6.09; $p < .05$) and cognitive (Chi-square = 4.67; $p = .05$) responders at the low-dose medication condition. Furthermore, this proportion more than doubled for both behavioural (Chi-square = 18.12; $p < .01$) and cognitive (Chi-square = 15.56; $p < .01$) responder classifications at the high-dose medication condition.

Validity of Cognitive Responder Classification

Table 5 shows the association between the number of participants in each classification category within the ADHD group for the low- and high-dose medication conditions. No significant association was found between classifications of behavioural and cognitive response for the low-dose condition (Chi-square = 3.76; $p = .10$); however, the majority of cases in this analysis (57%) were classified as behavioural and cognitive non-responders. For the high-dose condition, there was a significant association between classification as a behavioural *and* cognitive responder (Chi-square = 9.55; $p = .01$) with the majority of cases (71%) classified as responders across both domains.

The magnitude of cognitive response observed in individual children with ADHD (as defined by the cognitive z-composite scores) was strongly associated with the magnitude of behavioural response in the same children (as defined by the behavioural z-composite score) for both the low- ($r = .72$; $p < .01$; see Figure 1) and high-dose conditions ($r = .85$; $p < .01$; see Figure 2). Finally, despite being too subtle to yield many cases of treatment response, the magnitude of cognitive response at the low-dose was associated significantly with the magnitude of cognitive response at the high-dose medication condition ($r = .83$; $p < .01$; see Figure 3) for the children with ADHD.

Table 3. Number of participants demonstrating improvement on behavioural and cognitive measures

Measure		Group	Low-dose (Assessment two)	High-dose (Assessment four)
Behavioural				
	ACRS	Healthy	0	0
		ADHD	7**	8**
	CIS	Healthy	2	1
		ADHD	6	8**
	RGIS-H	Healthy	1	2
		ADHD	2	3
	z-composite	Healthy	0	0
		ADHD	3	11**
Cognitive				
	Detection	Healthy	0	0
		ADHD	6**	7**
	Identification	Healthy	0	0
		ADHD	3	4*
	Matching	Healthy	1	2
		ADHD	3	5
	Monitoring	Healthy	0	1
		ADHD	0	2
	Working memory	Healthy	0	0
		ADHD	2	2
	Sorting	Healthy	0	0
		ADHD	1	4*
	Learning	Healthy	0	0
		ADHD	1	1
	z-composite	Healthy	0	0
		ADHD	2	10**

Note. ACRS = Abbreviated Conners Rating Scale; CIS = Children's Impulsivity Scale; RGIS-H = Hyperactivity Composite score of the Rutter and Graham Interview Schedule; * = $p < .05$; ** = $p < .01$.

Table 4. Behaviour and cognitive responders for the ADHD group and their matched controls within each treatment condition

Group	Low-dose (assessment two)	High-dose (assessment four)
Behavioural measures		
Healthy	0	0
ADHD	5*	11**
Cognitive measures		
Healthy	0	0
ADHD	4*	10**

Note. * = $p < .05$; ** = $p < .01$.

Table 5. Summary of the number of children with ADHD in each classification category at both the low- and high-dose conditions

		Cognitive response	Cognitive non-response
Low-dose condition			
	Behavioural response	3	2
	Behavioural non-response	1	8
High-dose condition**			
	Behavioural response	10	1
	Behavioural non-response	0	3

Note. ** = $p < .01$.

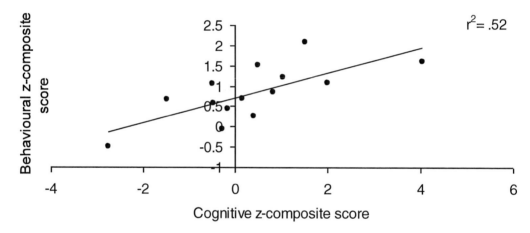

Figure 1. Scattergram of the z-composite scores for participants with ADHD in the low-dose condition.

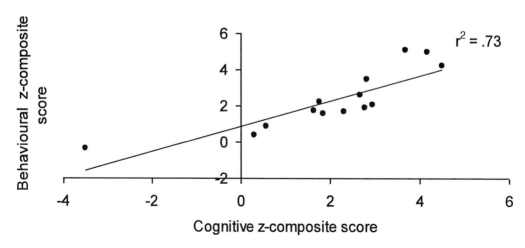

Figure 2. Scattergram of the z-composite scores for participants with ADHD in the high-dose condition.

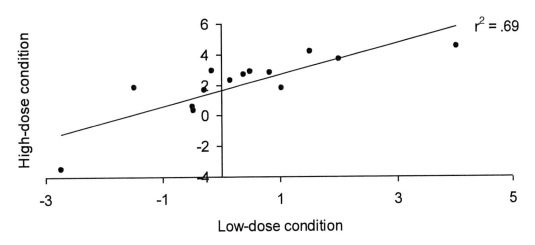

Figure 3. Scattergram of the cognitive z-composite scores for participants with ADHD at the low- and high-dose conditions.

Discussion

The results of this study indicate that the statistical method developed for defining cognitive and behavioural response to stimulant medication was able to identify individuals with ADHD who showed a true positive response without classifying any treatment response in healthy, non-treated children assessed at the same time intervals. By considering simultaneously both the criterion used to define change on each cognitive or behavioural measure as well as the number of measures given, the decision rule yielded a high sensitivity and high specificity to treatment response. Although preliminary, these results suggest that it is possible to develop an evidence-based approach to clinical decision-making about the response to medication of individual children with ADHD. Whereas clinicians are experienced in making decisions about the effectiveness of medication on the basis of change in behavioural symptoms, the utility of cognitive function as a measure of treatment response in individual patients is not clear. By taking what has been learned about cognitive tests and statistical methods from other contexts (e.g., detecting cognitive decline after coronary surgery, concussion management; Collie et al., 2002) and modifying them so that they are appropriate for investigating ADHD, the current study provides a reliable method for identifying treatment response in individual children.

In the children with ADHD, the cognitive and behavioural response to medication (as defined in this study) was remarkably consistent following the high dose of medication. Ten participants (71%) were classified as having responded to the medication on both the behavioural rating scales and cognitive performance measures. Furthermore, the three children who showed no behavioural response also showed no cognitive response (see Table 5). The one inconsistent classification at the high dose occurred when a child who was classified as a behavioural responder did not satisfy the criteria for a cognitive response. However, inspection of the data for this case indicated that the child's cognitive composite score was actually 1.64; hence, she failed narrowly to satisfy the classification criterion for cognitive response (as shown in Figure 2). Thus, in contrast to the work of Sprague and

Sleator (1977) the results of the current study suggest that a single dose of stimulant medication is effective in ameliorating behavioural *and* cognitive symptoms of ADHD.

Of the three children with ADHD who showed no behavioural or cognitive response at the high-dose, two also showed no improvement in cognitive and behavioural function at the low-dose of medication. Interestingly, the characteristics of these two non-responders were qualitatively similar to profiles of children with ADHD that are proposed to demonstrate a poor response to stimulant medication (for a review, see Gray and Kagan, 2000). For example, these two children had clinically significant ratings on the Anxious/Depressed subscale of the CBCL (i.e., scores of 75 and 76; M for responders = 57.75 ±4.90), as well as relatively low baseline ratings of behavioural disturbance on the ACRS (i.e., mean baseline ratings of 4 and 1.5; M for responders = 8.69 ±2.97). One of these children also had a co-morbid diagnosis of Generalised Anxiety Disorder. An increased severity of co-morbid internalizing symptoms such as anxiety and depression has been associated with a poor response to medication within both the cognitive and behavioural domains (DuPaul et al., 1994; Swanson et al., 1978; Tannock et al., 1995). In addition, children who display less severe behavioural symptoms of ADHD, including those measured by the Conners' rating scales (i.e., the ACRS), are less likely to demonstrate medication-related improvement (Buitelaar et al., 1995; Efron et al., 1997; Taylor et al., 1987). Therefore, the non-responsiveness of these two children is consistent with knowledge regarding variables that have been associated with poor response to stimulant medication in ADHD.

The third behavioural and cognitive non-responder at the high medication dose actually showed significant improvement in both cognitive and behavioural function following the low-dose. Although this finding is in contrast with the linear dose-response effects that are typically observed in ADHD (Greenhill, 2001), it also highlights the importance of considering inter-individual variation in medication response. For example, this may represent a case where treatment was optimised at a low-dose of stimulant medication. Previous studies have defined treatment response, albeit primarily based on behavioural symptoms, as occurring with improvement at *either* a low or high medication dose (e.g., Denney and Rapport, 1999). Had this same criterion been used in the current study, the number of true negatives would have been reduced to two and the number of true positive cognitive responders increased to 11 participants (79%).

When considered as a continuous variable, the magnitude of treatment response on the cognitive performance measures was associated in a strong, linear fashion with the magnitude of treatment response on the behavioural rating scales (see Figures 1 and 2). Furthermore, strong dose response relationships were evident in the linear association between cognitive response at the low and high doses (see Figure 3). These results highlight the internal consistency of the decision rules applied to define treatment response. Although factors that predict treatment response have not yet been elucidated (Gray and Kagan, 2000), further investigation of possible predictor variables could occur using the current experimental design and cognitive assessment methodology in a larger sample.

It is important to emphasise that the ability of the current methodology to identify treatment response in children with ADHD is not due only to the statistical criteria applied. The ability of the current study to detect cognitive change was based on a number of important methodological and statistical issues that were considered simultaneously in this study. For example, the performance measures to which the statistical rules were applied were selected because they yielded data that were normally distributed, did not suffer from floor or ceiling effects, and were not subject to range restriction (Mollica et al., 2005; Silbert et al., 2004). Furthermore, these performance measures were generated by a series of cognitive tasks that minimise the effects of practice, could be administered rapidly, and could be performed reliably by healthy children and children with ADHD (Mollica et al., 2005; Mollica et al., submitted). A study of groups of children has also indicated that the psychomotor function, attention, executive function and memory tasks were sensitive to both ADHD (when compared to healthy children) and the effect of stimulant medication (Mollica, et al., submitted). Finally, the current experimental design was developed to allow assessment of the effects of medication in children with ADHD without substantially disrupting clinical practice or requiring that medication be withheld for a substantial period of time. Thus, the development of the statistical approach used here could occur only after the foundations for data generation had been completed. Now that the methodology has been developed, it needs to be validated in other clinical settings using additional types of single-case investigations (e.g., Kent et al., 1999).

Although the current study was designed to identify cognitive change in individual children, the results are also relevant to aetiological models of ADHD. For example, it was surprising to find that the stability of cognitive performance was equivalent in unmedicated children with ADHD and healthy matched controls (see Table 2). There appears to be general consensus in the literature that the cognitive performance of individual children with ADHD is subject to greater variability than in children without the disorder (Douglas, 1999). This variability is detected most reliably on simple stimulus response tasks, and in fact some tasks deliberately entail long administration times to accentuate inconsistent performance within a single assessment session. The reason for this variability in performance is not understood completely. For example, it is reasonable to expect that children whose disorder is characterised by a difficulty in sitting still and remaining 'on task' show unreliable performance on cognitive tasks that are long, repetitive, and simple. Alternatively, the variability in performance could reflect true central nervous system disruption. Recent reviews by Castellanos, Tannock, and colleagues have lead to the hypothesis of a brain-behaviour model that links the performance variability shown by children with ADHD on clinical instruments to disordered temporal processing related to cerebellar dysfunction (see Castellanos and Tannock, 2002; Paule et al., 2000).

In our research involving children and adults with different psychiatric disorders, we have employed brief testing sessions in order to maximise compliance and thereby attain data with the greatest reliability (e.g., Cairney et al., 2001; Maruff et al., 1996; Wood et al., 1999). Consequently, the cognitive tasks used here were designed to keep administration time under 20 minutes. Therefore, in the ADHD group, the normal performance stability at baseline may have occurred because these children were not assessed for long enough, or, because seven different tasks were given within the 20 minute assessment, the duration of each task may have been too brief. Nevertheless, the current results suggest that short administration times did not decrease the sensitivity of the tasks to dysfunctions associated with ADHD. Moreover, these cognitive tasks provide reliable data with attractive statistical properties of normality and homogeneity of variance, which allow them to be investigated using the more powerful parametric statistical techniques.

The data set used here to explore the classification of treatment response in individuals has been previously employed to study medication-related improvement in cognitive function at a group level in ADHD (Mollica et al., submitted). In this previous study, the ADHD group showed significant improvement in performance on the detection, identification, monitoring and learning tasks following the high-dose of stimulant medication. It may have been expected that these tasks would also possess the greatest sensitivity to medication-related improvement in individual children with ADHD. This was not the case, as the tasks for which the greatest number of significant responses occurred were the detection, identification, matching and sorting tasks (see Table 3).

Interestingly, the cognitive composite score provided the greatest sensitivity to treatment response at both the low- and high-dose, while yielding no false positive classifications. In addition, the magnitude of response to treatment as defined by this cognitive composite score was correlated strongly between the low- and high-dose medication conditions. Although preliminary, these data suggests that response to stimulant medication may be characterised by subtle and generalised cognitive improvement rather than as a large improvement in specific aspects of cognitive function. These data are consistent with previous conclusions (Bergman et al., 1991) and highlight the heterogeneous nature of individual medication-response in ADHD. They also show how the performance of individual children with ADHD can be analysed in group studies of ADHD, at least in the form of post-hoc analyses. The statistical methods presented here would be appropriate for such analyses provided that the tests used to assess cognitive function possess the desired metric properties (as discussed above and in Mollica et al., 2005).

The findings of the current study are preliminary, and replication is warranted to address several limitations of this design. First, participants within the ADHD group were recruited from a Child and Adolescent Mental Heath Service and included children who required on-going case management. Consequently, these children were more likely to experience an increased rate of symptom severity and co-morbid difficulties. These clinical features may **have influenced the participant's cognitive** performance as well as their response to stimulant medication. In addition, these children were all diagnosed with ADHD-CT. Given the known differences in cognitive dysfunction across ADHD subtypes (Nigg et al., 2002), replication of this study using participants with primarily inattentive and hyperactive/impulsive subtypes would be of importance prior to clinical application of the methods described herein. Second, replication with a larger sample is warranted to add support to the inferences outlined above. Although clear trends in medication response were observed at the high-dose condition, it is of importance to validate these and the more general findings by using a larger group of participants and increasing the statistical power of analyses. In addition, the behavioural measures were specifically designed to be readministered at such brief intervals; the reliability and validity of these measures when used in this manner could be clarified with additional research. Finally, it may be useful to investigate the application of these cognitive assessment tasks and statistical methodologies across greater test-retest intervals. Although this brief medication trial was practical in minimising disruption to clinical management, increasing the time between assessments could be useful in aiding investigation into more long-term medication effects and potential side-effects. As outlined in a previous study, the integration of double-blind methodology is also important for future investigations to prevent the effects of experimenter bias or participant placebo effects (see Mollica et al., submitted). Overall, however, the statistical method explored within the current study appears to be a novel and effective way of empirically investigating the effects of stimulant medication on the cognitive and behavioural function of individual children with ADHD.

The next study will outline a novel "intensive design" method for assessing stimulant medication-related improvement in cognitive function in children with ADHD.

Study 2

Stimulant medication is currently the most effective pharmacological treatment for ADHD, although medication-responsiveness can vary according to many factors such as the severity of ADHD symptoms, medication dose, or the presence of co-morbid disorders (DuPaul et al. 1994; Buitelaar et al. 1995; Denney and Rapport 1999). The rate of true non-response is less than 25% (Elia et al. 1991; Santosh and Taylor 2000; Clarke et al. 2002).

In addition to the core symptoms of ADHD, impairment in cognitive function is reliably observed in ADHD and is an important aspect of the clinical presentation of this disorder (Denney and Rapport 2001; Castellanos and Tannock 2002). Further, improvement in cognitive function remains an important end point in studies assessing the effectiveness of novel and established therapies for ADHD (Gittelman-Klien and Klein 1975, 1976). Recent open-label and placebo-controlled trials demonstrate that stimulant medication improves cognitive function (de Sonneville et al. 1994; Efron et al. 1997; Bedard et al. 2003), although the majority of these studies base such conclusions on observations of improved performance on measures of speeded reaction time or inhibition. Although these improvements are reliable, their magnitude is generally moderate (an average effect size of 0.6) and less than the magnitude of improvement in ADHD symptoms (Greenhill 2001; Conners 2002). A small number of studies have measured the effect of stimulant medication on higher-level cognitive functions such as executive function or memory in prospective studies and reported equivocal results (Gittelman-Klien and Klein 1975, 1976; Tannock and Schachar 1992; Aman et al. 1998; Scheres et al. 2003). For example, some find no stimulant medication-related improvement in higher cognitive functions (Gittelman-Klien and Klein 1975; Scheres et al. 2003), or where improvement is detected, the magnitude is small (effect sizes <0.5; Gittelman-Klien and Klein 1976; Brodeur and Pond 2001).

There are a number of explanations for the different effects of stimulant medication observed across different cognitive functions. First, stimulant medication may exert greater effects on the simple cognitive functions required for tests such as the CPT, stop, and vigilance tasks compared to the more complex and multifactorial cognitive functions required for performance on executive function and memory tasks. Consequently, complex cognitive functions may require higher doses of medication to yield improvements of the same magnitude as that found for simple cognitive tasks (Swanson et al. 1991). Third, simple attention and motor tasks may be more sensitive to the measurement of cognitive change than tests of higher cognitive functions because these tests generally deliver more observations or trials per assessment, yield interval level data (e.g., reaction time) that are not constrained by floor or ceiling effects, and are less susceptible to the effects of practice than tests of higher cognitive function. For example, tasks such as the Wisconsin Card Sorting Task (WCST) and Tower of London (TOL) are commonly employed in ADHD research, yet they often yield restricted ranges of possible scores and may induce large practice effects when given repeatedly. These factors can reduce the sensitivity of such tests to subtle cognitive change (Basso et al. 1999; Collie et al. 2003). Finally, poor study designs can impede the detection of change. For example, inadequate control for the effects of repeated assessment, such as that found with natural history studies or studies without a crossover design, can produce data sets that obscure true change in cognitive function and promote erroneous inferences (McCaffrey and Westervelt 1995). Thus, to measure simultaneously the magnitude of stimulant medication-related improvement on tasks requiring either simple or complex cognitive functions, and to compare this to the magnitude of improvement in symptoms, it is necessary to control practice effects and to utilise tests that yield data that are optimal for the statistical analysis of change.

We have recently modified a series of tests of psychomotor, attentional, executive and memory function so that they can be given rapidly and repeatedly, do not give rise to practice effects once individuals are familiar with the task requirements, and generate optimal data for the assessment of change. These tests can be performed easily by children and are sensitive to cognitive changes associated with head injury, and fatigue and alcohol intoxication in adults (Mollica et al. 2005; Falleti et al. 2003). Therefore, these cognitive tests may be also useful for assessing stimulant-related cognitive change in children with ADHD.

Two experimental designs are typically used to study the effects of stimulant medication on cognitive function in ADHD: open-label and placebo controlled trials. In open-label trials, the participants, their parents and the assessors are aware of medication status, which can potentially influence performance or ratings. Also, the effect of repeated assessment cannot be determined because there is generally no appropriate unmedicated control group. Double-blind, placebo-controlled crossover trial designs are now common in ADHD research and this methodology does control the effects of expectancy and practice. However, such trials are expensive and require that children with ADHD have their medication withheld for a substantial period of time in order to complete the placebo condition (de Sonneville et al. 1994). The withholding of medication or disruption to children's conventional clinical management may be warranted when there is good *prima face* evidence for medication efficacy and broader clinical trials are required for regulatory approval. However, hypothesis-generating studies about the relationship between cognitive functions, symptoms, medication and the disorder may not warrant such disruption or expense. Given that the half-life of stimulant medication is generally short, it is possible to design a study in which practice effects are controlled, children are assessed on and off their medication at brief retest intervals, and reliable conclusions about the effect of medication on cognitive function can be made. Children with ADHD can act as their own controls by being assessed before and after taking medication on the same day (AB design). Repeating the cognitive assessments each day while the children are unmedicated, and also at the end of the trial, provides an estimate of any practice effects that may have arisen from the repeated assessment (ABA design). Furthermore, different doses of stimulant medication can be given on different days to determine whether dose-response relationships exist for any of the cognitive functions measured (ABACA design). Provided the assessments are brief, children and their parents are not inconvenienced and there will be minimal disruption to the child's conventional clinical management (Mollica et al. 2004). Uhlenhuth et al. (1977) and Klein (2008) strongly support such "intensive designs" to determine whether a given child is actually responding to a particular treatment.

The aim of the current study is to investigate the effect of low and high doses of stimulant medication on psychomotor, attentional, executive and memory function and compare this to the effect on symptoms in children with ADHD assessed using an ABACA experimental design. The null hypothesis is that there would be no differences in performance between the baseline and the low- or high-dose medication conditions in the ADHD group.

Method

Participants
The participants included a clinical group of 14 children with ADHD [12 males; 110.71 (22.38) months; full scale IQ 100.07 (8.44)] and a group of healthy children matched to the ADHD group for gender, age and full scale IQ [12 males; 111.50 (20.73); $t(26) = 0.10$, $p=.92$; full scale IQ 104.43 (5.67); $t(26) = 1.60$, $p=.12$]. All children with ADHD met DSM-IV criteria for ADHD, combined type (ADHD-CT), defined by a semi-structured clinical interview with the child's parent(s) [Anxiety Disorders Interview Schedule for Children (A-DISC) (Silverman and Albano 1996)] and a parent and/or teacher report assessing the core symptom domains of ADHD-CT being greater than 1.5 standard deviations above the mean for a given child's age and gender [Abbreviated Conners' Rating Scale (ACRS) score 21.48 (4.73) (Conners 1997)]. The children were all medication naïve and met the inclusion criteria of living in a family home (and not in an institution) and attending normal primary schools. Exclusion criteria for the ADHD group included: colour blindness, hearing impairment, a history of major neurological impairment, or a Full Scale Intelligence Quotient (FSIQ) below 80. Five participants had co-morbid psychiatric diagnoses (oppositional defiant disorder $n = 3$, generalised anxiety disorder $n = 2$).

Healthy control participants were recruited from a primary school and selected to match each ADHD participant for age, gender and full scale IQ. Exclusion criteria for the healthy participants included: colour blindness; hearing impairment; past or present psychiatric or neurological disorder; an estimated FSIQ below 80; or a T score of 60 or greater on any subscale of the *Child Behaviour Checklist- Parent and Teacher Form* (CBCL; Achenbach, 1991). No child in the healthy group received any medication during the study. No child withdrew or was excluded from the study. The study was approved by institutional ethics committees, and all children and their parents gave informed consent before beginning the study.

Measures
Intelligence testing. All of the children with ADHD completed the Wechsler Intelligence Scale for Children- Fourth Edition (WISC-IV) (Wechsler, 2004).

Clinical measures. The CBCL attention subscale T score was used to compare the participant groups. The ACRS was used to rate the children's behaviour within each testing session (approximately 30 minutes duration). The Children's Impulsiveness Scale (CIS; Vance and Barnett 2002) comprises 10 items that list and describe common features of behavioural impulsivity. The assessor is required to indicate the frequency of these behaviours as displayed by each child according to a four-point scale ranging from zero (*absence*) to three (*continuous*). The total CIS score was analysed in the present study. The *Revised Children's Manifest Anxiety Scale* (RCMAS; Reynolds and Richmond 1978) was used to gain subjective reports of anxiety. The total RCMAS score was analysed in the present study. The *Rutter and Graham Interview Schedule* (Rutter and Graham 1968) was employed to provide information about symptoms of anxiety and hyperactivity. Composite anxiety (RGIS-A) and hyperactivity (RGIS-H) ratings were analysed in the present study.

Cognitive measures. The cognitive assessment battery was presented on lap top computers complete with headphones. All tasks within the battery were adaptations of standard neuropsychological and experimental psychological tests, and assess a range of cognitive functions such as psychomotor speed, attention, decision-making and working memory. This battery required approximately 15-20 minutes to complete. It consisted of seven tasks in the form of card games that were presented in succession on a green background. In order to aid individuals with the task, written instructions were presented to the left of the screen to indicate the rule for each new task. Participants were then given an interactive demonstration and, once they had successfully completed a sufficient number of practice trials to demonstrate their awareness of the rules, the task began.

A grey keyboard resembling a computer keyboard appeared in the lower half of the computer screen and the cards associated with each task were presented in the upper half. Participants were only required to respond with two keys throughout the entire battery by using the 'D' or 'K' keys. The beginning of each new task was indicated with a shuffling of the cards. An error beep sounded when an individual pressed an incorrect key at any time. Each trial was time limited and the same error beep sounded if a response was not made within the required time. Participants were able to pause the test at any stage using the 'Escape' key. The dependent variables recorded for each task included reaction time (RT) and accuracy (i.e., the percentage of correct responses). The seven tasks included in the battery have been described previously (Faletti et al. 2003).

Procedure

A parent/guardian for each child completed the CBCL prior to testing. Each participant completed five testing sessions over three consecutive weekdays at 1000 hours and 1300 hours: two on day one, two on day two, and one on day three. Each testing session was completed within 30 minutes and participants had a two-hour break between sessions that were conducted on the same day. During this break, the children with ADHD had free time with a parent/guardian in a waiting room with a play area. The healthy children were tested at their primary school and participated in regular school activities between same-day testing sessions.

The children were tested individually in a small, quiet room. At the beginning of each session participants completed the RCMAS independently. Next, children were seated at a computer and instructed to position their headphones. They were informed that instructions for each task would appear to the left of the screen and that they were required to complete each task as quickly and as accurately as possible. They were also told that an error beep would sound each time they made a mistake and that they could briefly pause the test if needed. The experimenter was seated next to participants throughout each trial and remained silent once the administration had begun. The children were informed that the experimenter was unable to provide assistance with the task itself; however, assistance was provided if a child demonstrated difficulty in reading or comprehending the task instructions. Following completion of the computerised task, the experimenter completed the ACRS, CIS and the RGIS for each child based on his or her behaviour during the entire testing session.

Immediately after completion of the first and third sessions, the children with ADHD were administered a dosage of stimulant medication. They received 2.5mg of Dexamphetamine after session one and 7.5mg of Dexamphetamine following session three. The healthy children were not administered any medication throughout testing.

Statistical Analysis

Each participant's CBCL ratings were converted to standardised T scores. The healthy children's T scores were inspected to ensure that none exceeded 60. For each of the cognitive measures the number of correct responses was recorded and expressed as a percentage of the total trials. Arcsine transformations were then applied to these percentage scores in order to normalise the distributions of data in each group (Hopkins 2008). The reaction times (RTs) for each correct trial were identified and transformed using a log10 transformation. The mean log10 RT for each individual was computed for each test.

For each measure, performance on the initial baseline was excluded because it is known that familiarisation can occur from the first to second assessment in healthy children (Mollica et al. 2005). In order to keep the experimental design balanced, clinical data from the first assessment was also excluded. Data for the remaining two baseline assessments (assessments three and five) were then collapsed to form a single average baseline measure for each dependent variable. Exploratory analysis had indicated no difference between the baseline assessments for any cognitive measure.

Cognitive performance and clinical ratings were compared between-groups only for the average baseline condition, as both groups were not medicated at the time of assessments two and four. For each statistically significant between-group difference, the magnitude of that difference was calculated using an estimate of effect size (Cohen's d; Cohen 1988).

The main hypotheses in the study were tested by the presence of differences in performance between the baseline and the low- or high-dose medication conditions in the ADHD group. These were tested statistically by setting two orthogonal t-tests (average baseline versus low-dose medication condition, and average baseline versus high-dose medication condition) within a repeated measures analysis of variance (ANOVA). Inspection of any change in performance over the same assessment schedule in children without ADHD (using identical analyses) allowed us to determine the effect of repeated cognitive assessment independent of the state of disorder or medication. The magnitude of significant within-group differences was computed using a repeated measures effect size estimate (Dunlap's d; Dunlap et al. 1996). Data from the two groups were not compared directly in an ANOVA because this would have lead to an unbalanced design with medication status and group confounded.

Despite the relatively large number of statistical tests conducted in this study, the alpha level used to indicate statistical significance was maintained at $p \leq .05$. That criterion was selected because this is a relatively new approach to understanding the effects of medication on cognitive function and should therefore err toward generating hypotheses for future research. However, theoretical inferences were protected from false positive results because, even though many measures were used in the study, any change detected in the different cognitive and clinical measures following medication in the ADHD group was by definition correlated. In addition, by considering both the statistical significance and effect size of each comparison it was possible to identify any significant but meaningless differences ($d < 0.2$).

Results

Between-group Comparison on the Average Baseline Condition

Group means for the clinical and cognitive measures on the average baseline condition are displayed in Table 1, together with the results of *t*-tests and an estimate of effect sizes. The ADHD group received significantly higher ratings for all of the clinical measures of ADHD-related symptomatology. The magnitude of the group differences was large on each of these measures (d = 2.40 to 4.60). The groups did not differ on subjective or objective measures of anxiety. For measures of cognitive function, the performance speed of the ADHD group was significantly slower on the detection, identification and simple matching tasks with these group differences being large in magnitude ($ds > 1.00$).

Table 1. Between-group comparison of clinical and cognitive measures at baseline

Measures		Group	M	SD	t (26)	d
Clinical						
	ACRS	healthy	1.89	1.42	-6.29**	2.58
		ADHD	8.18	3.46		
	CIS	healthy	2.93	1.43	-6.18**	2.40
		ADHD	7.43	2.32		
	RGIS-H	healthy	2.46	0.82	-6.21**	2.47
		ADHD	5.46	1.61		
	RGIS-A	healthy	0.00	0.00	-1.79	0.96
		ADHD	0.18	0.37		
	CMAS	healthy	7.89	4.36	-1.19	0.46
		ADHD	10.57	7.22		
	CBCL-Att	healthy	52.64	3.52	-11.92**	4.60
		ADHD	73.07	5.36		
Cognitive						
	Detection	healthy	2.54	0.08	-3.94**	1.52
		ADHD	2.69	0.11		
	Identification	healthy	2.80	0.09	-2.63**	1.01
		ADHD	2.91	0.12		
	Simple matching	healthy	2.91	0.09	-3.56**	1.36
		ADHD	3.04	0.11		
	Monitoring	healthy	2.63	0.09	-0.64	0.24
		ADHD	2.66	0.12		
	Working memory	healthy	1.30	0.12	0.46	0.18
		ADHD	1.27	0.22		
	Complex matching	healthy	1.25	0.20	1.40	0.53
		ADHD	1.14	0.23		
	Learning	healthy	1.03	0.12	1.33	0.50
		ADHD	0.97	0.12		

Note. ACRS = Abbreviated Conners Rating Scale; CIS = Children's Impulsivity Scale; RGIS-H = Rutter and Graham Interviewing Schedule- Hyperactivity score; RGIS-A = Rutter and Graham Interviewing Schedule- Anxiety score; CMAS = Children's Manifest Anxiety Scale; ** indicates significant improvement from baseline, $p < .01$.

Within-group Analysis of Change on Clinical and Cognitive Measures

Group means for the clinical measures at each assessment are displayed in Table 2. A summary of the results of the *t*-tests used to compare performance between assessments and estimates of effect sizes are shown in Table 3. For the healthy children, no differences were observed between average baseline and the second or fourth assessments on any clinical measure. For the ADHD group, a significant reduction in symptom severity from the average baseline was found on the ACRS, CIS and RGIS-H. On each of these measures, the magnitude of symptom reduction from baseline to the high-dose condition was substantially greater than that observed for the low-dose condition. Figure 1 illustrates effect sizes for the change in the clinical measures of both groups across assessment conditions.

Table 2. Group means (SD) for the clinical and cognitive measures, within-group comparison across conditions and r^2 values for the linear trendline equation

Measures		Group	Average baseline	Assessment 2 (low dose)	Assessment 4 (high dose)	r^2
Clinical						
	ACRS	healthy	1.89 (1.42)	2.21 (1.93)	1.93 (1.44)	
		ADHD	8.18 (3.46)	4.86 (2.51)**	3.79 (1.63)**	0.92
	CIS	healthy	2.93 (1.43)	2.50 (1.34)	3.00 (1.30)	
		ADHD	7.43 (2.32)	5.71 (2.55)**	4.64 (1.91)**	0.98
	RGIS-H	healthy	2.46 (0.82)	2.71 (1.33)	2.64 (1.01)	
		ADHD	5.46 (1.61)	4.36 (1.78)**	3.21 (1.37)**	0.99
	RGIS-A	healthy	0.00	0.00	0.00	
		ADHD	0.18 (0.37)	0.21 (0.43)	0.07 (0.27)	0.65
	RCMAS	healthy	7.89 (4.36)	7.79 (3.79)	7.57 (4.01)	
		ADHD	10.57 (7.22)	12.00 (8.05)	10.64 (7.07)	0.01
Cognitive						
	Detection	healthy	2.54 (0.08)	2.56 (0.06)	2.55 (0.07)	
		ADHD	2.69 (0.11)	2.60 (0.08)**	2.57 (0.07)**	0.92
	Identification	healthy	2.80 (0.09)	2.81 (0.06)	2.84 (0.09)	
		ADHD	2.91 (0.12)	2.87 (0.14)	2.86 (0.14)*	0.89
	Simple matching	healthy	2.91 (0.09)	2.90 (0.09)	2.93 (0.10)	
		ADHD	3.04 (0.11)	3.03 (0.11)	3.00 (0.12)	0.92
	Monitoring	healthy	2.63 (0.09)	2.67 (0.12)	2.67 (0.08)	
		ADHD	2.66 (0.12)	2.67 (0.17)	2.60 (0.11)**	0.63
	Working memory	healthy	1.30 (0.12)	1.27 (0.18)	1.25 (0.15)	
		ADHD	1.27 (0.22)	1.18 (0.21)	1.28 (0.18)	0.01
	Complex matching	healthy	1.25 (0.20)	1.15 (0.23)#	1.20 (0.18)#	
		ADHD	1.14 (0.23)	1.05 (0.17)	1.17 (0.18)	0.06
	Learning	healthy	1.03 (0.12)	0.97 (0.11)##	1.01 (0.16)	
		ADHD	0.97 (0.12)	0.97 (0.16)	1.03 (0.14)*	0.75

Note. ACRS = Abbreviated Conners Rating Scale; CIS = Children's Impulsivity Scale; RGIS-H = Rutter and Graham Interviewing Schedule- Hyperactivity score; RGIS-A = Rutter and Graham Interviewing Schedule- Anxiety score; CMAS = Children's Manifest Anxiety Scale; ** indicates significant improvement from baseline, $p < .01$; * indicates significant improvement from baseline, $p < .05$; ## indicates significant decline from baseline, $p < .01$; # indicates significant decline from baseline, $p < .05$.

Group mean performance on each of the cognitive measures is shown in Table 2. A summary of the results of the *t*-tests used to compare performance between assessments and estimates of effect size are shown in Table 3. For the healthy children, no change in cognitive performance was identified between the average baseline and the second or fourth assessments. A significant decline in performance accuracy was observed for the second and fourth assessments on the complex matching task and the second assessment on the learning task.

Table 3. t-statistics and effects size estimates of the change in clinical and cognitive measures across assessment conditions within the healthy and ADHD groups

		Baseline v low		Baseline v high	
Measure	Group	t	d	t	d
Clinical					
ACRS	healthy	-0.87	0.18	-0.43	0.02
	ADHD	4.28**	1.06	5.12**	1.52
CIS	healthy	-0.96	0.31	0.35	-0.05
	ADHD	3.78**	0.70	4.63**	1.30
RGIS-H	healthy	0.80	-0.22	0.75	-0.19
	ADHD	3.08**	0.65	7.87**	1.48
RGIS-A	healthy	-	-	-	-
	ADHD	0.43	-0.09	-1.38	0.32
CMAS	healthy	-0.26	0.02	-0.82	0.07
	ADHD	1.57	-0.18	0.13	-0.01
Cognitive					
Detection	healthy	0.96	-0.20	0.49	-0.06
	ADHD	-4.27**	0.90	-5.82**	1.07
Identification	healthy	0.49	-0.11	1.61	-0.36
	ADHD	-1.39	0.33	-2.37*	0.41
Simple matching	healthy	-0.60	0.12	1.02	-0.25
	ADHD	-0.92	0.14	-1.51	0.33
Monitoring	healthy	1.53	-0.32	1.84	-0.43
	ADHD	0.37	-0.05	-3.08**	0.54
Working memory	healthy	0.90	-0.25	1.35	-0.40
	ADHD	1.78	-0.45	-0.13	0.04
Complex matching	healthy	2.27[#]	-0.47	2.11[#]	-0.31
	ADHD	-1.86	0.43	-0.63	0.17
Learning	healthy	2.98[##]	-0.50	0.53	-0.14
	ADHD	-0.17	0.03	-2.47*	0.45

Note. ACRS = Abbreviated Conners Rating Scale; CIS = Children's Impulsivity Scale; RGIS-H = Rutter and Graham Interviewing Schedule- Hyperactivity score; RGIS-A = Rutter and Graham Interviewing Schedule- Anxiety score; CMAS = Children's Manifest Anxiety Scale; ** indicates significant improvement from baseline, $p < .01$; * indicates significant improvement from baseline, $p < .05$; [##] indicates significant decline from baseline, $p < .01$; [#] indicates significant decline from baseline, $p < .05$.

The magnitude of these performance declines was low to moderate according to convention ($d = -0.13$ to -0.49; Cohen 1988). For the ADHD group, a significant improvement in performance on the detection task was observed from the average baseline to the low-dose *and* high-dose medication conditions. The magnitude of this improvement was greater for the high-dose condition (see Table 3). Improvements from the average baseline to the high-dose medication condition were also observed for performance speed on the identification and monitoring tasks, and performance accuracy on the learning task. The magnitude of improvements in performance for the ADHD group ranged from moderate to high ($d = 0.41$ to 1.04). Figure 1 displays the effect sizes of performance change on the cognitive measures for both groups across assessment conditions.

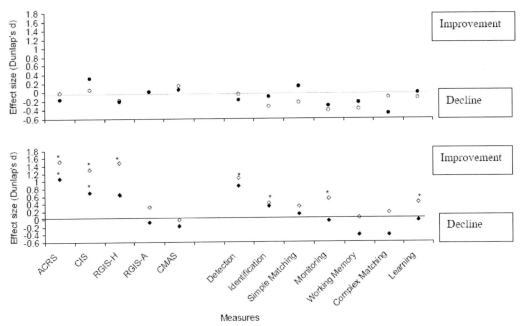

Note. • =lowdose-baseline comparison and ○=highdose-baseline comparison;
* = significant improvement in comparison to the average baseline condition.

Figure 1. Changes in the clinical and cognitive measures across conditions for the healthy (upper) and ADHD (lower) groups, as estimated by effect size.

Discussion

The results of the study indicate that a novel within-subjects design can be used to examine the short-term effects of stimulant medication on cognitive and behavioural functions in children with ADHD, without substantially altering their clinical management. First, as expected, children with ADHD showed high levels of disordered behaviour when unmedicated, although self-reported and objective ratings of anxiety were within normal range. With medication, the levels of ADHD symptoms and hyperactive behaviour were reduced in a dose-dependent fashion. When considered against the conventions for determining experimental effect sizes, the magnitude of improvement with the highest dose was large (Cohen, 1988). Despite this reduction however, there was still overlap between distributions of pre- and post-medication symptom ratings. Thus, stimulant medication did not ameliorate symptoms completely in children with ADHD, even at high doses.

Interestingly, the unmedicated children with ADHD performed worse than age-, IQ-, and gender-matched controls only in the speed of psychomotor function (detection task) and the speed of visual attentional functions (identification and simple matching tasks). The magnitude of these impairments ranged from moderate to large. Following the low dose of stimulant medication, performance improved significantly only in the speed of psychomotor function (detection task). However at the higher dose, improvements were observed in the speed of psychomotor function (detection task), visual attention (identification and monitoring) and in the accuracy of learning. Although the magnitude of these medication-related improvements was moderate to large, it was more subtle than that observed for the response of behavioural symptoms to stimulant medication in the same children (Figure 1).

The larger effect of stimulant medication on ADHD symptoms relative to cognitive function has been reported previously and described in research reviews; yet there is currently no explanation for this difference (Greenhill 2001; Conners 2002). It is possible that the cognitive sequelae of ADHD are more complex than the characteristic behavioural symptoms and are therefore more treatment resistant. Alternatively, the cognitive dysfunctions may be comparatively less severe than the behavioural problems, thus the magnitude of improvement required to normalise cognitive function is substantially less. The results of the current study support the latter proposal. First, the magnitude of group differences at baseline was substantially greater for the behavioural measures (see Table 1). In addition, observation of the means in Table 2 shows that during the high-dose medication condition, the ADHD group was comparable to the healthy group across most cognitive measures (Cohen's $d < 0.30$ for the detection, identification, working memory, complex matching and learning tasks) but not those of clinical function. Thus, the larger effect of stimulant medication on behavioural symptoms compared to cognitive function in children with ADHD may simply reflect a greater degree of dysfunction within the behavioural domain.

Although the interrelationships between cognitive function, behavioural symptoms and stimulant medication are central to current neurobiological theories of ADHD, surprisingly few studies have investigated these factors concurrently using a prospective design. Where these studies have been conducted, the results have varied between them. Some variation in research findings may reflect methodological differences, such as the types of tests used, the sample sizes studied, inclusion/exclusion criteria for ADHD, or the time period over which pre and post-medication assessment occurred. Despite this variability, there is consistent evidence that the speed of performance is improved following stimulant medication on relatively simple speeded response tasks (vigilance and reaction time tasks). This also occurred in the current study, where performance improved on the psychomotor and visual attention tasks following medication. Unfortunately, as few studies have reported the medication-related effect sizes (or provided statistics sufficient for their post-hoc calculation) it is not possible to determine whether the magnitude of improvement differs between experimental conditions. Alternatively, some studies report that the magnitude of improvement in performance on simple speeded psychomotor or visual attentional tasks shows a linear response relationship with the dose of stimulant medication (Barkley et al. 1991; O'Toole et al. 1997). Although within the current study improvements over baseline at the low and high medication dose were observed only on the measure of psychomotor function, linear trends were fitted to the mean cognitive performance of the ADHD group across the baseline, low and high dose conditions. These trends were significant for the detection, identification and simple matching tasks, with the variance in performance explained by medication approximately 90% for each task. Thus, robust dose response relationships were also detected. Medication was also found to facilitate performance accuracy on the learning task within the present study. Again, a significant linear trend was observed with approximately 75% of the variance in performance explained by medication dose. Similarly, other studies have reported improvement in memory and learning with low to moderate doses of stimulant medication in children with ADHD (Sprague and Sleator 1977; Evans et al.1986; Barkley et al. 1989). This consistency of outcome on learning task performance is also provocative, as memory impairment is not generally considered within current cognitive neuroscience models of ADHD.

Cross-sectional studies of cognitive function in ADHD consistently find impairment in executive function (for a review, see Pennington and Ozonoff, 1996; Kempton et al., 1999; Barnett et al., 2001, 2005; Shallice et al., 2002; Vance et al. 2003). However, we found no impairment on the tasks used to assess executive function in the current study. Although the performance accuracy of children with ADHD was lower on the tasks of executive function (working memory and complex matching tasks), these group differences did not reach statistical significance. The magnitude of impairment was, however, moderate ($ds < 0.54$), suggesting this may have been due to insufficient statistical power. The tasks that are typically employed to investigate executive dysfunction in children with ADHD vary greatly, but often include the WCST (Seidman et al. 1997; Houghton et al. 1999; Klorman et al. 1999), TOL (Aman et al. 1998; Nigg et al. 2002; Sonuga-Barke et al. 2002) or the spatial working memory task from the CANTAB battery (SWM; Barnett et al. 2005). We did not use such tests because we considered them inappropriate for the detection of change within a study that repeatedly assessed cognition using brief test-retest intervals. Future studies should examine newly developed measures of executive function for children that are suitable for repeated administration. This is important because executive dysfunction is known to be associated with increased academic and social difficulties (Clark et al. 2002).

One limitation of this study design was that the children, their parents and the assessor were always aware of medication status and the specific dose administered. In order to blind people completely to the medication status, the current design could be modified so that the drug (at different doses) or placebo were not identifiable and then randomised. Such measures were not taken within the current study, as we primarily sought to determine the practicality and statistical power of the design for assessing the effects of medication on symptoms and behaviour. The overall findings of this study demonstrate the design to be a novel, effective method for doing so.

References

Achenbach TM. (ed). *Manual for the Child Behaviour Checklist / 4-18 and 1991 Profile.* University of Vermont Department of Psychiatry: Burlington VT, 1991.

Aman CJ, Roberts RJ, Pennington BF. 1998. A neuropsychological examination of the underlying deficit in Attention Deficit Hyperactivity Disorder: Frontal lobe versus right parietal lobe theories. *Develop. Psychol.* 34: 956-969.

Aman MG, Turbott SH. 1991. Prediction of clinical response in children taking methylphenidate. *J. Autism Dev. Disord.* 21: 211-228.

American Psychiatric Association. (ed). *Diagnostic and Statistical Manual of Mental Disorders, Fourth edition.* Author: Washington DC, 1994.

Barkley RA, DuPaul GJ, McMurray MB. 1991. Attention Deficit Disorder with and without Hyperactivity: Clinical response to three dose levels of methylphenidate. *Pediatrics* 87: 519-531.

Barkley RA, McMurray MD, Edelbrock CS, Robbins BA. 1989. The response of aggressive and nonaggressive ADHD children to two doses of methylphenidate. *J. Am. Acad. Child. Adol. Psychiatry* 28: 873-888.

Barnett R, Maruff P, Vance A, Wood K, Costin J, Luk E. submitted. Executive and cognitive dysfunction in ADHD. *A. N. Z. J. Psychiatry*.

Barnett R, Vance A, Maruff P, Luk SL, Costin J, Pantelis C. 2001. Abnormal executive function in attention deficit hyperactivity disorder: The effect of stimulant medication and age on spatial working memory. *Psychol. Med.* 31: 1107-1115.

Basso M R, Bornstein R A, Lang J M. 1999. Practice effects of commonly used measures of executive function across twelve months. *Clin. Neuropsychol.* 13: 283-292.

Bedard AC, Ickowicz A, Logan GD, Hogg-Johnson S, Schachar R, Tannock R. 2003. Selective inhibition in children with Attention-Deficit Hyperactivity Disorder off and on stimulant medication. *J. Abnorm. Child Psychol.* 31: 315-327.

Bergman A, Winters L, Cornblatt B. 1991. Methylphenidate: Effects on sustained attention. In *Ritalin: Theory and Patient Management,* Greenhill L, Osman B (eds). Maryann Liebert: New York; 223-231.

Bland JM, Altman DG. 1996. Statistical notes: Measurement error. *Br. Med. J.* 313: 744.

Brodeur DA, Pond M. 2001. The development of selective attention in children with Attention Deficit Hyperactivity Disorder. *J. Abn. Child Psych.* 29: 229-239.

Buitelaar JK, van der Gaag RJ, Swaab-Barneveld H, Kuiper M. 1995. Prediction of clinical response to methylphenidate in children with Attention Deficit Hyperactivity Disorder. *J. Am. Acad. Child Adolesc. Psychiatry* 34: 1025-1032.

Cairney S, Maruff P, Vance A, Barnett R, Luk E, Currie J. 2001. Contextual abnormalities of saccadic inhibition in children with attention deficit hyperactivity disorder. *Exp. Brain Res.* 141: 507-518.

Castellanos FX, Tannock R. 2002. Neuroscience of Attention-Deficit/Hyperactivity Disorder: the search for endophenotypes. *Nat. Rev. Neurosci.* 3: 617-628.

Clark C, Prior M, Kinsella G. 2002. The relationship between executive function abilities, adaptive behaviour, and academic achievement in children with externalising behaviour problems. *J. Child Psychol. Psychiatry* 43: 785-796.

Clarke AR, Barry RJ, Bond D, McCarthy R, Selikowitz M. 2002. Effects of stimulant medications on the EEG of children with Attention-Deficit/Hyperactivity Disorder. *Psychopharm.* 164: 277-284.

Cohen J. 1988. Statistical power for the behavioural sciences 2nd ed., Hilsdale, NJ, Lawrence Erlbaum.

Collie A, Darby DG, Falleti MG, Silbert BS, Maruff P. 2002. Determining the extent of cognitive change after coronary surgery: a review of statistical procedures. *Ann. Thorac. Surg.* 73: 2005-2011.

Collie A, Maruff P, Makdissi M, McCrory P, McStephen M, Darby D. 2003. CogSport: reliability and correlation with conventional cognitive tests used in post-concussion medical evaluations. *Clin. J. Sport Med.* 13: 28-32.

Collie A, Maruff P, McStephen M, Darby DG. 2003. The effects of practice on the cognitive test performance of neurologically normal individuals assessed at brief test-retest intervals. *J. Int. Neuropsychol. Soc.* 9: 419-428.

Conners C K. 1997. *Conners' Rating Scales, revised.* New York, Multi-Health Systems.

Conners CK, Epstein JN, March JS, Angold A, Wells K, Klaric J, Swanson JM, Arnold LE, Abikoff HB, Elliot GR, Greenhill LL, Hechtman L, Hinshaw SP, Hoza B, Jensen PS, Kraemer HC, Newcorn JH, Pelham WE, Severe JB, Vitiello B, and Wigal T. 2001. Multimodal treatment of ADHD in the MAT: an alternative outcome analysis. *J. Am. Acad. Child Adolesc. Psychiatry* 40: 159-167.

Conners CK. 1985. Parent Symptom Questionnaire. *Psychopharmacology* 21: 816-822.

Conners CK. 2002. Forty years of methylphenidate treatment in Attention-Deficit/Hyperactivity Disorder. *Journal of Attention Disorders* 5: S17-S30.

de Sonneville LMJ, Njiokiktjien C, Bos H. 1994. Methylphenidate and information processing. Part 1: differentiation between responders and nonresponders; Part 2: efficacy in responders. *J. Clin. Exp. Neuropsychol.* 16: 877-897.

Denney C B, Rapport M D. 2001. Cognitive Pharmacology of Stimulants in children with ADHD. In: *Stimulant Drugs and ADHD.* Edited by Solanto MV, Arnsten AFT, Castellanos FX. Oxford, Oxford University Press, pp 283-302.

Denney CB, Rapport MD. 1999. Predicting methylphenidate response in children with ADHD: Theoretical empirical and conceptual models. *J. Am. Acad. Child Adolesc. Psychiatry* 38: 393-401.

Douglas VI, Barr RG, Desilets J, Sherman E. 1995. Do high doses of stimulants impair flexible thinking in Attention-Deficit Hyperactivity Disorder? *J. Am. Acad. Child Adolesc. Psychiatry* 34: 877-885.

Douglas VI, Barr RG, O'Neill ME, Britton BG. 1986. Short term effects of methylphenidate on the cognitive learning and academic performance of children with Attention Deficit Disorder in the laboratory and the classroom. *J. Child Psychol. Psychiatry* 27: 191-211.

Douglas VI. 1999. Cognitive control processes in Attention-Deficit Hyperactivity Disorder. In *Handbook of Disruptive Behaviour Disorders*, Quay HC, Hogan AE. (eds). Plenum Press: New York; 105-138.

Dunlap WP, Cortina JM, Vaslow JB, Burke MJ. 1996. Meta-Analysis of experiments with matched groups or repeated measures designs. *Psychol. Methods* 1: 170-177.

DuPaul G, Barkley RA, McMurray MB. 1994. Response of children with ADHD to methylphenidate: Interaction with internalizing symptoms. *J. Am. Acad. Child Adolesc. Psychiatry* 33: 894-903.

Efron D, Jarman F, Barker M. 1997. Methylphenidate versus dexamphetamine in children with Attention Deficit Hyperactivity Disorder: A double-blind crossover trial. *Pediatrics* 100: E6.

Elia J, Borcherding BG, Rapoport JL, Keysor CS. 1991. Methylphenidate and dextroamphetamine treatments of hyperactivity: Are there true nonresponders? *Psychiatry Res.* 36: 141-155.

Elliott R, Sahakian BJ, Matthews K, Bannerjea A, Rimmer J, Robbins TW. 1997. Effects of methylphenidate on spatial working memory and planning in healthy young adults. *Psychopharm.* 131: 196-206.

Evans RW, Gualtieri CT, Amara I. 1986. Methylphenidate and memory: dissociated effects in hyperactive children. *Psychopharm.* 90: 211-216.

Falleti MG, Maruff P, Collie A, Darby DG, McStephen, M. 2003. Qualitative similarities in cognitive impairment associated with 24 hours of sustained wakefulness and a blood alcohol concentration of 0.05%. *J. Sleep Res.* 12: 265-274.

Gittelman-Klein R, Klein DF. 1975. Are behavioural and psychometric changes related in methylphenidate-treated, hyperactive children? *Int. J. Men Health* 4: 182-198.

Gittelman-Klein R, Klein DF. 1976. Methylphenidate effects in learning disabilities. *Arch Gen. Psychiatry* 33: 655-664.

Gray JR, Kagan J. 2000. The challenge of predicting which children with Attention Deficit-Hyperactivity Disorder will respond positively to methylphenidate. *J. Appl. Dev. Psychol.* 21: 471-489.

Greenberg LM, Waldman ID. 1993. Developmental normative data on the test of variables of attention (T.O.V.A.). *J. Child Psychol. Psychiatry* 34: 1019-1030.

Greenhill LL. 2001. Clinical effects of stimulant medication in ADHD. In *Stimulant drugs and ADHD*, Solanto MN, Arnsten AFT, Castellanos FX (eds). Oxford University Press: New York; 31-71.

Hoagwood K, Kelleher KJ, Feil MMS, Comer DM. 2000. Treatment services for children with ADHD: A national perspective. *J. Am. Acad. Child Adolesc. Psychiatry* 39: 198-206.

Hopkins WG. 2003. A new view of statistics (On-line). Available: www.sportsci.org/resource/stats/index.html. Accessed December 2003.

Houghton S, Douglas G, West J, Whiting K, Wall M, Langsford S, Powell L, Caroll A. 1999. Differential patterns of executive function in children with attention-deficit hyperactivity disorder according to gender and subtype. *J. Child Neurol.* 14: 801-805.

Ingraham LJ, Aitken CB. 1996. An empirical approach to determining criteria for abnormality in test batteries with multiple measures. *Neuropsychology* 10: 120-124.

Jacobson NS, Traux P. 1991. Clinical significance: a statistical approach to defining meaningful change in psychotherapy research. *J. Consult. Clin. Psychol.* 59: 12-19.

Kempton S, Vance A, Maruff P, Luk E, Costin E, Pantelis C. 1999. Executive function and attention deficit hyperactivity disorder: stimulant medication and better executive function performance in children. *Psychol. Med.* 29: 527-538.

Kent MA, Camfield CS, Camfield PR. 1999. Double-blind methylphenidate trials: practical useful and highly endorsed by families. *Arch Pediatr Adolesc. Med.* 153: 1292-1296.

Klein DF. 2008. The loss of serendipity in psychopharmacology. *JAMA* 299: 1063-1065.

Klorman R, Hazel-Fernandez LA, Shaywitz SE, Fletcher JM, Marchione KE, Holahan JM, Stuebing KK, Shaywitz BA. 1999. Executive functioning deficits in attention-deficit/hyperactivity disorder are independent of oppositional defiant or reading disorder. *J. Am. Acad. Child Adol. Psychiatry* 38: 1148-1155.

Maruff P, Currie J, Pantelis C, Smith D. 1996. Deficits in the endogenous control of covert attention in chronic schizophrenia. *Neuropsychologia* 34: 1079-1084.

McCaffrey R J, Westervelt H J. 1995. Issues associated with repeated neuropsychological assessments. *Neuropsych. Rev.* 5: 203-221.

Mehta M, Goodyear IM, Sahakian BJ. 2004. Methylphenidate improves working memory and set shifting in ADHD: relationships to baseline memory capacity. *J. Child Psychol. Psychiatry* 45: 293-305.

Mollica CM, Maruff P, Collie A, Vance A. 2005. The effects of age and practice on the consistency of performance on a computerized assessment of cognitive change in children. *J. Child Neuropsychol.* 11: 303-310.

Mollica CM, Maruff P, Vance A, Collie A, submitted. The effects of stimulant medication on the cognitive and behavioural function of children with attention-deficit/hyperactivity disorder- combined type (ADHD-CT). *A. N. Z. J. Psychiatry*

Nigg JT, Blaskey LG, Huang-Pollock CL, Rappley MD. 2002. Neuropsychological executive functions and DSM-IV ADHD subtypes. *J. Am. Acad. Child Adolesc. Psychiatry* 41: 59-66.

O'Toole K, Abramowitz A, Morris R, Duclan M. 1997. Effects of methylphenidate on attention and nonverbal learning in children with Attention-Deficit Hyperactivity Disorder. *J. Am. Acad. Child Adol. Psychiatry* 36: 531-538.

Paule MG, Rowland AS, Ferguson SA, Chelonis JJ, Tannock R, Swanson JM, Castellanos FX. 2000. Attention Deficit/Hyperactivity Disorder: Characteristics interventions and models. *Neurotoxicol. Teratol.* 22: 631-351.

Pennington BF, Ozonoff S. 1996. Executive functions and developmental psychopathology. *J. Child Psychol. Psychiatry* 37: 51-87.

Pietrzak RH, Mollica CM, Maruff P, Snyder PJ. 2006. Cognitive effects of immediate release methylphenidate in children with ADHD. *Neurosci. Biobehav. Rev.* 30: 1225-1245.

Rapport MD, Denney CB. 2000. Attention Deficit Hyperactivity Disorder and methylphenidate: assessment and prediction of clinical response. In *Ritalin: Theory and Practice*, Greenhill LL, Osman BB (eds). Mary Ann Liebert: Larchmont NY; 45-70.

Rapport MD, Kelly KL. 1993. Psychostimulant effects on learning and cognitive function. In *Handbook of Hyperactivity in Children*, Matson JL (ed). Allyn and Bacon: Boston; 97-136.

Rasmussen LS, Larsen K, Houx P, Skovgaard LT, Hanning CD, Moller JT. 2001. The assessment of postoperative cognitive function. *Acta Anaesthesiol. Scand.* 45: 275-289.

Reynolds CR, Richmond BO. 1978. What I Think and Feel: a revised measure of the children's manifest anxiety. *J. Abn. Child Psychol.* 6: 271-280.

Rodgers J, Marckus R, Kearns P, Windebank K. 2003. Attentional ability among survivors of leukaemia treated without cranial irradiation. *Arch Dis. Child* 88: 147-150.

Rutter M, Graham P. 1968. The reliability and validity of the psychiatric assessment of the child: I. Interview with the child. *Br. J. Psychiatry* 114: 563-579.

Santosh PJ, Taylor E. 2000. Stimulant Drugs. *Eur. Child Adol. Psychiatry* 9: I/27 - I/43.

Scheres A, Oosterlaan J, Swanson J, Morein-Zamir S, Meiran N, Schut H, Vlasveld L, Sergeant JA. 2003. The effect of methylphenidate on three forms of response inhibition in boys with AD/HD. *J. Abn. Child Psychol.* 31: 105-120.

Seidman LJ, Biederman J, Faraone SV, Weber W, Oullette C. 1997. Toward defining a neuropsychology of attention deficit-hyperactivity disorder: performance of children and adolescents from a large clinically referred sample. *J. Consul. Clin. Psychol.* 65: 150-160.

Shallice T, Marzocchi GM, Coser S, Del Savio M, Meuter RF, Rumiatia RI. 2002. Executive function profile of children with Attention Deficit Hyperactivity Disorder. *Dev. Neuropsych.* 21: 43-71.

Silbert BS, Maruff P, Evered LA, Scott DA, Kalpokas M, Martin KJ, Lewis MS, Myles PS. 2004. Detection of cognitive decline after coronary surgery: A comparison of computerised and conventional tests. *Br. J. Anaesth.* 92: 814-820.

Silverman WK, Albano AM. 1996. *Anxiety Disorders Interview Schedule for Children (DSM-IV)*. Texas, Graywind.

Sonuga-Barke EJ, Dalen L, Daley D, Remington B. 2002. Are planning, working memory, and inhibition associated with individual differences in preschool ADHD symptoms? *Dev. Neuropsychol.* 21: 255-272.

Sprague RL, Sleator EK. 1977. Methylphenidate in hyperkinetic children: Differences in dose effects on learning and social behaviour. *Science* 198: 1274-1276.

Swanson JM, Cantwell D, Lerner M, McBurnett K, Hanna G. 1991. Effects of stimulant medication on learning in children with ADHD. *J. Learning Disab.* 24: 219-230.

Swanson JM, Kinsbourne M, Roberts W, Zucker K. 1978. Time-response analysis of stimulant medication on the learning ability of children referred for hyperactivity. *Pediatrics* 61: 21-29.

Swanson JM, McBurnett K, Wigal T, Pfiffner LJ, Williams L, Christian DL. 1993. Effects of stimulant medication on children with attention deficit disorder: A 'review of reviews'. *Exceptional Children* 60: 154-162.

Tannock R, Ickowicz A, Schachar R. 1995. Differential effects of methylphenidate on working memory in ADHD children with and without comorbid anxiety. *J. Am. Acad. Child Adolesc. Psychiatry* 34: 886-896.

Tannock R, Schachar R. 1992. Methylphenidate and cognitive perseveration in hyperactive children. *J. Child Psychol. Psychiatry* 33: 1217-1228.

Taylor E, Schachar R, Thorley G, Wieselberg HM, Everitt B, Rutter M. 1987. Which boys respond to stimulant medication? A controlled trial of methylphenidate in boys with disruptive behaviour. *Psychol. Med.* 17: 121-143.

Uhlenhuth EH, Turner DA, Purchatzke G, Gift T, Chassan J. 1977. Intensive design in evaluating anxiolytic agents. *Psychopharm (Berl)* 52: 79-85.

Vance A, Barnett R. 2002. *Children's Impulsiveness Scale*. Melbourne, Australia, Alfred Hospital.

Vance A, Maruff P, Barnett R. 2003. Attention Deficit Hyperactivity Disorder, combined type (ADHD-CT): better executive function performance with longer-term psychostimulant medication. *A. N. Z. J. Psychiatry* 37: 570-576.

Wechsler D. 2004. *Wechsler Intelligence Scale for Children, 4th ed.* San Antonio TX, Psychological Corporation.

Winer BJ (ed). *Statistical Principles in Experimental Design.* McGraw-Hill: London, 1971.

Wood K, Maruff P, Farrow M, Levy F, Hay D. 1999. Covert orienting of visual attention in children with attention deficit disorder: Does comorbidity make a difference? *Arch Neuropsychol.* 14: 179-189.

In: Encyclopedia of Neuroscience Research
Editors: Eileen J. Sampson and Donald R. Glevins

ISBN 978-1-61324-861-4
© 2012 Nova Science Publishers, Inc.

Chapter XXIII

Novel Therapies for Alzheimer's Disease: Potentially Disease Modifying Drugs

Daniela Galimberti[*]*, Chiara Fenoglio and Elio Scarpini*
Dept. of Neurological Sciences, "Dino Ferrari" Center,
University of Milan, Milan, Italy

Abstract

The two major neuropathologic hallmarks of Alzheimer's disease (AD) are extracellular Amyloid beta (Aβ) plaques and intracellular neurofibrillary tangles (NFTs). Several additional pathogenic mechanisms likely play a role in the pathogenesis of the disease, including inflammation, oxidative damage, ion disregulation and cholesterol metabolism. A number of compounds have been developed, trying to interfere with the above mentioned altered mechanisms. Conversely to symptomatic drugs available to date, these new compounds are supposed to modify pathological steps leading to AD, thus acting on the evolution of the disease. Some of them are under clinical testing, others are in preclinical phases of development. In this chapter, the main pathogenic steps leading to neurodegeneration will be discussed, together with an update of potentially disease-modifying drugs under testing.

1. Introduction

Alzheimer's disease (AD) is the most common cause of dementia in the elderly, with a prevalence of 5% after 65 years of age, increasing to about 30% in people aged 85 years or older. It is characterized clinically by progressive cognitive impairment, including impaired judgement, decision-making and orientation, often accompanied, in later stages, by

---

[*] Corresponding Author: IRCCS Fondazione Ospedale Maggiore Policlinico, Via F. Sforza 35, 20122, Milan, Italy. phone ++ 39.2.55033847, FAX ++ 39.2.50320430. e-mail: daniela.galimberti@unimi.it.

psychobehavioural disturbances as well as language impairment. Mutations in genes encoding for Amyloid Precursors Protein or presenilins 1 and presenilin 2 genes (*APP*, *PSEN1* and *PSEN2*, respectively) account for about 3% of cases, characterized by an early onset (before 65 years of age). So far, 31 different mutations have been described in the *APP* gene in 83 families, together with 175 mutations in *PSEN1* and 14 mutations in *PSEN2* (http://molgen-www.uia.ac.be).

The two major neuropathologic hallmarks of AD are extracellular Amyloid beta (Aβ) plaques and intracellular neurofibrillary tangles (NFTs). The production of Aβ, which represents a crucial step in AD pathogenesis, is the result of cleavage of APP, that is overexpressed in AD [1]. Aβ forms highly insoluble and proteolysis resistant fibrils known as senile plaques (SP). NFTs are composed of the tau protein. In healthy subjects, tau is a component of microtubules, which represent the internal support structures for the transport of nutrients, vesicles, mitochondria and chromosomes within the cell. Microtubules also stabilize growing axons, which are necessary for the development and growth of neurites [1]. In AD, tau protein is abnormally hyperphosphorylated and forms insoluble fibrils, originating deposits within the cell.

A number of additional pathogenic mechanisms, possibly overlapping with Aβ plaques and NFTs formation, have been described, including inflammation, oxidative damage, iron disregulation, cholesterol metabolism. In this chapter, these mechanisms will be discussed and treatments under development to interfere with these pathogenic steps presented.

2. Pathogenic Mechanisms at the Basis of Alzheimer's Disease

2.1. Role of Amyloid Protein

The APP plays a central role in AD pathogenesis and in AD research, as it is the precursor of Aβ, which is the heart of the amyloid cascade hypothesis of AD.

The human *APP* gene was first identified in 1987 by several laboratories independently. The two *APP* homologous, *APLP1* and *APLP2*, were discovered several years later. APP is a type I membrane protein. Two predicted cleavages, one in the extracellular domain (β-secretase cleavage) and another in the transmembrane region (γ-secretase cleavage) are necessary to release Aβ from the precursor protein. Notably, *APP* is located on chromosome 21, and this provided an immediate connection to the invariant development of AD pathology in trisomy 21 (**Down's syndrome**) individuals. The first mutations demonstrated to be causative of inherited forms of familial AD were identified in the *APP* gene [2], providing an evidence that APP plays a central role in AD pathogenesis. Notably, only *APP* but not its homologous *APLP1* and *APLP2* contain sequences encoding the Aβ domain.

Full-length APP undergoes sequential proteolytic processing. It is first cleaved by α-secretase (the so-called non-amyloidogenic pathway) or β-secretase (amyloidogenic pathway) within the luminal domain, resulting in the shedding of nearly the entire ectodomain and generation of α- or β-C-terminal fragments (CTFs). The major neuronal β-secretase, named BACE1 (β-site APP cleaving enzyme), is a transmembrane aspartyl protease which cleaves

APP within the ectodomain, generating the N-terminus of Aβ [3]. However, several zinc metalloproteinases such as TACE/ADAM17, ADAM9, ADAM10 and MDC-9, and the aspartyl protease BACE2, can cleave APP at the α-secretase site [4] located within the Aβ domain, thus precluding the generation of intact Aβ.

The second proteolytic event in APP processing involves intramembranous cleavage of α- and β-CTFs by γ-secretase, which liberates a 3kDa protein (p3) and Aβ peptide into the extracellular milieu. The minimal components of γ-secretase include presenilin (PS)1 or PS2, nicastrin, APH-1 and PEN-2 [5]. Protein subunits of the γ-secretase assemble early during biogenesis and cooperatively mature as they leave the endoplasmic reticulum. Biochemical evidence is consistent with PS1 (or PS2) as the catalytic subunit of the γ-secretase. APH-1 and PEN-2 are thought to stabilize the γ-secretase complex, and nicastrin to mediate the recruitment of APP CTFs to the catalytic site of the γ-secretase. Major sites of γ-secretase cleavage correspond to positions 40 and 42 of Aβ.

Amyloidogenic processing is the favoured pathway of APP metabolism in neurons, due to the greater abundance of BACE1, whereas non-amyloidogenic pathway predominates in other cell types.

It appears that none of the above mentioned secretases have unique substrate specificity towards APP. Besides APP, a number of other transmembrane proteins undergo ectodomain shedding by enzymes with α-secretase activity. Regarding BACE1, its low affinity for APP lead to the hypothesis that APP is not its sole physiological substrate. Similarly, PS1 and PS2 play a crucial role in intramembranous γ-secretase cleavage of several type I membrane proteins other than APP, including Notch1 receptors and its ligands [6]. A number of functional domains have been mapped to the extra- and intracellular region of APP, including metal (copper and zinc) binding, extracellular matrix components (heparin, collagen and laminin), neurotrophic and adhesion domains. Thus far, a thropic role for APP has been suggested, as it stimulate neurite outgrowth in a variety of experimental settings. The N-terminal heparin-binding domain of APP also stimulates neurite outgrowth and promotes synaptogenesis. In addition, an "RHDS" motif near the extralumenal portion of APP likely promotes cell adhesion, possibly acting in an integrin-like manner. Similarly, APP colocalizes with integrins on the surface of axons at sites of adhesion [7,8].

Despite APP was initially proposed to act as a cell surface receptor, the evidence supporting this hypothesis has been unconvincing. Only recently, aside of from interactions with extracellular matrix proteins, a candidate ligand has been proposed. In was in fact reported that F-spondin, a neuronal secreted signalling glycoprotein that may function in neuronal development and repair, binds to the extracellular domain of APP as well as of APLP1 and APLP2 [9]. This binding reduces β-secretase cleavage of APP, suggesting therefore that F-spondin binding may regulate APP processing.

APP-deficient animals are a useful model to better understand the role of APP. Deficient APP mice did not show major phenotypic abnormalities [10]. However, *APLP2$^{-/-}$/APLP1$^{-/-}$* and *APP$^{-/-}$/APLP2$^{-/-}$* mutants, but not *APP$^{-/-}$/APLP1$^{-/-}$* animals, showed early postnatal lethality, indicating that members of the APP gene family are essential genes, which exhibit partial overlapping functions. Deficiency of all the *APP* genes lead to death shortly after birth. The majority of animals studied showed cortical dysplasia suggestive of migrational abnormalities of the neuroblasts and partial loss of cortical Cajal Retzius cells [11]. Taken together, these findings presented a convincing picture that members of the *APP* family play essential roles in the development of the nervous system related to synapse structure and function as well as in neuronal migration.

Given the trophyc properties of APP, it would be natural to predict that overexpression of APP would lead to phenotypes related to the enhanced neurite outgrowth and cell growth, which indeed was demonstrated [12]. However, convincing negative phenotypes, in which APP does not act as trophyc factor, has been reported as well. For example, over-expression of APP in cells induced to differentiate into neurons lead to cell death [13]. Genetic in-vivo engineering to over-express APP carrying various familial AD mutations in transgenic mice resulted in the development of Aβ deposition and Aβ associated changes in the brain, including loss of synaptic markers, thus confirming the pathogenic nature of these mutations [14].

2.2. Tau and Alzheimer's Disease

Tau is relatively abundant in neurons but is present in all nucleated cells and functions physiologically to bind microtubules and stabilize microtubule assembly for polymerization. Tau encoding gene (*MAPT: Microtubule Associating Protein Tau*) consists of 16 exons. In the adult brain, alternative splicing of tau nuclear RNA transcribed on exons 2, 3, and 10, results is six tau isoforms, having either three or four peptide repeats of 31 or 32 residues in the C terminal region encoded on exon 10, comprising the microtubule binding domain or differing in the expression of zero, one or two inserts encoded on exon two and three. During neurodegeneration, tau is abnormally phosphorylated. The profile of alternative splicing differs among phatological phenotypes, such that tau accumulation in AD is a mixture of 3R and 4R tau, Pick disease tends to be 3R tau, corticobasal degeneration and progressive supranuclear palsy tends to be 4R tau, and so-called argyrophilic grain disease accumulates small inclusions comprised of 3R tau [15].

2.3. Role of Inflammation in Alzheimer's Disease

The fibrillar deposition of extracellular Aβ is closely associated with a neuro-inflammatory response, which includes a local up-regulation of acute-phase proteins, complement fragments, cytokines and other inflammatory mediators [16]. So far, epidemiological studies suggested that inflammatory processes play a role in the pathogenesis of AD. Prospective case-cohort studies showed that higher serum levels of certain acute-phase proteins are a risk factor for the development of AD [17-19]. Moreover, epidemiological studies indicate that longstanding use of non-steroidal anti-inflammatory drugs can prevent or delay the development of AD [20].

Microglial cells are the major producers of inflammatory factors. During the early stages of AD pathogenesis, activated microglia were clustered within classic (dense-cored) plaques in the AD neocortex [21]. These plaques showed strong immunostaining for complement factor C1q and serum amyloid P component (SAP). Plaque-associated factors C1q and SAP may trigger microglia to secrete high levels of proinflammatory cytokines [22].

Activated microglial cells colocalize with Aβ, and in vitro studies demonstrated that Aβ induces the production of Tumor Necrosis Factor (TNF)α in such cells [23]. This cytokine is a pleiotropic factor acting as an important mediator of inflammatory responses in a variety of tissues. Levels of TNFα in CSF from AD patients are 25-fold higher than in CSF from age-matched controls [24], suggesting a role for inflammation in neurodegeneration. Nevertheless, other findings demonstrated a protective effect of TNFα, as it likely protects neurons against Aβ triggered citotoxicity [25].

In AD, an increased production of IL-1 has been demonstrated by immuno-histochemistry. In particular, it is expressed by microglia localized around amyloid deposits, possibly participating to plaque formation [26].

Conflicting results have been reported with regard to IL-6 levels in serum and CSF of AD patients. However, it has been shown that its mRNA levels are increased in the entorhinal cortex and the superior temporal gyrus of AD patients [26].

Additional cytokines of the IL-6 family are IL-11 and leukaemia inhibitory factor (LIF) [27]. Interleukin-11 mean levels were significantly increased in AD and Frontotemporal Lobar Degeneration (FTLD) as compared with controls, whereas CSF LIF levels were not detectable either in patients or controls [28]. In accordance with previous results [29], in AD patients, a significantly positive correlation between Mini Mental State Examination (MMSE) scores and IL-11 CSF concentration was observed [28].

In contrast with the previously described cytokines, Transforming Growth Factor beta (TGF-β) has mainly an anti-inflammatory action. Several data show that its levels are increased in the brain of AD patients, as well as in plasma and CSF. In addition, TGF-β was also found both in amyloid plaques and tangles [26].

As a general comment, microglial-produced "inflammatory" cytokines have neurophatic as well as neuroprotective actions. For instance, whereas excess levels of TNFα might cause neurotoxicity, low-dose TNFα could, alternatively, trigger the neuroprotective and/or anti-apoptotic genes [30]. The role of glial cells is to support and sustain proper neuronal function and microglia are no exception to this general principle. In acutely injured central nervous system (CNS) microglia have a neuroprotective and pro-regenerative role [31]. Therefore, the primary mode of action of microglia seems to be the protection of the central nervous systems. Nevertheless, upon excessive or sustained activation, microglia could significantly contribute to chronic neuropathologies, leading to neurotoxicity [22].

Chemokines are low molecular weight chemotactic cytokines that have been shown to play a crucial role in early inflammatory events. Based on the arrangement of cysteine residues, they are divided into two main groups: CXC or α-chemokines, i.e Interferon-γ-inducible Protein-10 (IP-10) and Interleukin-8 (IL-8), responsible for attracting neutrophils, and CC or β-chemokines, i.e Monocyte Chemotactic Protein-1 (MCP-1), and Macrophage Inflammatory Protein-1α and β (MIP-1α and β), which act basically on monocytes [32].

Upregulation of a number of chemokines has been associated with AD pathological changes [33]. IP-10 immunoreactivity was markedly increased in reactive astrocytes in AD brains, as well as the level of its expression. Astrocytes positive for IP-10 were found to be associated with senile plaques and showed an apparently coordinated upregulation of MIP-1β [34,35]. Significant increased IP-10 levels were observed in CSF from patients with mild AD as compared with severe AD. Similarly to mildly impaired AD, IP-10 increased levels were also found in subjects with amnestic MCI [29]. Regarding MCP-1 and IL-8, significantly higher levels were found in all AD patients as compared with healthy subjects, and highest peaks were observed in mild AD and Mild Cognitive Impairment (MCI) [29].

With regard to a possible use of chemokine to easily predict evolution from MCI to AD, few investigation in serum have been so far carried out, despite a growing body of evidence supporting the hypothesis that some peripheral biochemical modifications also occur very early during AD pathogenesis. For instance, serum MCP-1 levels have been demonstrated to be increased in MCI subjects, similarly to findings described in the CSF [36]. Conversely, IP-10 serum levels were not increased in AD patients, but were found to correlate with aging [37].

2.4. Role of Oxidative Damage

Oxidative stress is supposed to play a relevant role in the pathogenesis of several neurodegenerative diseases, including AD. Aβ and other lesion-associated proteins are a major source of Reactive Oxygen Species (ROS) and other toxic radicals [38]. Increasing evidence supports a role of oxidative stress and impaired energy metabolism in the pathogenesis of the disease: an increase in DNA, lipid and protein oxidation metabolites has been observed in blood as well as post-mortem brain samples from AD patients compared with healthy subjects [39]. Free radicals are produced by mitochondria, as a side product, during the reduction of molecular oxygen. The production of radicals is thought to be higher in cerebral tissue, particularly vulnerable to free radical damage because of its low content of antioxidants, high content of polyunsaturated fatty acids in neuronal membranes and high oxygen requirements for its metabolic process [40]. Further observations indicate reduced cerebral metabolism in AD [40] as well as reduced activities of specific mitochondrial enzyme complexes, such as cytochrome oxidase [41-43]. Alterations in these key enzymes can favour the aberrant production of ROS. Intracellular oxidative balance is tightly regulated and, therefore, an upregulation of antioxidant compensatory mechanisms would be expected in AD. The induction of Cu/Zn superoxide dismutase, catalase, glutathione peroxidase (GSHPx), glutathione reductase (GSSG-R), peroxiredoxins and a number of heat shock proteins [44] suggests that vulnerable neuronal cells mobilize antioxidant defence in the face of increased oxidative stress [38]. On the other hand, the Total Antioxidant Capacity (TAC; including glutathione, ascorbic acid, uric acid and bilirubin) was shown to be reduced by 24% in plasma samples from AD patients [45]. A link between oxidative stress and hyperhomocysteinemia, which is a known risk factor for the development of AD [46], has been hypothesized, as homocysteine (Hcy) influences DNA repair, promoting the accumulation of DNA damage caused by oxidative stress [47]. Recent in vitro studies demonstrate that Hcy increases levels of thiobarbituric acid reactive substances, which represent an index of peroxidation, and decreases levels of total-trapping antioxidant potential in a model of rat hippocampus [48]. High tHcy levels are at present considered one of the major risk factors for the development of AD as a strong, graded association between tHcy levels and the risk of dementia and AD has been demonstrated [46]. In this regard, there are evidences that tHcy levels are increased in late onset AD (LOAD; disease onset >65 years), but not in early onset AD (EOAD; disease onset ≤65 years), suggesting an influence on this parameter of other pathological conditions, mainly vascular diseases, which often co-occur with LOAD [49].

Similarly to inflammation, emerging evidence indicates that oxidative damage to neuronal RNA and protein is an early event in AD pathogenesis [50]. In this regard, oxidative imbalance is likely to be present in subjects with MCI. Both in MCI and in AD patients, plasma mean levels of non-enzymatic antioxidants and lower activity of antioxidant enzymes appeared to be lower than in controls, with no parallel induction of antioxidant enzymes [51]. In this regard, it has been recently shown that subjects with MCI have plasma, urine and cerebrospinal fluid (CSF) levels of the isoprostane 8,12-iso-iPF2a -VI, which is a marker of in vivo lipid peroxidation, higher than healthy subjects [52]. This evidence clearly indicates that oxidative imbalance and subsequent oxidative stress are early events in AD evolution, and are probably secondary to other mechanisms specific to AD but not present in other neurodegenerative diseases [53].

On the basis of these studies, ROS, tHcy, and TAC were evaluated in samples from patients with AD, MCI and Vascular dementia (VaD), compared with age-matched healthy subjects. Total Hcy levels were significantly increased in AD as well as in VaD patients compared with controls. Notably, tHcy levels slightly increased were found in MCI patients compared with controls. As regards ROS levels, no significant differences were shown between patients and controls. TAC was significantly lower in AD patients than in either healthy subjects or VaD patients. No correlation between ROS and TAC levels in each subject was observed [54]. In conclusion, an alteration of some biochemical factors involved in oxidative stress occurs in AD patients. Both tHcy and TAC modifications seem to be early events in the pathogenesis of AD, whereas ROS levels appear to be correlated with age rather than with a specific dementing disorder. This consideration leads to the hypothesis that oxidative imbalance observed in AD is mainly due to a decreased TAC rather than to an increased production of ROS [54].

2.5. Role of Ions

It was first observed in 1994 that Aβ becomes amyloidogenic upon reaction with stoichiometric amounts of Zn^{2+} and Cu^{2+} [55]. Aβ is rapidly precipitated by Zn^{2+}. Cu^{2+} and Fe^{3+} also induce marked Aβ aggregation, but only under mildly acidic conditions [56], such as those believed to occur in AD brain. The precipitation of Aβ by these ions is reversible with chelation [57], in contrast with fibrillization, which is irreversible. Cu, Fe and Zn play more of a role than merely assembling Aβ. When binding Cu^{2+} or Fe^{3+}, Aβ reduces the metal ions and produces H_2O_2 by double electron transfer to O_2. In addition, Aβ promotes the Cu-mediated generation of toxic lipid oxidation product 4-hydroxynoneal (see [56] for review).

2.6. Role of Cholesterol and Vascular-related Risk Factors

It has been repeatedly shown that Apolipoprotein (Apo)E *ε4* carriers have a higher risk to develop AD. Since ApoE is the major cholesterol transporter in the central nervous system, a link between cholesterol and AD is suggested. The brain is the most cholesterol-rich organ of the body, which is synthesized by astrocytes. Additional vascular-related risk factors for AD include hypertension, atrial fibrillation, hyperhomocysteinemia, atherosclerosis, stroke (see [58] for review).

Hypertension is the strongest risk factor for AD and VaD when these conditions are considered together [59]. The penetrating arteries in the circle of Willis are particularly sensitive to the effects of hypertension and suffer early and selective damage during chronic hypertension [60]. Hypertension is closely associated with atherosclerosis and vascular function, and in the brain this results in hypoperfusion and ischemic conditions of the nucleus basalis Meynert. Targeting molecular mechanisms and using dietary methods and therapies are grounded in reducing free radicals and and associated oxidative stress related damage initiating hypertension [61].

3. Disease Modifying Drugs for Alzheimer's Disease Treatment

3.1. Drugs Interfering with Aβ Deposition

3.1.1. Anti-Amyloid Aggregation Agents

A number of anti-Aβ aggregation agents are currently in clinical testing. Despite their biological mechanisms of action are not completely understood, they are believed to prevent fibril formation and to facilitate soluble Aβ clearance. The most studied is named tramiprosate (Alzhemed™, Neurochem, Inc.), a glycosaminoglycan (GAG) mimetic. GAGs binds to soluble Aβ, promoting fibril formation and deposition of amyloid plaques. GAG mimetics compete for GAG-binding sites, thus blocking fibril formation and reducing soluble Aβ [62]. In transgenic mice, tramiprosate reduces plaque burden and decreases CSF Aβ levels, but cognitive and behavioural outcomes in this animal model have not been reported [63]. A phase I study in healthy adults demonstrated the drug is well tolerated. A 3-month phase II study was subsequently conducted in 58 patients with mild to moderate AD, who were randomized to tramiprosate 50 mg, 100 mg or 150 mg twice a day or placebo. Patients who completed the study were eligible for a 21-month open-label extension with 150 mg twice daily. Baseline CSF Aβ levels declined by up to 70% after 3 months for patients randomly assigned to the 100-mg or 150-mg twice-daily group. However, no differences were observed in cognitive functions between the tramiprosate and placebo groups [64]. A phase III study was then carried out in the US in 1052 patients with AD to test tolerability, efficacy and safety of the drug. The study failed to show any significant effect. Multiple factors likely contributed to the failure of the study. Overall, variability among the 67 clinical sites in the trial overwhelmed the observed treatment effects. In particular, changes in people's concomitant treatment with cognitive-enhancing drugs including cholinesterase inhibitors, memantine and antidepressants affected the results for the primary cognitive endpoints based on neuropsychological testing. Unexpected problems also arose in the control group, confounding the interpretation of Alzhemed efficacy. Thirty percent of the control group did not decline in cognition over the 18-month trial period, whereas a portion of this group unexpectedly showed a significant improvement in cognition. Another similar trial conducted in Europe has been discontinued. In addition, recent data suggest that tramiprosate promotes an abnormal aggregation of the tau protein in neuronal cells [65], emphasizing the importance of testing on both types of pathology (amyloid and tau) the potential drugs to be used for the treatment of AD.

Another molecule under testing is named colostrinin. It is a proline-rich polypeptide complex derived from sheep colostrum (O-CLN; ReGen Therapeutics), which inhibits Aβ aggregation and neurotoxicity in cellular assays and improves cognitive performance in animal models [66]. A 3-week phase I study in patients with AD demonstrated it is well tolerated [67]. A subsequent phase II trial demonstrated modest improvements in MMSE scores for patients with mild AD over a treatment period of 15 months, but this beneficial effect was not sustained during 15 additional months of continued treatment [68].

In 2000, McLaurin et al. [69] described a compound, named *scyllo*-inositol, which is able to stabilize oligomeric aggregates of Aβ and to inhibit Aβ toxicity. *Scyllo*-inositol (AZD103) dose-dependently rescued long-term potentiation in mouse hippocampus from the inhibitory effects of soluble oligomers of cell-derived human Aβ [70]. This compound is in Phase II clinical trials (Transition Therapeutics/Elan).

3.1.2. Amyloid Removal: Vaccination

In 1999 Schenk et al. [71] demonstrated that immunization with Aβ as an antigen attenuated AD-like pathology in transgenic mice over-expressing the *APP* gene by removing amyloid from the central nervous system. This transgenic mouse model of AD progressively develop several neuropathological features of the disease in an age-related and brain-region-dependent manner. Immunization of young animals with Aβ prevents the development of plaque formation, neuritic dystrophy and astroglyosis, whereas in older animals, vaccination reduces extent and progression of AD-like pathologies. Given these preclinical results, a multicenter, randomized, placebo-controlled, phase II double-blind clinical trial using active immunization with Aβ42 plus adjuvant was started in 2001 on 300 patients using the pre-aggregated Aβ peptide AN1792. However, following reports of aseptic meningo-encephalitis in 6% of treated patients, the trial was halted after 2-3 injections. Of the 300 patients treated, 60% developed antibody response. The final results of the trial were published in 2005 [72].

Double-blind assessment were maintained for 12 months, demonstrating no significant differences in cognition between antibody responders and placebo group for ADAS-Cog, Disability Assessment for Dementia (DAS), Clinical Dementia Rating, Mini Mental State Examination and Clinical Global Impression of Change. In a small subset of patients, CSF tau levels were decreased in antibody responders but Aβ levels were unchanged.

A quite disappointing observation was the finding of greater brain volume decrease and great ventricular enlargement at MRI in responders than in placebo patients [73]. Nevertheless, this brain atrophy was not associated with worsening of cognitive performances. A possible explanation is that the brain volume changes observed may result from an association between amyloid removal and intracerebral fluid shifts.

Long-term follow-up of treated patients and further analysis of autopsy data modified and moderated the negative impact of the first results, encouraging additional clinical attempts. Subsequent observations on AN1792 vaccinated patients or transgenic models and on brain tissue derived from mice and humans using a new tissue amyloid immunoreactivity (TAPIR) method suggested that antibodies against Aβ-related epitopes are capable of slowing the progression of neuropathology in AD. Hock and Nitsch [74] followed for four years 30 patients who received a prime and booster immunization over the first year after vaccination, providing further support to continue investigation of antibody treatment in AD.

In 2008 a paper was published describing the relation between Aβ42 immune response, degree of plaque removal and long-term clinical outcomes [75]. In June 2003, 80 patients (or their caregivers), who had entered the phase I AN1792 trial in 2000, gave their consent for long-term clinical follow-up and post-mortem neuropathological examination. In patients who received immunisation, mean Aβ load was lower than in the placebo group. Despite this observation, however, no evidence of improved survival or an improvement in time to severe dementia was observed in such patients. Therefore, plaque removal is not enough to halt progressive neurodegeneration in AD, prompting some intriguing challenges to the amyloid hypothesis.

Although severe adverse events occurred in the first AN1792 trial and cognitive results were unclear, immunization was not abandoned, but the treatment was modified from active into passive in order to avoid excessive activation of the T-cell response and thus prevent complications. The humanized monoclonal anti-Aβ antibody Bapineuzumab (Wyeth and Elan) has been tested in a phase II trial in 200 patients with mild to moderate AD. The 18-month, multi-dose, one-to-one randomization trial was conducted at about 30 sites in the US. It was designed to assess safety, tolerability and standard efficacy endpoints (ADAS-Cog, Neuropsychological Test Battery, DAS) of multiple ascending doses of Bapineuzumab in patients. On May 21, 2007, Elan and Wyeth announced their plans to start a phase III clinical trial of Bapineuzumab. The decision to launch phase III studies prior to the conclusion of the ongoing phase II was based on the totality of the accumulated clinical data from phase I, phase II and a 4.5-year follow-up study of those patients involved in the original AN1792 trial.

3.1.3. γ-Secretase Inhibition

Several compounds which inhibit γ-secretase activity in the brain have been identified. Nevertheless, γ-secretase has many biologically essential substrates [76]. One of the most physiologically important γ-secretase substrate in the Notch signaling protein, which is involved in the differentiation and proliferation of embryonic cells, T cells, gastrointestinal goblet cells and splenic B cells. Experience with transgenic mice showed that the administration of a γ-secretase inhibitor in doses sufficient to remove Aβ concentrations interferes with lymphocyte differentiation and alters the structure of intestinal goblet cells [77]. Therefore, safety is a very important consideration for this kind of compounds.

A nonselective γ-secretase inhibitor named LY450139 (Eli Lilly) has been evaluated in a phase I placebo-controlled study in 37 healthy adults (at doses ranging from 5 to 50 mg). Aβ CSF levels were reduced in both active treatment and placebo groups, but differences were not statistically significant. Transient gastrointestinal adverse effects (bleeding, abdominal pain) were reported by 2 subjects treated with 50 mg [78]. A subsequent phase II randomized, controlled trial was carried out in 70 patients with AD. Patients were given 30 mg for 1 week followed by 40 mg for 5 weeks. Treatment was well tolerated. No significant changes in plasma and CSF Aβ40 and Aβ42 were observed [79].

Subsequently, a multicenter, randomized, double-blind, dose-escalation, placebo-controlled trial was carried out. Fifty-one patients with mild to moderate AD were randomized to receive placebo, or LY450139 (100 mg or 140 mg). The LY450139 groups received 60 mg/day for two weeks, then 100 mg/day for six weeks, then either 100 or 140 mg/day for six additional weeks. Primary outcomes included safety, tolerability and CSF/plasma Aβ levels; secondary outcome was neuropsychological testing. LY450139 was generally well tolerated at doses of up to 140 mg/day for 14 weeks. However, adverse events were seen, including 3 possible drug rashes, 3 reports of hair color change and 3 adverse event-related discontinuation, therefore a close clinical monitoring will be needed in future studies. Plasma Aβ, but not CSF, levels were reduced in treated patients, consistent with inhibition of γ-secretase. No differences were seen in cognitive or functional measures [80].

3.1.4. Selective Aβ42-Lowering Agents (SALAs)

Tarenflurbil is the first compound in this new class of drugs, which modulate γ-secretase activity without interfering with Notch or other γ-secretase substrates [81]. It binds to a γ-secretase site other than the active/catalytic center of relevance to production of Aβ42, thereby altering the conformation of γ-secretase and shifting production away from Aβ42 without interfering with other physiologically essential γ-secretase substrates.

Tarenflurbil (MPC-7869; Myriad Pharmaceuticals; Flurizan™) is the pure R-enantiomer of flurbiprofen. It shifts cleavage of APP away from Aβ42, leading to the production of shorter non-toxic fragments [82,83]. In contrast with S-flurbiprofen or other non-steroidal anti-inflammatory drugs (NSAIDs), it does not inhibit cyclo-oxygenase (COX) I or COX 2 and it is not associated with gastrointestinal toxicity [84]. In mice, treatment with tarenflurbil reduces amyloid plaque burden and prevents learning and behavioural deterioration [85].

A 3-week, placebo-controlled, phase I pharmacokinetic study of tarenflurbil (twice-daily doses of 400, 800 or 1600 mg) in 48 healthy, older volunteers, showed that the drug is well tolerated, with no evidence of renal or gastrointestinal toxicity. CSF was collected at baseline and after 3 weeks. The compound penetrated the blood brain barrier in a dose-dependent manner. No significant changes of Aβ42 CSF levels were shown after treatment. However, in plasma, higher drug concentrations were related to statistically significant lower Aβ levels [86].

Myriad conducted a large, placebo-controlled Phase II trial for Flurizan of 12 month-duration in 210 patients with mild to moderate AD (MMSE score: 15-26). Patients were randomly assigned to receive Tarenflurbil twice per day (400 mg or 800 mg or placebo) for 12 months. Primary outcome measures for the trial were the rate of change (slope of decline) of: activities of daily living, **quantified by the Alzheimer's Disease Cooperative Study—**Activities of Daily Living inventory (ADCS-ADL), global function, measured by the Clinical Dementia Rating-sum of boxes (CDR-sb), cognitive function, **measured by the Alzheimer's** disease Assessment Scale-cognitive subscale (ADAS-cog). In a 12-month extended treatment phase, patients who had received tarenflurbil continued to receive the same dose, and patients who had received placebo were randomly assigned to tarenflurbil at 800 mg or 400 mg twice a day.

A preliminary analysis revealed that patients with mild AD (MMSE: 20-26) and moderate AD (MMSE: 15-19) responded differently to tarenflurbil in the ADAS-Cog and the ADCS-ADL, therefore these groups were analyzed separately. Patients with mild AD in the 800 mg tarenflurbil group had lower rates of decline than did those in the placebo group in the activities of daily living, whereas slowing of cognitive decline did not differ significantly. In patients with moderate AD, 800 mg tarenflurbil twice per day had no significant effects on

ADCS-ADL and ADAS-Cog and had a negative effect on CDR-sb. The most common adverse events included diarrhoea, nausea and dizziness. Patients with mild AD who were in the 800 mg tarenflurbil group for 24 months had lower rates of decline for all three primary outcomes than did patients who were in the placebo group for months 0-12 and a tarenflurbil group for months 12-24 [87].

Given these results, two phase III study were carried out, in US and in Europe. ActEarliAD trial was started in 2007 all over Europe. It is a 18-month, multinational, randomized, double-blind, placebo controlled study in over 800 patients with AD. Patients enrolled in the trial were treated with 800 mg twice a day of either Flurizan or placebo and attended periodic physician visits for analysis of their performance on memory, cognition and behavioral tests. The two primary clinical endpoints of the trial were the change in cognitive decline and function, as measured by the ADAS-cog, and changes in activity of daily living, as measured by the ADCS-ADL. A secondary endpoint of the trial was the change in overall function, measured by the CDR-sb. Additional exploratory outcome measures were designed to assess the psychological, physical and financial impact of this disease on caregivers and medical resources. The trial was designed to meet the requirements of the European Agency for the Evaluation of Medicinal Products (EMEA) for marketing of Flurizan in Europe. The global endpoints in this trial were identical to those in the US trial. As was the case with the phase II trial, all patients in the phase III studies are allowed to take currently standard of care medicines in addition to Flurizan or placebo, provided their dose has been stable for 6 months.

Disappointingly, on July 2, 2008, the sponsor of Flurizan announced that this γ-secretase-modulating agent had fallen flat in its definitive Phase III trial and was finished as a development product (www.alzforum.org). In fact, on both primary efficacy endpoints, the ADAS-Cog and the ADCS activities of daily living scales, the treatment and placebo curves overlapped almost completely, and there was no effect whatsoever in the group as a whole. In addition, while the overall side effect profile was similar between placebo and treatment groups, anemia, infections and gastrointestinal ulcers appeared more often in people on Flurizan than in the placebo group.

3.2. Drugs Interfering with Tau Deposition

A phase II trial of a tau-blocking compound named methyl thioninium chloride (MTC) is ongoing (TauRx Therapeutics, Rember[TM]). This is a reducing agent better known as methylene blue, a deep blue dye used in analytical chemistry, as a tissue stain in biology and in various industrial products. MTC interfere with tau aggregation by acting on self-aggregating truncated tau fragments [88]. The company conducted a phase II trial randomizing 321 patients with mild or moderate AD to treatment with either placebo or one of three oral doses of MTC: 30 mg, 60 mg or 100 mg three times a day. Patients were not taking acetylcholinesterase inhibitors or memantine. Primary outcomes were to compare the effect of MTC to placebo on cognitive abilities measured by the ADAS-Cog at 24 weeks. Preliminary results were presented at the 2008 International Conference on Alzheimer's disease [89]. The 100mg dose was found to have a formulation defect limiting release of the therapeutic form of MTC, therefore this arm was discontinued. A significant improvement relative to placebo of -5.4 ADAS-cog units in CDR-moderate subjects at the 60mg dose was shown. There was no placebo decline in CDR-mild AD over the first 24 weeks preventing initial efficacy analysis. Significant efficacy was demonstrated separately in mild and moderate subgroups [89]. A problem with the use of this drug is that urine become blue, resulting in a lack of blinding.

An interesting approach to block tau deposition is to inhibit kinases responsible for tau hyperphosporylation. Despite the large number of tau phosphorylation sites and the ability of multiple kinases to phosporylate individual sites, Glycogen Synthase Kinase 3 (GSK3β) has emerged as potential therapeutic target (see [90] for review). The most studied compound able to inhibit GSK3 is lithium, but several other compounds are under development (reviewed in [91]).

3.3. Anti-inflammatory Drugs

A large body of epidemiologic evidence suggested that long-term use of NSAIDs protects against the development of AD [20,92]. Nevertheless, prospective studies of Rofecoxib, Naproxen or Diclofenac failed to slow progression of cognitive decline in patients with mild to moderate AD [93-95], as well as celecoxib [96], dapsone [97], hydroxychloroquine [98], nimesulide [99]. In contrast, Indomethacin may delay cognitive decline in this subset of patients, but gastrointestinal toxicity is treatment-limiting [100,101]. Because of general concerns about lack of efficacy, gastrointestinal toxicity, myocardial infarction and stroke, NSAIDs are not considered to be viable treatment options for patients with AD.

A rapid improvement in verbal fluency and aphasia following perispinal etanercept administration was described [102]. Etanercept is a TNFα inhibitor, which acts by blocking the binding of this cytokine to its receptor. It was tested in 12 patients with mild to severe AD, at a dose of 25-50 mg weekly for six months, showing improvement in a number of neuropsychological testing, particularly in verbal fluency [102].

3.4. Drugs Preventing Oxidative Damage

Ginkgo biloba extracts have antioxidant activity, and counteract the aggregation or deposition of Aβ in vitro and in animal models. These promising preclinical hints lead to the development of a large trial. The GEM study is a Randomised, double-blind, placebo-controlled trials designed to test whether EGb 761, a commercial extract contained in many over-the-counter ginkgo preparations, at a dose of 120 mg twice a day, could delay the onset of AD in older adults. Participants included 2587 cognitively normal elderly volunteers, and 482 with MCI. Half subjects received the treatment, half the placebo. All of them were followed for an average of 6.1 years. The measured endpoints was onset of dementia of any cause, or AD. No differences were found in the incidence of all-cause dementia, or AD in particular, between the ginkgo takers and the placebo group [103].

Additional potential antioxidants include mitoquinone (Antipodian Pharmaceuticals), vitamin E, and natural polyphenols such as green tea, wine, blueberries and curcumin. Clinical trial with vitamin E and omega-3 fatty acids did not show beneficial effects in AD patients (see [104] for review).

A trial to determine whether reduction of homocysteine levels with high-dose high-dose folate, vitamin B(6), and vitamin B(12) supplementation can slow the rate of cognitive decline in subjects with AD has been tried in a multicenter, randomized, controlled clinical trial named VITAL (High Dose Supplements to Reduce Homocysteine and Slow the Rate of Cognitive Decline in Alzheimer's Disease). Four hundred and nine individuals with mild to moderate AD (MMSE between 14 and 26, inclusive) and normal folic acid, vitamin B(12), and Hcy levels were included. Participants were randomly assigned to 2 groups of unequal size (60% treated with high-dose supplements [5 mg/d of folate, 25 mg/d of vitamin B(6), 1 mg/d of vitamin B(12)] and 40% treated with identical placebo); duration of treatment was 18 months. The main outcome measure was the change in the cognitive subscale of the Alzheimer Disease Assessment Scale (ADAS-cog). A total of 340 participants completed the trial. Although the vitamin supplement regimen was effective in reducing Hcy levels, it had no beneficial effect on the primary cognitive measure, rate of change in ADAS-cog score during 18 months or on any secondary measures [105].

3.5. Drugs Interfering with Metals

PBT2 was designed to modify the course of AD by preventing metal-dependent aggregation, deposition, and toxicity of Aβ. PBT2 acts at three levels of the "amyloid cascade": it inhibits the redox-dependent formation of toxic soluble oligomers, prevents deposition of Aβ as amyloid plaques, and promotes clearance by mobilizing and "neutralizing" Aβ from existing deposits [106]. PBT2 has been recently tested in a phase II trial. Seventy-eight patients with mild AD were randomly assigned to PBT2 50mg, PBT2 250mg or placebo (in addition to acetylcholinesterase inhibitors) for 12 weeks. No serious adverse events were reported by patients on PBT2. Patients treated with PBT2 250 mg had a dose-dependent and significant reduction in CSF Aβ42 concentration compared with those treated with placebo [107]. Cognitive efficacy was however restricted to two measures only, therefore future larger and longer trials are needed to test the efficacy of this drug on cognition.

The parent compound clioquinol (PBT1, Prana Biotechnology) was tested in a clinical trial for AD, showing a reduction in the rate of cognitive decline in the subgroup of more severely affected patients only [108]. According to Cochrane Collaborative study, it was not clear from this trial that clioquinol showed any positive clinical result. The two statistically significant positive results were seen for the more severely affected subgroup of patients; however, this effect was not maintained at the 36-week end- point, and this group was small (eight treated subjects). The sample size was small. Details of randomization procedure or blinding were not reported [109].

3.6. Statins

Epidemiological studies indicated that patients treated for cardiovascular disease with cholesterol-lowering therapy (statins) showed a decreased prevalence of AD [110].

Simvastatin is a pro-drug, hydrolyzed in vivo to generate mevinolinic acid, an active metabolite that is structurally similar to HMG-CoA. This metabolite competes with HMG-

CoA for binding HMG-CoA reductase, a hepatic microsomal enzyme. Simvastatin metabolites are high affinity HMG-CoA reductase inhibitors, reducing the quantity of mevalonic acid, a precursor of cholesterol.

CLASP is an ongoing randomized, double-blind, pacebo controlled, pararallel assignment, phase III trial to investigate the safety and effectiveness of simvastatin to slow the progression of AD. The clinical trial include the treatment of patients with mild to moderate AD, and the objective is to evaluate the safety and efficacy of simvastatin to slow the progression of AD, as measured by ADAS-cog. Measures of clinical global change (ADCS-CGIC), mental status, functional ability, behavioral disturbances, quality of life and economic indicators will be made also. Sample size will include 400 participants enrolled from approximately 40 sites. Study medication will be as follows: 20mg of simvastatin or matching placebo to be given for 6 weeks, followed by 40mg of simvastatin or matching placebo for the remainder of the 18-month study period.

The Lipitor's Effect in Alzheimer's Dementia (LEADe) study tests the hypothesis that a statin (atorvastatin 80mg daily) will provide a benefit on the course of mild to moderate AD in patients receiving background therapy of a cholinesterase inhibitor (donepezil 10mg daily). An international, multicenter, double-blind, randomized, parallel-group study with a double-blind randomized withdrawal phase of patients with mild to moderate AD (Mini-Mental State Examination [MMSE] score, 13 to 25) was started. Inclusion criteria included age 50 to 90 years, receiving donepezil 10mg for at least 3 months before randomization, and low-density lipoprotein cholesterol levels (LDL-C) 2.5 to 3.5 mmol/L (95 to 195 mg/dL). Co-primary end points are changes in ADAS-cog and ADCS-CGIC scale scores. A confirmatory end point is rate of change in whole brain and hippocampal volumes in patients who were enrolled in the magnetic resonance imaging substudy. Enrollment of 641 subjects is complete [111].

Acknowledgments

Part of data presented have been carried out thanks to grants from Associazione "Amici del Centro Dino Ferrari", Monzino Foundation, IRCCS Fondazione Ospedale Maggiore Policlinico, and Ing. Cesare Cusan.

References

[1] Griffin WS. Inflammation and neurodegenerative diseases. *Am. J. Clin. Nutr.* 83 (suppl), 470S-4S (2006)

[2] Hardy J. Amyloid, the presenilins and Alzheimer's disease. *Trends Neurosci.* 20(4), 154-9 (1997)

[3] Vassar R. BACE1: the beta-secretase enzyme in Alzheimer's disease. *J. Mol. Neurosci.* 23 (1-2), 105-14 (2004)

[4] Allinson TM, Parkin ET, Turner AJ, Hooper NM. ADAMs family members as amyloid precursor protein alpha-secretases. *J. Neurosci. Res.* 74(3), 342-52 (2003)

[5] Edbauer D, Winkler E, Regula JT, et al. Reconstitution of gamma-secretase activity. *Nat. Cell Biol.* 5(5): 486-8 (2003)

[6] Koo EH, Kopan R. Potential role of presenilin-regulated signaling pathways in sporadic neurodegeneration. *Nat. Med.* 10 Suppl, S26-33 (2004)

[7] Storey E, Spurck T, Pickett-Heaps J, et al. The amyloid precursor protein of Alzheimer's disease is found on the surface of static but not activity motile portions of neurites. *Brain Res.* 735(1), 59-66 (1996)

[8] Yamazaki T, Koo EH, Selkoe DJ. Cell surface amyloid beta-protein precursor colocalizes with beta 1 integrins at substrate contact sites in neural cells. *J. Neurosci.* 17 (3), 1004-10 (1997)

[9] Ho A, Südhof TC. Binding of F-spondin to amyloid-beta precursor protein: a candidate amyloid-beta precursor protein ligand that modulates amyloid-beta precursor protein cleavage. *Proc. Natl. Acad. Sci. USA* 101(8), 2548-53 (2004)

[10] Zheng H, Jiang M, Trumbauer ME, et al. beta-Amyloid precursor protein-deficient mice show reactive gliosis and decreased locomotor activity. *Cell* 81(4), 525-31 (1995)

[11] Herms J, Anliker B, Heber S, et al. Cortical dysplasia resembling human type 2 lissencephaly in mice lacking all three APP family members. *EMBO J.* 23(20), 4106-15 (2004)

[12] Leyssen M, Ayaz D, Hébert SS, et al. Amyloid precursor protein promotes post-developmental neurite arborization in the Drosophila brain. *EMBO J.* 24(16), 2944-55 (2005)

[13] Yoshikawa K, Aizawa T, Hayashi Y. Degeneration in vitro of post-mitotic neurons overexpressing the Alzheimer amyloid protein precursor. *Nature* 359(6390), 64-7 (1992).

[14] Stokin GB, Lillo C, Falzone TL, et al. Axonopathy and transport deficits early in the pathogenesis of Alzheimer's disease. *Science* 307(5713), 1282-8 (2005)

[15] Castellani RJ, Nunomura A, Lee H, et al. Phosphorylated tau: toxic, protective, or none of the above. *J. Alzheimers Dis.* 14, 377-83 (2008)

[16] Akiyama H, Barger S, Barnum S, et al. **Inflammation and Alzheimer's disease.** *Neurobiol. Aging* 21, 383-421 (2000)

[17] Yaffe K, Lindquist K, Penninx BW, et al. Inflammatory markers and cognition in well-functioning African-American and white elders. *Neurology* 61(1), 76-80 (2003)

[18] Engelhart MJ, Geerlings MI, Meijer J, et al. Inflammatory proteins in plasma and the risk of dementia : the Rotterdam study. *Arch Neurol.* 61, 668-72 (2004)

[19] Dik MG, Jonker C, Hack CE, et al. Serum inflammatory proteins and cognitive decline in older persons. *Neurology* 64, 1371-7 (2005)

[20] McGeer PL, Schulzer M, Mc Geer EG. Arthritis and anti-inflammatory agents as possible protective factors **for Alzheimer's disease:** a review of 17 epidemiologic studies. *Neurology* 47, 425-32 (1996)

[21] Veerhuis R, Van Breemen MJ, Hoozemans JM, et al. Amyloid beta plaque-associated proteins C1q and SAP enhance the Abeta 1-42 peptide-induced cytokine secretion by adult human microglia in vitro. *Acta Neuropathol.* 105, 135-44 (2003)

[22] Hoozemans JJM, Veerhuis R, Rozemuller JM, Eikelenboom P. Neuroinflammation and regeneration in the early stages **of Alzheimer's disease** pathology. *Int. J. Neuroscience* 24, 157-65 (2006)

[23] Meda L, Cassatella MA, Szendrei GI, et al. Activation of microglial cells by beta-amyloid protein and interferon-gamma. *Nature* 374, 647-50 (1995)

[24] Tarkowski E, Wallin A, Blennow K, Tarkowski A. Intracerebral production of tumor necrosis factor-alpha, a local neuroprotective agent, in Alzheimer disease and vascular dementia. *J. Clin. Immunol.* 19, 223-30 (1999)

[25] Barger SW, Hörster D, Furukawa K, et al. Tumor necrosis factors alpha and beta protect neurons against amyloid beta-peptide toxicity: evidence for involvement of a kappa B-binding factor and attenuation of peroxide and Ca2+ accumulation. *Proc. Natl. Acad. Sci. USA* 92, 9328-32 (1995)

[26] Cacquevel M, Lebeurrier N, Chéenne S, Vivien D. Cytokines in neuroinflammation and Alzheimer's disease. *Current Drug Targets* 5, 529-34 (2004)

[27] Taga T, Kishimoto T. Gp130 and the interleukin-6 family of cytokines. *Annu. Rev. Immunol.* 15, 797-819 (1997)

[28] Galimberti D, Venturelli E, Fenoglio C, et al. Intrathecal levels of IL-6, IL-11 and LIF in Alzheimer's disease and Frontotemporal Lobar Degeneration. *J. Neurol.* 255(4), 539-44 (2008)

[29] Galimberti D, Schoonenboom N, Scheltens P, et al. Intrathecal chemokine synthesis in mild cognitive impairment and Alzheimer disease. *Arch Neurol.* 63(4), 538-43 (2006)

[30] Rozemuller JM, van Muiswinkel FL. Microglia and neurodegeneration. *Eur. J. Clin. Invest* 30, 469-70 (2000)

[31] Streit WJ. Microglia as neuroprotective, immunocompetent cells of the CNS. *Glia* 40, 133-9 (2002)

[32] Scarpini E, Galimberti D, Baron P, et al. IP-10 and MCP-1 levels in CSF and serum from multiple sclerosis patients with different clinical subtypes of the disease. *J. Neurol. Sci.* 195, 41-6 (2002)

[33] Xia MQ, Hyman BT. Chemokines/chemokine receptors in the central nervous system and Alzheimer's disease. *J. Neurovirol.* 5, 32-41 (1999)

[34] Xia MQ, Qin SX, Wu LJ, et al. Immunohistochemical study of the beta-chemokine receptors CCR3 and CCR5 and their ligands in normal and Alzheimer's disease brains. *Am. J. Pathol.* 153, 31-7 (1998)

[35] Xia MQ, Bacskai BJ, Knowles RB, et al . Expression of the chemokine receptor CXCR3 on neurons and the elevated expression of its ligand IP-10 in reactive astrocytes: in vitro ERK1/2 activation and role in Alzheimer's disease. *J. Neuroimmunol.* 108, 227-35 (2000)

[36] Galimberti D, Fenoglio C, Lovati C, et al. Serum MCP-1 levels are increased in mild cognitive impairment and mild Alzheimer's disease. *Neurobiol. Aging* 27(12), 1763-8 (2006)

[37] Galimberti D, Venturelli E, Fenoglio C, et al. IP-10 serum levels are not increased in Mild Cognitive Impairment and Alzheimer's disease. *Eur. J. Neurol.* 14(4), e3-e4 (2007)

[38] Zhu X, Raina AK, Lee HG, et al. Oxidative stress signaling in Alzheimer's disease. *Brain Res.* 1000, 32-9 (2004)

[39] Markesbery WR, Carney JM. Oxidative alterations in Alzheimer's disease. *Brain Pathol.* 9, 133-46 (1999)

[40] Christen Y. Oxidative stress and Alzheimer's disease. *Am. J. Clin. Nutr.* 71, 621S-9S (2000)

[41] Gibson GE, Sheu FK, Blass JP. Abnormalities of mitochondrial enzymes in Alzheimer's disease. *J. Neural. Transm.* 105, 855-70 (1998)

[42] Maurer I, Zierz S, Moller HJ. A selective defect of cytochrome c oxidase is present in brain of Alzheimer's disease patients. *Neurobiol. Aging* 21, 455-62 (2000)
[43] Cottrell DA, Blakely EL, Johnson MA, et al. Mitochondrial enzyme-deficient hippocampal neurons and choroidal cells in AD. *Neurology* 57, 260-4 (2001)
[44] Aksenov MY, Tucker HM, Nair P, et al. The expression of key oxidative stress-handling genes in different brain regions in Alzheimer's disease. *J. Mol. Neurosci.* 11, 151-64 (1998)
[45] Repetto MG, Reides CG, Evelson P, et al. Peripheral markers of oxidative stress in probable Alzheimer's patients. *Eur. J. Clin. Invest.* 29, 643-9 (1999)
[46] Seshadri S, Beiser A, Selhub J, et al. Plasma homocysteine as a risk factor for dementia and Alzheimer's disease. *New Engl. J. Med.* 346, 476-83 (2002)
[47] Kruman II, Kumaravel TS, Lohani A, et al. Folic acid deficiency and homocysteine impair DNA repair in hippocampal neurons and sensitize them to amyloid toxicity in experimental models of Alzheimer's disease. *J. Neurosci.* 22, 1752-62 (2002)
[48] Streck EL, Vieira PS, Wannmacher CM, et al. In vitro effect of homocysteine on some parameters of oxidative stress in rat hippocampus. *Metab. Brain Dis.* 18, 147-54 (2003)
[49] Guidi I, Galimberti D, Venturelli E, et al. Influence of the Glu298Asp polymorphism of NOS3 on age at onset and homocysteine levels in AD patients. *Neurobiol. Aging* 26(5), 789-94 (2005)
[50] Nunomura A, Perry G, Aliev G, et al. Oxidative damage is the earliest event in Alzheimer disease. *J. Neuropathol. Exp. Neurol.* 60(8), 759-67 (2001)
[51] Rinaldi P, Polidori MC, Metastasio A, et al. Plasma antioxidants are similarly depleted in mild cognitive impairment and in Alzheimer's disease. *Neurobiol. Aging* 24, 915-9 (2003)
[52] Praticò D, Clark CM, Liun F, et al. Increase of brain oxidative stress in mild cognitive impairment: a possible predictor of Alzheimer's disease. *Arch Neurol.* 59, 972-6 (2002)
[53] Praticò D, Sung S. Lipid peroxidation and oxidative imbalance: early functional events in Alzheimer's disease. *J. Alzheimers Dis.* 6, 171-5 (2004)
[54] Guidi I, Galimberti D, Lonati S, et al. Oxidative imbalance in patients with mild cognitive impairment and Alzheimer's disease. *Neurobiol. Aging* 27, 262-9 (2006)
[55] Bush AI, Pettingell WH, Multhaup G, et al. Rapid induction of Alzheimer Aβ amyloid formation by zinc. *Science* 265, 1464-7 (1994)
[56] Bush AI. Drug development based on the metals hypothesis of Alzheimer's disease. *J. Alzheimers Dis.* 15, 223-40 (2008)
[57] Cherny RA, Atwood CS, Xilinas ME, et al. Treatment with a copper-zinc chelator markedly and rapidly inhibits β-amyloid accumulation in Alzheimer's disease transgenic mice. *Neuron* 30, 665-76 (2001)
[58] Hooijmans CR, Kiliaan AJ. Fatty acids, lipid metabolism and Alzheimer pathology. *Eur. J. Pharmacol.* 585, 176-96 (2008)
[59] Skoog I, Lernfelt B, Landahl S, et al. 15-year longitudinal study of blood pressure and dementia. *Lancet* 347, 1141-5 (1996)
[60] Fox NC, Warrington EK, Seiffer AL, et al. Presymptomatic cognitive deficits in individuals at risk of familial Alzheimer's disease. A longitudinal prospective study. *Brain* 121(Pt9), 1631-9 (1998)

[61] Vasdev S, Gill V, Singal P. Role of advanced glycation end products in hypertension and atherosclerosis: therapeutic implications. *Cell Biochem. Biophys.* 49, 48-63 (2007)

[62] Gervais F, Chalifour R, Garceau D, et al. Glycosaminoglycan mimetics: a therapeutic approach to cerebral amyloid angiopathy. *Amyloid* 8(Suppl.1), 28-35 (2001)

[63] Gervais F, Paquette J, Morissette C, et al. Targeting soluble Abeta peptide with Tramiprosate for the treatment of brain amyloidosis. *Neurobiol. Aging* 28(4), 537-47 (2007)

[64] Aisen PS, Saumier D, Briand R, et al. A phase II study targeting amyloid-beta with 3APS in mild-to-moderate Alzheimer disease. *Neurology* 67(10), 1757-63 (2006)

[65] Santa-Maria I, Hernández F, Del Rio J, et al. Tramiprosate, a drug of potential interest for the treatment of Alzheimer's disease, promotes an abnormal aggregation of tau. *Mol. Neurodegener.* 2(1), 17 (2007)

[66] Gladkevich A, Bosker F, Korf J, et al. Proline-rich polypeptides in Alzheimer's disease and neurodegenerative disorders - therapeutic potential or a mirage? *Prog. Neuropsychopharmacol. Biol. Psychiatry* 31(7), 1347-55 (2007)

[67] Leszek J, Inglot AD, Janusz M, et al. Colostrinin: a proline-rich polypeptide (PRP) complex isolated from ovine colostrum for treatment of Alzheimer's disease. A double-blind, placebo-controlled study. Arch Immunol Ther Exp (Warsz) 47(6), 377-85 (1999)

[68] Bilikiewicz A, Gaus W. Colostrinin (a naturally occurring, proline-rich, polypeptide mixture) in the treatment of Alzheimer's disease. *J. Alzheimer's Dis.* 6, 17-26 (2004)

[69] McLaurin J, Golomb R, Jurewicz A, et al. Inositol stereoisomers stabilizes an oligomeric aggregate of Alzheimer amyloid beta peptide and inhibit Abeta-induced toxicity. *J. Biol. Chem.* 275, 18495-502 (2000)

[70] Townsend M, Cleary JP, Mehta T, et al. Orally available compound prevents deficits in memory caused by the Alzheimer Amyloid-β oligomers. *Ann. Neurol.* 60, 668-76 (2006)

[71] Schenk D, Barbour R, Dunn W. Immunization with amyloid-beta attenuates Alzheimer-disease-like pathology in PDAPP mouse. *Nature* 400, 173-7 (1999)

[72] Gilman S, Koller M, Black RS. Clinical effects of Aβ immunization (AN1792) in patients with AD in an interrupted trial. *Neurology* 64, 1553-62 (2005)

[73] Fox N, Black RS, Gilman S. Effects of Aβ immunization on MRI measures of cerebral volume in Alzheimer's disease. *Neurology* 64, 1563-72 (2005)

[74] Hock C, Nitsch R. Clinical observations with AN1792 using TAPIR analyses. *Neurodeg. Dis.* 2, 273-76 (2005)

[75] Holmes C, Boche D, Wilkinson D, et al. Long-term effect of Aβ$_{42}$ immunisation in Alzheimer's disease: follow-up of a randomised, placebo-controlled phase I trial. *Lancet* 372, 216-23 (2008)

[76] Pollack SJ, Lewis H. Secretase inhibitors for Alzheimer's disease: challenges of promiscuous protease. *Curr. Opin. Investig. Drugs* 6, 35-47 (2005)

[77] Wong GT, Manfra D, Poulet FM, et al. Chronic treatment with the γ-secretase inhibitor LY-411,575 inhibit β-amyloid peptide production and alters lymphopoiesis and intestinal cell differentiation. *J. Biol. Chem.* 279, 12876-82 (2004)

[78] Siemers E, Skinner M, Dean RA, et al. Safety, tolerability, and changes in in amyloid β concentrations after administration of a γ-secretase inhibitor in volunteers. *Clin. Neuropharmacol.* 28, 126-32 (2005)

[79] Siemers E, Quinn J, Kaye J, et al. Effects of a γ-secretase inhibitor in a randomized study of patients with Alzheimer disease. *Neurology* 66, 602-4 (2006)

[80] Fleisher AS, Raman R, Siemers ER, et al. Phase 2 safety trial targeting amyloid beta production with a gamma-secretase inhibitor in Alzheimer disease. *Arch Neurol.* 65(8), 1031-8 (2008)

[81] Weggen S, Eriksen JL, Das P, et al. A subset of NSAIDs lower amyloidogenic Abeta42 independently of cyclooxygenase activity. *Nature* 414(6860), 212-6 (2001)

[82] Beher D, Clarke EE, Wrigley JD, et al. Selected nonsteroidal anti-inflammatory drugs and their derivatives target γ-secretase at a novel site: evidence for an allosteric mechanism. *J. Biol. Chem.* 279, 43419-26 (2004)

[83] Lleo A, Berezovska O, Herl L, et al. Non steroidal anti-inflammatory drugs lower Aβ42 and change presenilin1 conformation. *Nat. Med.* 10, 1065-6 (2004)

[84] Townsend KP, Praticò D. Novel therapeutic opportunities for Alzheimer's disease: focus on nonsteroidal anti-inflammatory drugs. *FASEB J.* 19(12), 1592-601 (2005)

[85] Kukar T, Prescott S, Eriksen JL, et al. Chronic administration of R-flurbiprofen attenuates learning impairments in transgenic amyloid precursor protein mice. *BMC Neurosci.* 8, 54 (2007)

[86] Galasko DR, Graff-Radford N, May S, et al. Safety, tolerability, pharmacokinetics, and Aβ levels after short-term administration of R-flurbiprofen in healthy elderly individuals. *Alzheimer Dis. Assoc. Disord.* 21(4), 292-9 (2007)

[87] Wilcock GK, Black SE, Hendrix SB, et al. Efficacy and safety of tarenflurbil in mild to moderate Alzheimer's disease: a randomised phase II trial. *Lancet Neurol.* 7(6), 468-9 (2008)

[88] Wischik CM, Edwards PC, Lai RY, et al. Selective inhibition of Alzheimer disease-like tau aggregation by phenothiazines. *Proc. Natl. Acad. Sci. USA* 93(20), 11213-8 (1996)

[89] Wischik CM, Bentham P, Wischik DJ, Seng KM. Tau aggregation inhibitor (TAI) therapy with Rember™ arrests disease progression in mild and moderate Alzheimer's disease over 50 weeks. *Alzheimers Dement* 4(4), Suppl.2, T167 (2008)

[90] Balaraman Y, Limaye AR, Levey AI, Srinivasan S. Glycogen synthase kinase 3β and Alzheimer's disease: pathophysiological and therapeutic significance. *Cell Mol. Life Science* 63, 1226-35 (2006)

[91] Martinez A, Perez DI. GSK-3 inhibitors: a ray of hope for the treatment of Alzheimer's disease? *J. Alz. Dis.* 15, 181-91 (2008)

[92] Szekely CA, Thorne JE, Zandi PP, et al. Nonsteroidal anti-inflammatory drugs for the prevention of Alzheimer's disease: a systematic review. *Neuroepidemiology* 23, 159-69 (2004)

[93] Reines SA, Block GA, Morris JC, et al. Rofecoxib: no effect on Alzheimer's disease in a 1-year, randomized, blinded, controlled study. *Neurology* 62, 66-71 (2004)

[94] Aisen PS, Schafer KA, Grundman M, et al. Effects of rofecoxib or naproxen vs placebo on Alzheimer's disease progression: a randomized controlled trial. *JAMA* 289, 2819-26 (2003)

[95] Scharf S, Mander A, Ugoni A, et al. A double-blind, placebo controlled trial of diclofenac/misoprostol in Alzheimer's disease. *Neurology* 53, 197-201 (1999)

[96] Soininen H, West C, Robbins J, Niculescu L. Long-term efficacy and safety of celecoxib in Alzheimer's disease. *Dement Geriatr. Cogn. Disord.* 23(1), 8-21 (2007)

[97] Eriksen JL, Sagi SA, Smith TE, et al. NSAIDs and enantiomers of flurbiprofen target gamma-secretase and lowe Abeta 42 in vivo. *J. Clin. Invest.* 112(3), 440-9 (2003)

[98] Aisen PS, Marin DB, Brickman AM, et al. Pilot tolerability studies of hydroxychloroquine and colchicine in Alzheimer disease. *Alz. Dis. Assoc. Disord.* 15 (2), 96-101 (2001)

[99] Aisen PS, Schmeidler J, Pasinetti GM. Randomized pilot study of nimesulide treatment in Alzheimer's disease. *Neurology* 58(7), 1050-4 (2002)

[100] Rogers J, Kirby LC, Hempelman SR, et al. Clinical trial of indomethacin in Alzheimer's disease. *Neurology* 43, 1609-11 (1993)

[101] Tabet N, Feldman H. Indomethacin for the treatment of Alzheimer's disease patients. *Cochrane Database Syst. Rev.* CD003673 (2002)

[102] Tobinick EL, Gross H. Rapid improvement in verbal fluency and aphasia following perispinal etanercept in Alzheimer's disease. *BMC Neurology* 8, 27 (2008)

[103] DeKosky ST, Williamson JD, Fitzpatrick AL, et al. Ginkgo biloba for prevention of dementia: a randomized controlled trial. *JAMA* 300(19), 2253-62 (2008)

[104] Barten DM, Albright CF. Therapeutic strategies for Alzheimer's disease. *Mol. Neurobiol.* 37, 171-86 (2008)

[105] Aisen PS, Schneider LS, Sano M, et al. High-dose B vitamin supplementation and cognitive decline in Alzheimer disease: a randomized controlled trial. *JAMA* 300(15): 1774-83 (2008)

[106] Cherny RA, Atwood CS, Xilinas ME, et al. Treatment with a copper-zinc chelator markedly and rapidly inhibits beta-amyloid accumulation in Alzheimer's disease transgenic mice. *Neuron* 30(3), 665-76 (2001)

[107] Lannfelt L, Blennow K, Zetterberg H, et al. Safety, efficacy, and biomarker findings of PBT2 in targeting Abeta as a modifying therapy for Alzheimer's disease: a phase IIa, double-blind, randomized, placebo-controlled trial. *Lancet Neurol.* 7(9), 779-86 (2008)

[108] Ritchie CW, Bush AI, Mackinnon A, et al. Metal-protein attenuation with iodochlorhydroxyquin (Clioquinol) targeting Aβ amyloid deposition and toxicity in Alzheimer disease. *Arch Neurol.* 60, 1685-91 (2003)

[109] Jenagaratnam L, McShane R. Clioquinol for the treatment of Alzheimer's Disease. *Cochrane Database Syst. Rev.* 25;(1), CD005380 (2006)

[110] Jick H, Zornberg GL, Jick S, Seshadri S. Statins and the risk of dementia. *Lancet* 356, 1627-31 (2000)

[111] Jones RW, Kivipelto M, Feldman H, et al. The Atorvastatin/Donepezil in Alzheimer's Disease Study (LEADe): design and baseline characteristics. *Alzheimers Dement* 4(2), 145-53 (2008)

Table. Potentially disease-modifying drugs tested in clinical trials in patients with AD

	Name	N. of patients	Duration (months)	Mode of Action	Clinical trial phase	Results	Ref.
Drugs influencing Aβ deposition							
Anti-aggregants	Tramiprosate (Alzhemed™)	1052	18	GAG mimetic	Phase III	No efficacy (definitive)	
	Colostrinin	105	15+15 extension	Inhibits Aβ aggregation	Phase II	Modest improvement in MMSE (not definitive)	[68]
	Scyllo-inositol (AZD103)	340*	18	Inhibits Aβ aggregation	Phase II	Ongoing	
Vaccination	AN1792	300	halted after ≅12 months	Aβ removal (active immunisation)	Phase II	Unclear cognitive results-severe adverse events (definitive)	[71,72]
	Bapineuzumab	200	18	Aβ removal (passive immunisation)	Phase III	Ongoing	
γ-secretase inhibitors	LY450139	51	3	Inhibits γ-secretase	Phase II	No changes in cognitive measures Phase III trial ongoing	[80]
SALAs	Tarenflurbil (Flurizan™)	210	12+12 extension	Inhibits γ-secretase	Phase II	No effect on cognition	[87]
		>800	18		Phase III	No effect on cognition (definitive)	
Drugs influencing tau deposition							
	MTC (Rember™)	321	6	Interferes with tau aggregation	Phase II	Improvement in cognition (not definitive)	[89]
Anti-inflammatory drugs							
	Rofecoxib	351	12	NSAID, inhibits COX2	Phase III	No efficacy (definitive)	[93,94]
		692	12		Phase III	No efficacy (definitive)	

Name	N. of patients	Duration (months)	Mode of Action	Clinical trial phase	Results	Ref.
Naproxene	351	12	Non-selective NSAID	Phase III	No efficacy (definitive)	[94]
Diclofenac	41	25 weeks	NSAID	Phase II	No efficacy (definitive)	[95]
Celecoxib		52 weeks	NSAID	Phase II	No efficacy (definitive)	[96]
Hydroxychloroquine	20	3	NSAID	Phase II	No efficacy (definitive)	[98]
Nimesulide	40	6	NSAID, inhibits COX2	Phase II	No efficacy (definitive)	[99]
Indomethacin		6	NSAID	Phase II	No efficacy-toxicity (definitive)	[100]
Etanercept	12	6	Inhibits TNF-α	Open study	Improvement in cognition (not definitive)	[102]
Drugs preventing oxidative damage						
Folate/B6/B12	340	18	Reduction of Hcy	Phase III	No effects	[105]
Drugs interfering with metals						
PBT2	78	3	Metal-protein attenuation	Phase II	Improvement in cognition (not definitive)	[107]
Clioquinol	36	9	Inhibits zinc and copper from binding to Aβ	Phase II	Reduction in cognitive decline in more severely affected patients only (definitive)	[108]
Statins						
Simvastatin	400*	18	Cholesterol reduction	Phase III	Ongoing	
Atorvastatin	641	20	Cholesterol reduction	Phase III	Ongoing	[111]

*estimated.

In: Encyclopedia of Neuroscience Research
Editors: Eileen J. Sampson and Donald R. Glevins
ISBN 978-1-61324-861-4
© 2012 Nova Science Publishers, Inc.

Chapter XXIV

Cognitive Interventions to Improve Prefrontal Functions

Yoshiyuki Tachibana*, Yuko Akitsuki and Ryuta Kawashima
Department of Functional Brain Imaging, IDAC, Tohoku University,
Sendai, Japan

Abstract

The human prefrontal cortex (PFC) plays major roles in higher cognitive functions necessary for maintaining a healthy social life. Psychological and psychiatric problems are often associated with cognitive impairments associated with PFC. Thus, previous cognitive intervention studies have been conducted to improve the functions associated with PFC. In this article, first we describe the functions associated with PFC and its importance in cognitive intervention studies. Then, we describe recent advancements in cognitive intervention methods, particularly interventions to prevent cognitive decline in healthy older adults and those to enhance their emotional control and resilience in preschool children. We also discuss on transfer effects of previous cognitive intervention, which are often observed. Finally, we discuss on the unresolved issues on the mechanism underlying the effect of cognitive intervention. We consider that a deeper understanding of the effect of cognitive intervention will greatly contribute to human welfare and education for all generations. Further multidisciplinary research will be required to achieve this ultimate goal.

Keywords: prefrontal cortex, cognitive intervention, executive function, transfer effects, older adults, early childhood

---

* Correspondence should be addressed to: Yoshiyuki Tachibana, Seiryo-machi 4-1, Aoba-ku Sendai, 980-8575, Japan. Tel/Fax: +81-022-217-7988. tatibana@idac.tohoku.ac.jp.

1. Functions of Human Prefrontal Cortex

The human prefrontal cortex (PFC) plays major roles in higher cognitive functions necessary for maintaining a healthy social life. For example, patients with PFC lesions have difficulties in planning and controlling their emotions. Such patients, however, have no cognitive dysfunction of language, memory, and sensory (Damasio, 1995). The first seminal case of a prefrontal brain lesion was that of Phineas Gage (Damasio et al., 1994). His frontal lobes were damaged by a railway construction accident in 1848. Although he retained normal memory, and speech and motor skills, his personality changed after the accident. He became irritable, quick-tempered, and impatient. Owing to these changes, his friends described him as "no longer Gage." He was a capable and efficient worker before the accident. After the accident, however, he was unable to complete multiple tasks. His case became the focus of attention in studies of the importance of PFC functions.

Previous neuroimaging studies showed that dysfunction of PFC is related to various psychiatric disorders, such as, schizophrenia (Lawrie et al., 2008; Weinberger et al., 2001); depression (Drevets, 2000); bipolar disorder (Adler et al., 2006; Monkul et al., 2005); mild cognitive impairment (Petrella et al., 2006; Prvulovic et al., 2005); and Attention Deficit / Hyperactivity Disorder (ADHD) (Valera et al., 2007). Furthermore, decrease in PFC volume was observed in people exposed to repeated stressors (Liston et al., 2006; Radley et al., 2006), suicide victims (Rajkowska, 1997), criminals diagnosed as sociopaths (Pridmore et al., 2005; Yang et al., 2008), and drug addicts (Wilson et al., 2004; Yucel and Lubman, 2007). It is also suggested that neurological representations of various psychological stases are associated with PFC, such as, feeling of guilt or remorse (Kiehl, 2006; Martens, 2001); interpretation of reality (Abraham and von Cramon, 2009; Blackwood et al., 2001); and lying (Spence et al., 2008; Spence and Kaylor-Hughes, 2008).

2. Functions Associated with PFC as Target of Intervention Studies

As mentioned above, PFC is deeply associated with a healthy social life. Thus, previous intervention research was conducted to improve the functions associated with PFC (Details will be provided later in this article). It is also noteworthy that cognitive intervention aimed at improving the functions associated with PFC is considered to be useful for primary prevention of psychiatric or psychological disorders (Blair and Diamond, 2008). In particular, interventions for older adults and children have been the focus of much attention owing to their importance as preventive strategies; namely, to prevent dementia in older adults and to enhance healthy development in children.

It should be stressed that the target functions associated with PFC in cognitive intervention studies differ across age groups. This is because PFC is one of the last regions to reach maturity during the course of development in humans (Fuster, 2002). Thus, the developmental state of PFC differs with age (Lemaitre et al., 2005). For example, gray matter volume does not reach adult levels in the dorsolateral prefrontal cortex until at least the end of adolescence (Whitford et al., 2007). Moreover, myelination of the dorsolateral prefrontal cortex continues into the 20s or possibly 30s (Huttenlocher and Dabholkar, 1997).

On the basis of these developmental characteristics of PFC, most of the cognitive interventions in early childhood are aimed at improving executive function; executive function is vastly developed in early childhood between 3 and 5 years old (Garon et al., 2008). During the preschool years, enhancement of executive function is regarded as one of the most important factor for healthy development (Blair and Diamond, 2008). The details of the importance of executive function in cognitive intervention for children will be discussed later.

By contrast, cognitive interventions for older adults (aged more than 60 years) have targeted a broader range of functions associated with PFC; PFC has already reached maturity by this age (Hubert et al., 2007; Olesen et al., 2007). For older adults, interpretations in many studies aimed to maintain various functions in daily life to prevent dementia (Fillit et al., 2002; Kramer et al., 2005).

3. Transfer Effects of Cognitive Interventions

Transfer effects were observed in cognitive intervention studies, which means that the effects (i.e., improvements) of a particular cognitive aspect directly targeted by a training program generalize to either related cognitive constructs or behaviors associated with the trained construct (Thorell et al., 2009). For example, Thorell et al. (2009) observed that children trained in working memory showed transfer effects to attention. Moreover, Olesen et al. (2004) showed that training in working memory tasks increase brain activity in multimodal areas of the prefrontal and parietal cortices. This might be related to the observation that certain areas of the frontal and parietal lobes seem to be multimodal, rather than being associated with any one type of sensory stimulus, and are activated by both hearing and vision working memory tasks (Klingberg and Roland, 1997). Thus, the authors claimed that developing a multimodal area would probably be more useful than developing an area that is only associated with one function, for example, hearing. That is building up bottleneck areas through intervention would be beneficial in enhancing many functions.

In this article, we will describe some transfer effects of intervention studies targeting the functions associated with PFC. We, however, think it is beyond the scope of this review to cover all the cognitive interventions for older adults and children in every detail. Instead, we describe selected studies with highly relevant findings in this review.

4. Recent Advancements in Cognitive Intervention Methods

1) Interventions for Older Adults

Kawashima et al. (2005) developed a cognitive intervention program named "Learning Therapy" for senile dementia. The concept of the program was based on the findings of neuroimaging and clinical studies. Learning Therapy has been developed to stimulate the cognitive functions of the dorsolateral PFC, as well and the temporal and parietal association cortices. Learning Therapy has two characteristics that distinguish it from previous methods of cognitive rehabilitation. One is that this method is based on the findings of previous neuroscience research—that is, solving arithmetic problems and reading aloud bilaterally activate the dorsolateral PFC of humans (Kawashima et al., 2004; Miura et al., 2003). The second is that it aims to mediate the transfer of different cognitive functions within the dorsolateral PFC—that is, from reading aloud and solving arithmetic problem functions toward general cognitive functions such as communication, independence, and conceptualization. The materials for the training program were two tasks in arithmetic and Japanese language, which were systematized basic problems in reading and arithmetic. This is based on the findings of previous brain imaging studies indicating that reading sentences or words aloud and simple arithmetic operations activate PFC (Kawashima et al., 2004; Miura et al., 2003; Miura et al., 2005). On the other hand, both reading aloud and solving arithmetic problems require working memory. Thus, prefrontal stimulation by reading aloud and solving arithmetic problems may lead to the positive transfer effects on other cognitive functions (Kawashima et al., 2005). It is noteworthy that the authors developed the training program such that the reading aloud and solving arithmetic problems are very simple and easy tasks that even people with senile dementia can understand, perform, and continue the tasks. Thirty-two people were recruited from a nursing home in Japan. The participants were randomly assigned to these two groups: sixteen participants in the experimental group and sixteen in the control group. All of the participants in both groups were clinically diagnosed as having senile dementia of the Alzheimer type on the basis of the criteria of the American Psychiatric Association's Diagnostic and Statistical Manual of Mental Disorders, 4th edition (DSM-IV). The participants in the experimental group were instructed to perform a training program using learning tasks in reading and arithmetic for 2-6 days a week. The function of the frontal cortex of the subjects was assessed using the Frontal Assessment Battery (FAB) test at the bedside. After 6 months of training, the FAB score of the experimental group decreased slightly over the 6-month period, and the difference in the score between the experimental and control groups was statistically significant. They also observed the restoration of communication skills and independence in the experimental group. Their results indicate that the learning tasks of reading aloud and arithmetic calculation can be used for cognitive rehabilitation of dementia patients.

Uchida and Kawashima (2008) then applied a similar daily cognitive training program to a study of 124 community-dwelling healthy seniors aged 70 to 86 years (mean age = 76.2 years). The aim was to evaluate the effects of training on cognitive functions, particularly on the functions associated with PFC. The study design was a single-blind randomized controlled trial. In this study, neuropsychological measures were determined prior to and six months after the intervention (post-test) by the following tests: mini-mental state examination (MMSE), FAB test at bedside, and digit-symbol substitution test (DST) of WAIS-R. Interestingly, the results showed that the speed of processing (as measured by DST) and executive function (as measured by FAB) showed a statistically significant improvement in the post-test compared with the pretest. It is noteworthy that these properties are not directly tied to the contents of the intervention. Moreover, such an improvement was maintained up to six months of follow-up tests in only the experimental group. The results of their investigations indicate that the transfer effects of the cognitive intervention by reading and solving arithmetic problems on nontargeted cognitive functions was demonstrated in this study. The results also demonstrated that cognitive training has the beneficial effects of maintaining and improving cognitive functions.

On the other hand, Ball et al. (2002) and Jobe et al. (2001) performed a very large scale intervention study of independently living vital older adults. Two thousand eight hundred and thirty-two participants aged 65 to 94 years (mean age = 73.6 years) were recruited from senior housing, community centers, and hospital/clinics in six metropolitan areas in the United States. The subjects were randomly assigned to one of four groups: memory training group (verbal episodic memory: n = 711), reasoning training group (ability to solve problems that follow a serial pattern: n = 705), speed of processing training group (visual search and identification: n = 712), and control condition group. For the three treatment groups, booster training was provided to 60% of randomly selected initially trained subjects 11 months after the initial training was provided. Functional activities were measured by performance-based and self-reported questionnaires on cognitive function and cognitively demanding everyday functioning (See Table. 1) immediately after the intervention and at 2- and 5-year follow-ups. The results indicated that each intervention improved the targeted cognitive ability compared with the baseline before the intervention started. Ball et al. determined the percentage of participants who showed reliable improvement in each training group. A participant was classified as having improved reliably on a particular measure of his or her performance when the measure at a follow-up exceeded the baseline measure of the performance by 1 standard error of measurement (Dudek, 1979). The percentages of participants in the speed of processing, reasoning, and memory interventions who showed reliable cognitive improvement after the interventions were 87%, 74%, and 26%, respectively. Moreover, booster training enhanced training gains in speed and reasoning interventions, which were maintained at 2-year follow-up. Training effects were of a magnitude equivalent to the amount of decline expected in older adults without dementia over 7- to 14-year intervals. Willis et al. (2004) assessed the long-term effects of this training program. Their assessment using the self-reported instrumental activities of daily living (IADL) test indicated that reasoning training resulted in less functional decline. Each intervention produced immediate improvement in the cognitive ability, which was retained for five years. At year five, participants in all three intervention groups reported less difficulty in performing the IADL test compared with the control group. However, this effect was significant only for the reasoning training group. Neither speed of processing training nor memory training had a significant effect on IADL.

Another intervention study was carried out by Edwards et al, (2002). They evaluated the extent to which a standardized speed of processing training transfers to similar and dissimilar speeded cognitive measurements as well as to other domains of cognitive functioning in older adults. Ninety-seven older adults aged 60 to 87 years (mean age = 73.7 years) were randomly assigned to either the intervention group or the control group. The subjects of the intervention group received ten 1-hour sessions of speed of processing training for six week. They were administered a battery of cognitive tests (See Table. 1) and the timed IADL Test. Their intelligence, memory, attention, verbal fluency, visual-perceptual ability, speed of processing, and functional abilities were assessed by the tests. The subjects of the intervention group showed some improvement in speed of processing measures, including performance of the IADL test. However, the effects did not transfer to other domains of cognitive functioning.

Table 1. Cognitive interventions for older adults

	Contents of the interventions	Subjects	Target of the intervention	Methods of measurement	Results
Kawashima et al., 2005	Learning Therapy: Solving systematized basic problems in reading and arithmetic every day for 6 months	Community-dwelling, healthy older adults. N = 124 (80 males and 44 females) Age range = 70 to 86 years (mean age = 78.7 years, SD = 4.0)	Functions of dorsolateral prefrontal cortex	- Mini mental state examination - Frontal assessment battery at bedside	- Digit symbol substitution test of WAIS-R Transference effects in MMSE, FAB, DST, not directly tied to the interventions were shown
Willis et al., 2006; Ball et al., 2002; Jobe et al., 2001	ACTIVE (Advanced Cognitive Training for Independent and Vital Elderly) trial: Memory, reasoning, training, and processing training in small group settings in ten 60- to 75- minute sessions over 5- to 6-week periods	Older adults recruited from senior housing, community centers, and hospital/clinics in 6 metropolitan areas in the United States. N = 2832 (706 males and 2126 females) Age range = 65 to 94 years (mean age = 73.6 years, SD = 5.9)	Memory, reasoning, speed of processing	- Hopkins Verbal Learning Test - Auditory Verbal Learning Test - Rivermead Behavioral Memory Test - Reasoning (Word series, Letter series, Letter sets - Useful Field of View	Besides the improvements in the abilities for the trained domain, reasoning training was revealed to improve functional declines in self-reported instrumental activities of daily living (IADL)
Edwards et al., 2002	Standardized speed of processing training: Speed processing intervention in ten 1-hour sessions for six weeks	Community-dwelling, healthy older adults N = 91 (42 males and 53 females) Age range = 61 to 195 years (mean age = 73.7 years)	Evaluation of the extent to which standardized speed of processing training transfers to other domains of cognitive functioning	- Useful Field of View - Road Sign Test - Timed Instrumental Activities of Daily Living - Finding As - Identical Pictures - Letter/Pattern Comparison - Digit Symbol - Stroop test	Speed of processing training enhanced the abilities of performing instrumental activities of daily living
Mahncke et al., 2006	Brain plasticity-based training program: Computer-based cognitive training for 1 hour per day 5 days a week for 8 to 10 weeks	Community-dwelling, healthy older adults. N = 182 (91 males and 91 females) Age range = 60 to 87 years (mean age = 70.9 years)	Processing speed Working memory	- Repeatable Battery for the Assessment of Neuropsychological Status (including six tests of auditory cognition, list learning, story memory, digit span forward, delayed free list recall, delayed free story recall)	Not only task-specific performance but also non-related standardized neuropsychological measures of memory were improved
Stuss et al., 2007; Levine et al., 2007; Craik et al., 2007	Cognitive rehabilitation program in the elderly: The program, consisted of the three distinct 4-week modules mentioned above, applied for 12 weeks in a small-group format to which individuals as a group met weekly in relatively short (3 hr), highly interactive sessions 1. memory training 2. psychosocial training 3. goal management seminar	Community-dwelling, healthy older adults, not fitting the criteria of mild cognitive impairment or dementia N = 49 (22 males and 27 females) Age range = 71 to 87 years (mean age = 78.7 years, SD = 3.9)	General strategies abilities in ways that would be expressed in a broad range of functional domains in daily life	- Mini mental state examination - National Adult Reading Test-Revised - Digit Span - Logical Memory, immediate recall of two short stories - Judgement of Line Orientation - Wisconsin Card Sorting Test	- Boston Naming Test - Beck Anxiety Index - Geriatric Depression Scale - Alpha Span Test - Brown-Peterson test - Hopkins Verbal Learning Test-Revised - Simulated real life tasks - Dysexecutive questionnaire - Memory, psychosocial and goal management abilities were enhanced by the program - Untrained memory (secondary measures, and simulated IADL tasks were improved
Oswald et al. 1996	SIMA project: 1) Competence-training program 2) Memory-training program 3) Psychomotor-training program 4) Combined competence and psychomotor training program 5) Combined memory- and psychomotor training program The trainings were conducted once a week over 30 sessions, each lasting 2 to 3 hours for nine months	Community-dwelling, healthy older adults. N = 375 Age range = 75 to 89 years (mean age = 79.5 years, SD = 3.5)	Determination of the importance of memory and reinforcement factor in cognitive training	Global Constructs (e.g., cognitive status) Subconstructs (e.g., fluid and crystallized cognitive performance) Individual Measurement Parameters (e.g., speed of information processing in the ZVT-G)	The memory plus psychomotor training group showed transfer effects on the composite cognitive index, and led to an improvement of psychomotor performance

Cognitive intervention studies, with highly relevant findings, to improve prefrontal functions of older adults are described in this table. These are the summaries of the papers, which is described in this review.

Table 2. Cognitive Interventions for children

	Contents of the intervention	Subjects	Target of the intervention	Measurement	Results
Diamond et al., 2007	"Tools of the Mind (Tools)" • Preschool curriculum based on Vygotsky's theory • Forty Executive Function (EF) promoting activities, including "self-regulatory private speech", dramatic play, memory, and attention training. • Every day curriculum for a year	Children in a low-income urban school district preschool (All came from low-income families 78% with yearly income < $25,000) N = 147, the numbers of boys and girls were unknown Age range = 4 to 5 years (mean of 11 years, SD = 0.34)	Improvement of cognitive control	• Dots task • Flanker task • Social Skills Rating Scale (SSRS) • Peabody Picture Vocabulary Test—III (PPVT) • IDEA Oral Language proficiency Test • The Wechsler Preschool and Primary Scales of Intelligence	• Woodcock Johnson Test of Achievement, Letter-Word Identification Test • Woodcock Johnson Test of Achievement, Applied Problem Test • Get Ready To Read • Expressive One-Word Picture Vocabulary Test	The intervention enhanced the children's EF
Domitrovich et al., 2007	Promoting Alternative Thinking Strategies (PATHS) Curriculum • Preschool curriculum in Head Starts Lessons on compliments, basic and advanced feelings, a self-control strategy and problem solving • Thirty "PATHS" curriculum lessons weekly and extension activities for 9 months	Children in socioeconomically low district preschool (All came from Head Starts) N = 246, 120 boys and 126 girls Age range = 4 to 5 years (mean of 4.3 years, SD = 0.55) African-American: 47% European-American: 38% Hispanic: 0% The mean actual income of families was $7005	Improvement of social competence and reduction of problem behaviors	• Kusche Emotions Inventory (KEI) • Assessment of Children's Emotions Scales (ACES) • Denham Puppet Interview • Day/Night task • Luria's tapping test		Children's emotion knowledge skills were enhanced
Bierman et al., 2007	REDI (Research-Based, Developmentally Informed) Program • Preschool curriculum in Head Starts • Preliteracy skills such as learning the alphabet and learning to manipulate the sounds of presented letters	Children in socioeconomically low district preschool N = 356, 184 boys and 172 girls African American: 27% European American: 54% Hispanic: 7% Mean age = 4.5 years old (SD = 0.31)	Enhancement of learning skills and school readiness	• Word span test • Peg tapping test • Change card sort • Walk-a-line slowly • Task orientation • Expressive One-Word Picture Vocabulary Test • Blending and Elision Scales of the Test of Preschool Early Literacy (TOPEL)	• Print Knowledge Scale of TOPEL-assessed children's familiarity with written text • Social Competence Scale • Teacher Observation of Child Adaptation-Revised	Besides enhancing learning skills, impulsivity, aggression, and attention problems were improved.
Kawashima et al., 2009	NOUTRE playing program (NOUTRE means "brain training" in Japanese) • Home-based program between mother and child • Playing program conducted at home 5 days a week for 3 months	Healthy children with typical development N = 227, 107 boys and 120 girls Age range = 4 to 5 years (mean of 5.1 years, SD = 0.59)	Enhancement of children's and maternal mental health Enhancement of children's cognitive development	• Parenting Stress Index • S-5 Intelligence test revised version • Goodenough draw-a-man intelligence test • Questionnaire for children's lifestyle	• Child and maternal attachment was enhanced • Working memory and performance IQ were enhanced

Table 2. (Continued)

Rueda et al., 2005	Computer-based executive attention training for preschool • 9 sessions for 4-year-old children and 10 sessions for 6-year-old children for 5 days	Healthy children with typical development • 4-year-old children N = 49, 25 boys and 24 girls Mean age = 4.3 years SD = 0.18 • 6-year-old children N = 24, 12 boys and 12 girls Mean age: 6.4 years SD = 0.27 • All participants were recruited from a database of births in the Eugene-Springfield, OR, area • Healthy normal developed children	Enhancement of executive attention	• Children's version of Attention Network Test (Child ANT) • Kaufman Brief Intelligence Test • Children's Behavior Questionnaire (CBQ) • Event-related potential recording during attention network test performance • Genotyping (Cheek swab)	• Both 4- and 6-year-old children showed more mature performance after the training than the control groups • Their study indicated that the executive attention network appears to develop under strong genetic control but that it is subject to educational interventions during development
Thorell et al., 2009	Working memory training: • Computerized visuospatial working memory training for preschool • Containing inhibition training: Go/no-go task, stop-signal task, flanker task. • Fifteen minutes per day for 5 weeks	Normal healthy developed children Age range = 4 to 5 years (mean of 4.7 years; SD = 0.43) 1. Working memory training group N = 17, 8 boys and 9 girls Mean age = 4.5 years 2. Inhibition training group N = 18, 9 boys and 9 girls Mean age = 4.5 years 3. Active control group N = 14, 7 boys and 7 girls Mean age = 4.8 years 4. Passive control group N = 16, 7 boys and 9 girls Mean age = 5.0 years	Assessment of training and transfer effects of executive functions by working memory training	• Span board • Word Spans • Stroop-like task • Go/no-go • Auditory CPT • Block design	• Working memory training has significant effects on the performance of nontrained working memory tasks within both the spatial and the verbal domains • It also has significant transfer effects on laboratory measures of attention • Training inhibitory control did not have any significant effects relative to the control group • In at least some of the trained tasks, the children's inhibitory control improved

Cognitive intervention studies, with highly relevant findings, to improve prefrontal functions of children are described in this table. These are the summaries of the papers, which is described in this review.

Their results indicated that the speed of processing training can transfer to the performance of the IADL test. Furthermore, Edwards et al., (2005) extended their investigation by assigning the subjects into the social- and computer-contact control group. The results were consistent with those of the authors' previous study of the transfer of a speed of processing intervention (Edwards et al., 2002).

Mahncke et al., (2006) developed a computer-based program for healthy adults. Most older adults experience nonpathological losses in cognitive function, frequently called "age-related cognitive decline (ARCD)" (Bischkopf et al., 2002; Fillit et al., 2002; Ritchie et al., 2001). Their study targeted ARCD. One hundred and eighty-two community-dwelling, healthy older adults aged 60 to 87 years (mean age = 70.9 years) participated in the study. They were assigned to the computer-based intervention program or the no-contact control group. The program focused on processing speed and working memory. In the intervention program, the participants performed the exercises for one hour per day five days a week for eight to ten weeks. After the intervention period, not only task-specific performance but also nonrelated standardized neuropsychological measures of memory were improved. The data at three-month follow-up were compared with those at pretraining, and the results indicated that the enhancement of the performance of the digit span forward task was maintained in the intervention group.

Stuss et al. (2007) performed a randomized control trial to evaluate a cognitive rehabilitation program. The purpose of their study was to enhance the general strategic abilities in ways that would be expressed in a broad range of functional domains. On the other hand, Craik et al. (2007) examined 49 elderly adults who were cognitively normal and participated in a twelve-week intervention program. The training program consisted of three distinct modules, each requiring four consecutive weeks. The participants met once a week in three-hour sessions. The modules were presented in a fixed order: memory training, goal management seminars, and psychosocial training (Craik et al., 2007; Stuss et al., 2007). The authors used a cross-over design; the participants were assigned to either the "early-training group (ETG)" or "late-training group (LTG)." During the intervention periods, the participants were assigned a homework for one hour every day. The authors used the alpha span test (Craik, 1986), Brown-Peterson test (Floden et al., 2000, modification), and Hopkins verbal learning test-revised to measure the effects of the intervention on memory. There was no improvement in most outcome measures of working memory, primary memory, or recognition memory after the intervention nor at any of the additional two assessments for three months after follow-up. Memory enhancement in the tests was restricted to the ETG, possibly because the LTG lost their motivation as a consequence of their delayed training (Craik et al., 2007). Furthermore, improvement in the performance of an untrained simulated IADL task, in which participants had to set up carpools or assign people to swimming lessons, was observed immediately after the intervention. However, six months later, only the ETG group maintained the effects.

Oswald et al. (1996) conducted an intervention study in the SIMA (Maintaining and Supporting Independent Living in Old Age) project. Three hundred and seventy-five community-dwelling older adults aged to 75 to 89 years (mean age = 79.5 years) participated in the study. Their study aimed at identifying the importance of memory and psychomotor factor in cognitive training. Three different intervention programs were developed and applied for nine months to the following groups: 1) group for the competence-training program, aimed at conveying general strategies of coping with age-related changes and everyday problems of older adults; 2) group for the memory-training program based on a multi-storage model of memory and aging of memory; 3) group for the psychomotor-training program aimed at fostering coordination and security of motoricity taking age-related psychomotoric decreases into consideration; 4) group for the combined competence- and psychomotor-training program; and 5) group for the combined memory- and psychomotor-training program (Oswald et al., 1996). The participants were randomly assigned to one of the five groups. The interventions were held once a week over thirty sessions, each lasting two to three hours for nine months. The specific intervention measures were achieved in several small groups of 15 to 20 participants. The measures were recorded by two group supervisors. The training measures were evaluated hierarchically at different levels: global constructs (e.g., cognitive status), subconstructs (e.g., fluid and crystallized cognitive performance), and individual measurement parameters (e.g., speed of information processing). Both of the memory training group and memory plus psychomotor training group had higher scores than the control group after the intervention in processing speed, attention, reasoning, and memory. The memory plus psychomotor training group showed the largest effects on the composite cognitive index across the five-year follow-up period. In addition, only the memory plus psychomotor training led to an improvement of psychomotor performance. Neither psychomotor training alone nor memory training alone resulted in the amelioration of symptoms of dementia.

2) Interventions for Children

i) Importance of Executive Function (EF)

Prefrontal functions are deeply associated with children's social and emotional development (Monk, 2008). In particular, executive function (EF) has been revealed to play major roles in prefrontal functions (Elliott, 2003). Recently, it has also been suggested that EF has three components: shifting of mental sets, monitoring and updating of working memory, and inhibition of prepotent responses (Miyake et al., 2000). To clarify the profiles of these three components, Miyake and colleagues had the subjects perform a set of frequently used executive tasks: the Wisconsin Card sorting Test (WCST), Tower of Hanoi (TOH), random number generation (RNG), operation span, and dual tasking. Confirmatory factor analysis indicated that the three executive functions moderately correlated with one another and were clearly separable. Many recent executive function studies are based on their model (Bull and Scerif, 2001; Garon, Bryson, and Smith, 2008).

It has been suggested that inhibitory control plays a central role in fostering self-regulation, by delaying responses, which enables one to have behavioral flexibility and make a strategic selection of alternative behaviors (Barkley, 2001; Miyake et al., 2000). Inhibitory control may also play a critical role in academic learning, by promoting children's capacity to think about multiple dimensions of perspectives on a problem (Blair and Razza, 2007; Vivas et al., 2007). Working memory may also foster the acquisition of academic knowledge by allowing children to hold more information for a longer period of time, engage in mental rehearsal, and thereby increase opportunity for consolidation of information into long-term memory (Bull and Scerif, 2001).

On the other hand, some researchers use the term "EF skills" to refer to a complex but well-defined set of cognitive regulatory processes (Friedman et al., 2006) that underlie adaptive, goal-directed responses to novel or challenging situations (Hughes and Graham, 2002). EF capacities appear to play a central role in fostering the focused and rule-governed behavior that supports both cognitive and social-emotional adjustments (Bierman et al., 2008).

The neural areas that underlie EF skills include structures in the dorsolateral prefrontal, anterior cingulate, and parietal cortices, which have extensive interconnections with the ventral medial frontal and limbic brain structures associated with emotional reactivity and regulation. In combination, these cortical and limbic structures support the emotion-arousal system associated with behavioral inhibition in reaction to punishment or novelty and those associated with behavioral responses to a reward; they are also involved in responses to threat and stress (Blair et al., 2005). That is, the self-regulation processes that develop in the prefrontal cortex serve to modulate, either by enhancing or inhibiting, the reactive state of this arousal system (Derryberry and Rothbart, 1997). As such, the executive regulatory system directly affects and is affected by emotional and autonomic responses to stimulation. It plays a central role in the development of the ability to regulate attention, emotion, and behavior during the preschool years (Rueda et al., 2005).

Regarding the features of EF development, EF develops rapidly from three to five years old (Zelazo et al., 2003). This rapid development of EF enables children to think and behave more flexibly, to decrease their impulsivity, and to be more self-regulated (Rueda et al., 2005). The development of EF from three to five years old is quite important and thus interventions during this period have **a large effect on children's EF development** (Garon et al., 2008). Dysfunction of EF causes emotional regulation and behavioral problems, including aggression, depression, and attention disorders (Barkley, 2001). On the basis of these findings, many preventive interventions were developed focusing on supporting improved emotion regulation and problem-solving skills in which EF and the functions of the prefrontal lobe play a central role (Greenberg, 2006).

ii) Interventions for Children's EF Enhancement

Diamond et al. (2007) developed a preschool program to improve cognitive control. Their curriculum **named "The Tools of the Mind (Tools)" was designed to improve the EF of preschoolers in regular classrooms with regular teachers at a minimal cost.** The Tools curriculum is based on Vygotsky's theory related to EF and its development. Its core is 40 EF-**promoting activities, such as telling oneself out loud what one should do ("self-**regulatory private speech"), **dramatic play,** and aids to facilitate memory and attention. The teachers of the Tools curriculum spend up to 80% of day classes promoting EF skills. As a control condition, they used a **District's version of the Balanced Literacy curriculum (dBL).** The curriculum developed by the school district was based on balanced literacy and included thematic units. Although Tools and dBL covered the same academic contents, dBL did not address EF development. One hundred and forty-seven preschoolers aged 4 to 5 years (mean age = 5.1 years) participated in the study. Sixty-two children received the dBL curriculum, and eighty-five **children received the Tools curriculum. The children's cognitive abilities** were assessed by neuropsychological tests for EF. The children in the Tools curriculum significantly performed better on measures of EF than those in the dBL curriculum. As the Tools curriculum uses mature, dramatic plays to help improve EFs, Diamond et al. (2007) suggested that such plays are essential for EF development.

Domitrovich and colleagues developed a program named Promoting Alternative Thinking Strategies (PATHS) (Greenberg and Kusche, 1998) for preschool-aged children in "Head Start" (Domitrovich et al., 2007). Head Start is a federally sponsored early childhood education program in U.S.A. developed to reduce socioeconomic disparities in school readiness (Currie and Thomas, 1995). In the Head Start, the PATHS program was planned as a universal, teacher-taught social-emotional curriculum to improve children's social competence and reduce problem behavior. Dormitrovich and colleagues performed a randomized clinical trial to evaluate the efficacy of a preschool version of the PATHS curriculum and determined whether the curriculum improved children's social competence and reduce problem behavior. Twenty classrooms in two Pennsylvania communities participated in the study. The participants in the study were two hundred and forty-six children aged 4 to 5 years (mean age = 4.3 years). Forty-seven percent of the children were African-American, 38% were European-American, and 10% were Hispanic. Seventy-two percent of the primary caregivers were biological mothers. The mean annual income of families was $7039. The curriculum was divided into thematic units that included lessons on compliments, basic and advanced feelings, a self-control strategy, and problem solving. The self-control strategy was a revised version of the "Turtle Technique" (Robin, Schneider et al., 1976). The primary objectives of the curriculum were to (1) develop the children's awareness of and the ability to communicate their own and others' emotions; (2) develop self-control of arousal and behavior; (3) promote positive self-concept and peer relations; (4) develop children's problem solving skills by fostering the integration of their self-control, affect recognition, and communication skills; and (5) create a positive classroom atmosphere that supports social-emotional learning. The project was conducted over a three-year period. Teachers in the ten intervention classrooms implemented weekly lessons and extension activities during the 9-month period. Child assessments were collected at the beginning and end of the school year. The teachers' and parents' assessments of the children's behaviors were also collected at the same time. Analysis of covariance was used to control for baseline differences between the groups and pretest scores on each of the outcome measures. The results suggest that after exposure to PATHS, the intervention group showed higher emotion knowledge skills and was rated by parents and teachers as more socially competent than their peers. Furthermore, teachers rated children in the intervention group as less socially withdrawn at the end of the school year than the control group.

To enhance school readiness, Bierman et al. (2008) developed the Head Start REDI (Research-Based, Developmentally Informed) program. Their program included the preschool PATHS curriculum and components targeting language abilities and emergent literacy skills. In this study, the researchers compared the progress of students who received a conventional Head Start curriculum to those who received a curriculum that aimed to enhance the areas of social and emotional learning and prereading skills. The majority of Head Start programs in the U.S.A. use either the High/Scope curriculum, first developed in 1962 (Currie and Thomas, 1995), or the Creative Curriculum for Preschool, first developed in 1978 (Dodge, Colker, and Heroman, 2002). Both of these core curricula emphasize the importance of child-initiated learning supported by positive teacher-child relationships and strategic learning interactions. On the other hand, social-emotional competencies (in areas of prosocial behaviors, aggression control, emotional understanding, social problem-solving skills, and learning engagement) can be enhanced using systematic instructional approaches in the classroom (Elias et al., 1997). The study of Bierman et al. (2008) was carried out at forty-four Head Start centers in

Central Pennsylvania. The participants in the study included two cohorts of 356 4-year-old children (mean age = 4.49; 17% Hispanic, 25% African American). The centers were randomly assigned to either the intervention or comparison condition. Classrooms in the same center were always assigned to the same condition, to avoid inadvertent contamination of condition within centers. As compared with the children in the conventional Head Start program, the children in the REDI program scored higher on several tests of emotional and social development than the children in the traditional program. This included recognizing others' emotion and responding appropriately to situations involving a conflict. Moreover, the parents of children in the REDI group reported fewer instances of impulsivity, aggression, and attention problems than the parents of children in the conventional program. The children in the REDI program also scored higher than those in the conventional program in several tests of prereading skills (e.g., vocabulary, combining letter sounds together to form words, separating words into their component letter sounds, and in naming the letters of the alphabet). The interactions between emotion and cognition discussed in the report of Bierman et al. (2008) suggest that focusing on developing emotional, attention, and behavioral regulation in children at risk of school failure may be an effective strategy for promoting school success throughout the school years. It will be better than focusing only on acquiring academic content, such as exclusively focusing on teaching the basics of early literacy and math.

On the other hand, Kawashima and colleagues proposed an intervention play program named "NOUTORE play" (NOUTORE means brain training in Japanese) for mother and child (Tachibana et al., in press). Their intervention program had two purposes. One is to enhance attachment between mother and child. Another is to enhance children's cognitive development using specific games in the program. The games were selected from the tasks used in previous neuroimaging studies and thse tasks were demonstrated to be associated with PFC activation. In contrast to Diamond and Greenberg's program, Kawashima's intervention program was designed for healthy and nonsocioeconomically disadvantaged children. To identify the effects of this intervention on children's cognitive development and on maternal and child mental health, the researchers used a group randomized controlled trial. Two hundred and thirty-eight pairs of mother and child with typical development aged 4 to 6 (mean age = 5.1) from a kindergarten in Japan participated in this study. The pairs of mother and child were instructed to play at home for about ten minutes the five times a week for three months in accordance with the program schedule. The participants were randomly assigned to the intervention group or the control group. Parental Stress Index (PSI: for mothers), and the S-S intelligence test and Goodenough draw-a-man intelligence test (for children) were used before and six weeks after the intervention period. Post-preintervention changes in the test scores of the intervention group were compared with those of the control group. PSI indicated that the program helped mother and child to develop attachments between each other and enhanced maternal and child mental health. The results of the cognitive tests of children revealed that the program enhanced the children's working memory and performance intelligence quotient. Thus, the intervention program developed by Kawashima et al. enhanced not only children's cognitive development but also maternal and child mental health.

Rueda et al. (2005) studied an executive attention intervention for preschool children with typical development. They found that attention involves separable networks that compute different functions. One of these, the executive attention network, involves the anterior cingulate and lateral prefrontal areas and is activated strongly in situations that entail attentional control (Bush et al., 2000; Fan et al., 2005). Rueda et al. (2005) explored how a specific educational intervention targeting at the executive attention network might affect the development. Thus, the training program was designed to enhance executive attention. The program contained computerized games for executive attention. Forty-nine 4-year-old children (25 males; mean age, 52 months; SD, 2.2 months) and twenty-four 6-year-old children (12 males; mean age, 77 months; SD, 3.2 months) participated in the study. Twenty four 4-year-old children participated in Experiment (Exp.) 1, 25 4-year-old children in Exp. 2, and 24 6-year-old children in Exp. 3. For each experiment, the children were randomly divided into the experimental (to-be-trained, n = 12) group and the control (n = 12, n = 13 in Exp. 2 only) groups. Exp. 1 and Exp. 2 differed only in the control group. In Exp. 1, the 12 control children came to the laboratory only twice with no intervention. In Exp. 2, the control group underwent five sessions over a 2- to 3-week period during which they watched popular children's videos. The experimental and control groups were treated exactly the same as in Exp. 2, but Rueda et al. added one more exercise to complete the 5 training sessions. On the first day, the children were examined in terms of attention (children's version of the Attention Network Test (Child ANT), and intelligence (Kaufman Brief Intelligence Test, KBIT), and parent-reported children's temperament (Children's Behavior Questionnaire: CBQ). Then, the children underwent 5 days of training over a 2-to 3-week period. The children were presented with five pieces fish arranged horizontally on the monitor screen in the Child ANT. The task was to respond to the center fish on the screen by pressing a key in the direction in which the fish pointed. In congruent trials, the flanking fish pointed in the same direction as the center fish. In incongruent trials, the flanking fish pointed in the opposite direction. The conflict score was obtained by subtracting congruent reaction time from incongruent reaction time. On the final day the children underwent the same examinations as on day 1. On the final day, the caretakers were given the temperament questionnaire (Rothbart et al., 2001) to take home and return after filling it out on the basis of observations 2 weeks after the final session. Assessment sessions involved ERP recording during the performance of Child ANT. Cheek swabs were collected from most of the 6-year-old children and genotyping of DAT1 gene was performed. Marked improvements in executive attention and intelligence were found. After 5 days of executive attention training, the intervention group improved to a significantly greater extent than the control group in the KBIT. In addition, Rueda et al. found no significant training effects on a version of the flanker task (a type of inhibition test). The attention training program used by Rueda et al. did not improve inhibitory control in preschoolers, as measured on the basis of performance in the flanker task. Analysis of ERPs revealed that the training had a specific effect on the scalp distribution of ERPs, which was similar to the training effect on children's development. Their studies also showed evidence of generalization of the benefits of training to aspects of intelligence that were quite remote from the exercises. The improvement was small in overall intelligence and greatest in the subscale of abstracting reasoning skills. The matrices scale measures more culture-free aspects of intelligence such as simultaneous processing, nonverbal reasoning, and fluid thinking. Genetic analysis of the 6-year-old children revealed that the long form of DAT1 gene was associated with a stronger effortful control (inhibition ability) and less surgency (extroversion). Their findings suggest that the less outgoing and more controlled children may be less in need of attention training. The executive attention measure is related to executive attention during childhood (Posner et al., 2001). Posner et al. revealed that effortful control have a strong heritability and it is related DAT1 polymorphism. Rueda et al. indicated that the surgency difference may result from a greater control of action in children with the pure long allele.

Thorell et al. (2009) verified the effectiveness of their computer program training for executive function in preschool children. Their program was found effective for children with ADHD (Klingberg et al., 2005; Klingberg et al., 2002), for people recovering from a stroke (Westerberg et al., 2007), and for healthy adults (Olesen et al., 2004). The program enhanced the subjects' cognitive abilities including working memory. Children between the ages of 4 and 5 years (M = 56 months) participated in the study. They were assigned to the working memory (WM) training group (n = 17, nine boys), active control (inhibition training) group (n = 18, nine boys), or the passive control group (n = 16, seven boys). The WM training group children received computerized training of either visuospatial working memory or inhibition for 5 weeks. The active control group played commercially available computer games, and the passive control group took part in only before and after intervention period. Thorell et al. suggested that working memory training had significant effects on the performance of nontrained working memory tasks within both the spatial and the verbal domains. They also considered that working memory training has significant transfer effects on laboratory measures of attention. On the other hand, inhibitory control of the WM training group did not show any significant effects compared with the inhibitory control of the control group, although the children of the WM training group showed improved performance in at least some of the tasks they were trained on (the go/no go tasks, the flanker task, but not stop-signal tasks).

As reviewed in this article, cognitive interventions improved the children's cognitive functions and enhanced their executive functions. These findings are very important, because enhancing the children's executive function leads to reduction of their behavior problems, such as aggressive, disruptive, and withdrawn behaviors. It also develops their social-emotional competence (Domitrovich et al., 2007). Improving the children's cognitive functions, particularly executive function, can strengthen their resilience (Greenberg, 2006).

5. Future Directions of Cognitive Interventions

The intervention studies indicated that trainings of the functions associated with PFC improves not only the targeted abilities but also the untrained cognitive abilities. Not much is known, however, about the mechanism underlying this transfer effects, and the issue on which cognitive functions can be trained and to what extent the effect of cognitive training can be generalized to other cognitive functions and behavior are also unclarified. It is indicated that daily activities such as reading, playing board games, playing musical instruments, and dancing are associated with cognitive enhancement and prevent dementia (Verghese et al., 2003). This suggests that various daily activities might enhance cognitive abilities. Still, further neuroscientific studies should be performed to identify the effects of daily activities as cognitive trainings.

Recently, neuroimaging studies have revealed that cognitive trainings affect brain plasticity (Johansson, 2004; Watanabe et al., 2007). A recent positron emission tomography (PET) study indicated that a cognitive intervention in which participants perform working memory tasks with a difficulty level close to their individual capacity limit induces changes in cortical dopamine D1 receptor binding (McNab et al., 2009). This study showed that cognitive interventions can induce neuroendocrinal changes. However, only few neuroendocrinal studies on the effect of cognitive interventions have been conducted. Further interdisciplinary studies in various fields of neuroscience (e.g., collaboration with neuroendocrinal and neuroimaging researchers) should be conducted to clarify the transference effects of cognitive trainings/interventions.

6. Conclusions

We have reviewed previous studies on cognitive interventions that targeted the functions associated with PFC, as these functions are deeply associated with a healthy social life. Recently, these interventions have been reported to be useful for primary prevention of psychiatric or psychological problems (Busse et al., 2009; Holcomb, 2004). For example, cognitive interventions are suggested to prevent the development of dementia and cognitive decline (Uchida and Kawashima, 2008) and also to enhance a healthy social life in healthy adults (Willis et al., 2006). Moreover, cognitive interventions are indicated to prevent the development of psychological problems (Greenberg, 2006) and enhance school readiness and healthy minds in children (Blair, 2003). However, not much is known about the neurobiological/neuroendocrinal basis of cognitive interventions and the mechanism underlying the transfer effects, which are often observed in intervention studies. And the issue on which intervention (i.e., contents of intervention) is most effective remains unclarified. Further multidisciplinary investigations should be performed. We believe that a deeper understanding of the effects of cognitive interventions will contribute greatly to human welfare and education for all generations.

References

Abraham, A., and von Cramon, D. Y. (2009). Reality = relevance? Insights from spontaneous modulations of the brain's default network when telling apart reality from fiction. *PLoS ONE 4*, e4741.

Adler, C. M., DelBello, M. P., and Strakowski, S. M. (2006). Brain network dysfunction in bipolar disorder. *CNS Spectr 11*, 312-320; quiz 323-314.

Ball, K., Berch, D. B., Helmers, K. F., Jobe, J. B., Leveck, M. D., Marsiske, M., Morris, J. N., Rebok, G. W., Smith, D. M., Tennstedt, S. L., *et al.* (2002). Effects of cognitive training interventions with older adults: a randomized controlled trial. *Jama 288*, 2271-2281.

Barkley, R. A. (2001). The executive functions and self-regulation: an evolutionary neuropsychological perspective. *Neuropsychol. Rev. 11*, 1-29.

Bierman, K. L., Nix, R. L., Greenberg, M. T., Blair, C., and Domitrovich, C. E. (2008). Executive functions and school readiness intervention: impact, moderation, and mediation in the Head Start REDI program. *Dev. Psychopathol. 20*, 821-843.

Bischkopf, J., Busse, A., and Angermeyer, M. C. (2002). Mild cognitive impairment--a review of prevalence, incidence and outcome according to current approaches. *Acta Psychiatr. Scand. 106*, 403-414.

Blackwood, N. J., Howard, R. J., Bentall, R. P., and Murray, R. M. (2001). Cognitive neuropsychiatric models of persecutory delusions. *Am. J. Psychiatry 158*, 527-539.

Blair, C. (2003). Behavioral inhibition and behavioral activation in young children: relations with self-regulation and adaptation to preschool in children attending Head Start. *Dev. Psychobiol. 42*, 301-311.

Blair, C., and Razza, R. P. (2007). Relating effortful control, executive function, and false belief understanding to emerging math and literacy ability in kindergarten. *Child Dev. 78*, 647-663.

Blair, C., Zelazo, P. D., and Greenberg, M. T. (2005). The measurement of executive function in early childhood. *Dev. Neuropsychol. 28*, 561-571.

Bull, R., and Scerif, G. (2001). Executive functioning as a predictor of children's mathematics ability: inhibition, switching, and working memory. Dev Neuropsychol *19*, 273-293.

Bush, G., Luu, P., and Posner, M. I. (2000). Cognitive and emotional influences in anterior cingulate cortex. *Trends Cogn. Sci. 4*, 215-222.

Busse, J. W., Montori, V. M., Krasnik, C., Patelis-Siotis, I., and Guyatt, G. H. (2009). Psychological intervention for premenstrual syndrome: a meta-analysis of randomized controlled trials. *Psychother Psychosom. 78*, 6-15.

Craik, F.I.M. (1986). A functional account of age differences in memory. In F. Klix and H. H. Hagendorf (Eds.), Human memory and cognitive capabilities, mechanism, and performance (pp.409-422). Amsterdam: North-Holland.

Craik, F. I., Winocur, G., Palmer, H., Binns, M. A., Edwards, M., Bridges, K., Glazer, P., Chavannes, R., and Stuss, D. T. (2007). Cognitive rehabilitation in the elderly: effects on memory. *J. Int. Neuropsychol. Soc. 13*, 132-142.

Currie J. and Thomas D. (1995). Does Head Start make a difference? *The American Economic Review 85*(3), 341-364.

Damasio, A. R. (1995). On some functions of the human prefrontal cortex. *Ann. N. Y. Acad. Sci. 769*, 241-251.

Damasio, H., Grabowski, T., Frank, R., Galaburda, A. M., and Damasio, A. R. (1994). The return of Phineas Gage: clues about the brain from the skull of a famous patient. *Science 264*, 1102-1105.

Derryberry, D., and Rothbart, M. K. (1997). Reactive and effortful processes in the organization of temperament. *Dev. Psychopathol. 9*, 633-652.

Diamond, A., Barnett, W. S., Thomas, J., and Munro, S. (2007). Preschool program improves cognitive control. *Science 318*, 1387-1388.

Dodge, D.T., Colker, L.J., and Heroman, C. (2002). The creative curriculum for preschool (4th ed.). Washington, DC: Teaching Strategies.

Domitrovich, C. E., Cortes, R. C., and Greenberg, M. T. (2007). Improving young children's social and emotional competence: a randomized trial of the preschool "PATHS" curriculum. *J. Prim. Prev. 28*, 67-91.

Drevets, W. C. (2000). Neuroimaging studies of mood disorders. *Biol. Psychiatry 48*, 813-829.

Dudek, F.J. (1979). The continuing misinterpretation of the standard error of measurement. *Psychol. Bull. 86*, 335-337.

Edwards, J. D., Wadley, V. G., Myers, R. S., Roenker, D. L., Cissell, G. M., and Ball, K. K. (2002). Transfer of a speed of processing intervention to near and far cognitive functions. *Gerontology 48*, 329-340.

Edwards, J. D., Wadley, V. G., Vance, D. E., Wood, K., Roenker, D. L., and Ball, K. K. (2005). The impact of speed of processing training on cognitive and everyday performance. *Aging Ment Health 9*, 262-271.

Elias, M. J., Zins, J. E., Weissberg, R. P., Frey, K. S., Greenberg, M. T., Haynes, N. M., Norris M., Kessler, R., Scwab-Stone, M.E., Schriver, T.P. (1997). Promoting social and emotional learning: Guidelines for educators. Aleandria, VA: Association for Supervision and Curriculum Development.

Elliott, R. (2003). Executive functions and their disorders. *Br. Med. Bull. 65*, 49-59.

Fan, J., McCandliss, B. D., Fossella, J., Flombaum, J. I., and Posner, M. I. (2005). The activation of attentional networks. *Neuroimage 26*, 471-479.

Fillit, H. M., Butler, R. N., O'Connell, A. W., Albert, M. S., Birren, J. E., Cotman, C. W., Greenough, W. T., Gold, P. E., Kramer, A. F., Kuller, L. H., Perls, T.T., Shagan, B.G. (2002). Achieving and maintaining cognitive vitality with aging. *Mayo Clin. Proc. 77*, 681-696.

Floden, D., Stuss, D.T., Craik, F.I.M. (2000). Age differences in performance on two versions of the Brown-Peterson task. *Aging, Neuropsychology, and Cognition 7*, 245-259.

Friedman, N. P., Miyake, A., Corley, R. P., Young, S. E., Defries, J. C., and Hewitt, J. K. (2006). Not all executive functions are related to intelligence. *Psychol. Sci. 17*, 172-179.

Fuster, J. M. (2002). Frontal lobe and cognitive development. *J. Neurocytol. 31*, 373-385.

Garon, N., Bryson, S. E., and Smith, I. M. (2008). Executive function in preschoolers: a review using an integrative framework. *Psychol. Bull. 134*, 31-60.

Greenberg, M. T. (2006). Promoting resilience in children and youth: preventive interventions and their interface with neuroscience. *Ann. N. Y. Acad. Sci. 1094*, 139-150.

Greenberg, P., and Kusche, C. (1998). Preventive intervention for school-age deaf children: the PATHS curriculum. *J. Deaf Stud. Deaf Educ. 3*, 49-63.

Holcomb, H. H. (2004). Practice, learning, and the likelihood of making an error: how task experience shapes physiological response in patients with schizophrenia. *Psychopharmacology* (Berl) *174*, 136-142.

Hubert, V., Beaunieux, H., Chetelat, G., Platel, H., Landeau, B., Danion, J. M., Viader, F., and Desgranges, B. (2007). The dynamic network subserving the three phases of cognitive procedural learning. *Hum. Brain Mapp. 28*, 1415-1429.

Hughes, C. and Graham, A. (2002). "Measuring executive function in childhood: Problems and solutions?" *Child and adolescent mental health 7*: 131-142.

Huttenlocher, P. R., and Dabholkar, A. S. (1997). Regional differences in synaptogenesis in human cerebral cortex. *J. Comp. Neurol. 387*, 167-178.

Jobe, J. B., Smith, D. M., Ball, K., Tennstedt, S. L., Marsiske, M., Willis, S. L., Rebok, G. W., Morris, J. N., Helmers, K. F., Leveck, M. D., and Kleinman, K. (2001). ACTIVE: a cognitive intervention trial to promote independence in older adults. *Control Clin. Trials 22*, 453-479.

Johansson, B. B. (2004). Brain plasticity in health and disease. *Keio J. Med. 53*, 231-246.

Kawashima, R., Taira, M., Okita, K., Inoue, K., Tajima, N., Yoshida, H., Sasaki, T., Sugiura, M., Watanabe, J., and Fukuda, H. (2004). A functional MRI study of simple arithmetic--a comparison between children and adults. *Brain Res. Cogn. Brain Res. 18*, 227-233.

Kiehl, K. A. (2006). A cognitive neuroscience perspective on psychopathy: evidence for paralimbic system dysfunction. *Psychiatry Res. 142*, 107-128.

Klingberg, T., Fernell, E., Olesen, P. J., Johnson, M., Gustafsson, P., Dahlstrom, K., Gillberg, C. G., Forssberg, H., and Westerberg, H. (2005). Computerized training of working memory in children with ADHD--a randomized, controlled trial. *J. Am. Acad. Child Adolesc. Psychiatry 44*, 177-186.

Klingberg, T., Forssberg, H., and Westerberg, H. (2002). Training of working memory in children with ADHD. *J. Clin. Exp. Neuropsychol. 24*, 781-791.

Klingberg, T., and Roland, P. E. (1997). Interference between two concurrent tasks is associated with activation of overlapping fields in the cortex. *Brain Res. Cogn. Brain Res. 6*, 1-8.

Kramer, A. F., Colcombe, S. J., McAuley, E., Scalf, P. E., and Erickson, K. I. (2005). Fitness, aging and neurocognitive function. *Neurobiol. Aging 26 Suppl 1*, 124-127.

Lawrie, S. M., McIntosh, A. M., Hall, J., Owens, D. G., and Johnstone, E. C. (2008). Brain structure and function changes during the development of schizophrenia: the evidence from studies of subjects at increased genetic risk. *Schizophr. Bull. 34*, 330-340.

Lemaitre, H., Crivello, F., Grassiot, B., Alperovitch, A., Tzourio, C., and Mazoyer, B. (2005). Age- and sex-related effects on the neuroanatomy of healthy elderly. *Neuroimage 26*, 900-911.

Liston, C., Miller, M. M., Goldwater, D. S., Radley, J. J., Rocher, A. B., Hof, P. R., Morrison, J. H., and McEwen, B. S. (2006). Stress-induced alterations in prefrontal cortical dendritic morphology predict selective impairments in perceptual attentional set-shifting. *J. Neurosci. 26*, 7870-7874.

Mahncke, H. W., Connor, B. B., Appelman, J., Ahsanuddin, O. N., Hardy, J. L., Wood, R. A., Joyce, N. M., Boniske, T., Atkins, S. M., and Merzenich, M. M. (2006). Memory enhancement in healthy older adults using a brain plasticity-based training program: a randomized, controlled study. *Proc. Natl. Acad. Sci. USA 103*, 12523-12528.

Martens, W. H. (2001). Effects of antisocial or social attitudes on neurobiological functions. *Med. Hypotheses 56*, 664-671.

McNab, F., Varrone, A., Farde, L., Jucaite, A., Bystritsky, P., Forssberg, H., and Klingberg, T. (2009). Changes in cortical dopamine D1 receptor binding associated with cognitive training. *Science 323*, 800-802.

Miura, N., Iwata, K., Watanabe, J., Sugiura, M., Akitsuki, Y., Sassa, Y., Ikuta, N., Okamoto, H., Watanabe, Y., Riera, J.,Matsue, Y., Kawashima, R.(2003). Cortical activation during reading aloud of long sentences: fMRI study. *Neuroreport 14*, 1563-1566.

Miura, N., Watanabe, J., Iwata, K., Sassa, Y., Riera, J., Tsuchiya, H., Sato, S., Horie, K., Takahashi, M., Kitamura, M., and Kawashima, R. (2005). Cortical activation during reading of ancient versus modern Japanese texts: fMRI study. *Neuroimage 26*, 426-431.

Miyake, A., Friedman, N. P., Emerson, M. J., Witzki, A. H., Howerter, A., and Wager, T. D. (2000). The unity and diversity of executive functions and their contributions to complex "Frontal Lobe" tasks: a latent variable analysis. *Cogn. Psychol. 41*, 49-100.

Monk, C. S. (2008). The development of emotion-related neural circuitry in health and psychopathology. *Dev. Psychopathol. 20*, 1231-1250.

Monkul, E. S., Malhi, G. S., and Soares, J. C. (2005). Anatomical MRI abnormalities in bipolar disorder: do they exist and do they progress? *Aust. N. Z. J. Psychiatry 39*, 222-226.

Olesen, P. J., Macoveanu, J., Tegner, J., and Klingberg, T. (2007). Brain activity related to working memory and distraction in children and adults. *Cereb. Cortex 17*, 1047-1054.

Olesen, P. J., Westerberg, H., and Klingberg, T. (2004). Increased prefrontal and parietal activity after training of working memory. *Nat. Neurosci. 7*, 75-79.

Oswald, W. D., Rupprecht, R., Gunzelmann, T., and Tritt, K. (1996). The SIMA-project: effects of 1 year cognitive and psychomotor training on cognitive abilities of the elderly. *Behav. Brain Res. 78*, 67-72.

Petrella, J. R., Krishnan, S., Slavin, M. J., Tran, T. T., Murty, L., and Doraiswamy, P. M. (2006). Mild cognitive impairment: evaluation with 4-T functional MR imaging. *Radiology 240*, 177-186.

Posner, M. I., Rothbart, M. K., and Gerardi-Caulton, G. (2001). Exploring the biology of socialization. *Ann. N. Y. Acad. Sci. 935*, 208-216.

Pridmore, S., Chambers, A., and McArthur, M. (2005). Neuroimaging in psychopathy. *Aust. N. Z. J. Psychiatry 39*, 856-865.

Prvulovic, D., Van de Ven, V., Sack, A. T., Maurer, K., and Linden, D. E. (2005). Functional activation imaging in aging and dementia. *Psychiatry Res. 140*, 97-113.

Radley, J. J., Rocher, A. B., Miller, M., Janssen, W. G., Liston, C., Hof, P. R., McEwen, B. S., and Morrison, J. H. (2006). Repeated stress induces dendritic spine loss in the rat medial prefrontal cortex. *Cereb. Cortex 16*, 313-320.

Rajkowska, G. (1997). Morphometric methods for studying the prefrontal cortex in suicide victims and psychiatric patients. *Ann. N. Y. Acad. Sci. 836*, 253-268.

Ritchie, K., Artero, S., and Touchon, J. (2001). Classification criteria for mild cognitive impairment: a population-based validation study. *Neurology 56*, 37-42.

Robin, A.L., Schneider, M., Dolnick, M. (1976). The Turtle Technique: An extended case study of self control in the classroom. *Psychology in the Schools 13*, 449-453.

Rothbart, M. K., Ahadi, S. A., Hershey, K. L., and Fisher, P. (2001). Investigations of temperament at three to seven years: the Children's Behavior Questionnaire. *Child Dev. 72*, 1394-1408.

Rueda, M. R., Posner, M. I., and Rothbart, M. K. (2005). The development of executive attention: contributions to the emergence of self-regulation. *Dev. Neuropsychol. 28*, 573-594.

Rueda, M. R., Rothbart, M. K., McCandliss, B. D., Saccomanno, L., and Posner, M. I. (2005). Training, maturation, and genetic influences on the development of executive attention. *Proc. Natl. Acad. Sci. USA 102*, 14931-14936.

Spence, S. A., Kaylor-Hughes, C., Farrow, T. F., and Wilkinson, I. D. (2008). Speaking of secrets and lies: the contribution of ventrolateral prefrontal cortex to vocal deception. *Neuroimage 40*, 1411-1418.

Spence, S. A., and Kaylor-Hughes, C. J. (2008). Looking for truth and finding lies: the prospects for a nascent neuroimaging of deception. *Neurocase 14*, 68-81.

Stuss, D. T., Robertson, I. H., Craik, F. I., Levine, B., Alexander, M. P., Black, S., Dawson, D., Binns, M. A., Palmer, H., Downey-Lamb, M., and Winocur, G. (2007). Cognitive rehabilitation in the elderly: a randomized trial to evaluate a new protocol. *J. Int. Neuropsychol. Soc. 13*, 120-131.

Tachibana, Y., Fukushima, A., Akitsuki, Y., Saito, H., Yoneyama, S., Ushida, K., Yoneyama S., Kawashima, R. (in press). Encouraging Mother-Child Play Improves Maternal and Child Mental Health as well as Child Cognitive Development: A Randomized-Controlled Trial. Proceedings of the 56th Annual Meeting of American Academy of Child and Adolescent Psychiatry.

Thorell, L. B., Lindqvist, S., Bergman Nutley, S., Bohlin, G., and Klingberg, T. (2009). Training and transfer effects of executive functions in preschool children. Dev Sci *12*, 106-113.

Tsujimoto, S. (2008). The prefrontal cortex: functional neural development during early childhood. *Neuroscientist 14*, 345-358.

Uchida, S. and Kawashima R. (2008). Reading and solving arithmetic problems improve cognitive functions of normal aged people – A randomized controlled study. *Age 30*, 21-29.

Valera, E. M., Faraone, S. V., Murray, K. E., and Seidman, L. J. (2007). Meta-analysis of structural imaging findings in attention-deficit/hyperactivity disorder. *Biol. Psychiatry 61*, 1361-1369.

Verghese, J., Lipton, R. B., Katz, M. J., Hall, C. B., Derby, C. A., Kuslansky, G., Ambrose, A. F., Sliwinski, M., and Buschke, H. (2003). Leisure activities and the risk of dementia in the elderly. *N. Engl. J. Med. 348*, 2508-2516.

Vivas, A. B., Fuentes, L. J., Estevez, A. F., and Humphreys, G. W. (2007). Inhibitory tagging in inhibition of return: evidence from flanker interference with multiple distractor features. *Psychon. Bull. Rev. 14*, 320-326.

Watanabe, D., Savion-Lemieux, T., and Penhune, V. B. (2007). The effect of early musical training on adult motor performance: evidence for a sensitive period in motor learning. *Exp. Brain Res. 176*, 332-340.

Weinberger, D. R., Egan, M. F., Bertolino, A., Callicott, J. H., Mattay, V. S., Lipska, B. K., Berman, K. F., and Goldberg, T. E. (2001). Prefrontal neurons and the genetics of schizophrenia. *Biol. Psychiatry 50*, 825-844.

Westerberg, H., Jacobaeus, H., Hirvikoski, T., Clevberger, P., Ostensson, M. L., Bartfai, A., and Klingberg, T. (2007). Computerized working memory training after stroke--a pilot study. *Brain Inj. 21*, 21-29.

Whitford, T. J., Rennie, C. J., Grieve, S. M., Clark, C. R., Gordon, E., and Williams, L. M. (2007). Brain maturation in adolescence: concurrent changes in neuroanatomy and neurophysiology. *Hum. Brain Mapp. 28*, 228-237.

Willis, S. L., Tennstedt, S. L., Marsiske, M., Ball, K., Elias, J., Koepke, K. M., Morris, J. N., Rebok, G. W., Unverzagt, F. W., Stoddard, A. M., and Wright, E. (2006). Long-term effects of cognitive training on everyday functional outcomes in older adults. *Jama 296*, 2805-2814.

Wilson, S. J., Sayette, M. A., and Fiez, J. A. (2004). Prefrontal responses to drug cues: a neurocognitive analysis. *Nat. Neurosci. 7*, 211-214.

Yang, Y., Glenn, A. L., and Raine, A. (2008). Brain abnormalities in antisocial individuals: implications for the law. *Behav. Sci. Law 26*, 65-83.

Yucel, M., and Lubman, D. I. (2007). Neurocognitive and neuroimaging evidence of behavioural dysregulation in human drug addiction: implications for diagnosis, treatment and prevention. *Drug Alcohol. Rev. 26*, 33-39.

Zelazo, P. D., Muller, U., Frye, D., Marcovitch, S., Argitis, G., Boseovski, J., Chiang, J. K., Hongwanishkul, D., Schuster, B. V., and Sutherland, A. (2003). The development of executive function in early childhood. *Monogr. Soc. Res. Child Dev. 68*, vii-137.

In: Encyclopedia of Neuroscience Research
Editors: Eileen J. Sampson and Donald R. Glevins

ISBN 978-1-61324-861-4
© 2012 Nova Science Publishers, Inc.

Chapter XXV

Insights from Proteomics into Mild Cognitive Impairment, Likely the Earliest Stage of Alzheimer's Disease

*Renã A. Sowell1 and D. Allan Butterfield1, 2, 3 ***

1Department of Chemistry,
University of Kentucky, Lexington, USA
2 Sanders-Brown Center on Aging,
University of Kentucky, Lexington, USA
3 Center of Membrane Sciences,
University of Kentucky, Lexington, USA

Abstract

Mild cognitive impairment (MCI) is arguably the earliest form of Alzheimer's disease (AD). Better understanding of brain changes in MCI may lead to the identification of therapeutic targets to slow the progression of AD. Oxidative stress has been implicated as a mechanism associated with the pathogenesis of both MCI and AD. In particular, among other markers, there is evidence for an increase in the levels of protein oxidation and lipid peroxidation in the brains of subjects with MCI. Several proteins are oxidatively modified in MCI brain, and as a result individual protein dysfunction may be directly linked to these modifications (e.g., carbonylation, nitration, modification by HNE) and may be involved in MCI pathogenesis. Additionally, Concanavalin-A-mediated separation of brain proteins has recently led to the identification of key proteins in MCI and AD using proteomics methods. This chapter will summarize important findings from proteomics studies of MCI, which have provided insights into this cognitive disorder and have led to further understanding of potential mechanisms involved in the progression of AD.

---

* Correspondence to: D. Allan Butterfield, University of Kentucky, Lexington, KY 40506-0055, USA. Ph: (859)-257-3184, Fax: (859)-257-5876. e-mail: dabcns@uky.edu.

Keywords: Mild Cognitive Impairment, proteomics, oxidative modifications, Alzheimer's disease

1.0. Introduction

Mild cognitive impairment (MCI) can be considered as the earliest form of Alzheimer's disease (AD) existing as a transitional state between normal aging and AD [1-3]. MCI exists in two forms: amnestic MCI and nonamnestic MCI [2, 3]. Amnestic MCI patients are able to perform normal daily living activities and have no signs of dementia; however, they do have cognitive complaints that include bursts of episodic memory loss [1, 4]. In some cases, amnestic MCI patients can develop AD at a rate of ~10 to 15% annually, however in other cases, the patients revert back to normal conditions [5]. Pathologic characteristics of MCI are similar to those of AD. For example, MCI patients have hippocampal, entorhinal cortex (EC), and temporal lobe atrophy based on magnetic resonance imaging studies [6-8], synapse loss, neuronal loss, low cerebrospinal fluid (CSF)-resident β amyloid levels [6], genetic risk factors including preponderance in APOE4 allele [9, 10], and increased levels of oxidative stress [11-20].

Oxidative stress is one of the underlying indices associated with MCI, AD, and other neurodegenerative disorders such as Parkinson's disease and amyotrophic lateral sclerosis. Specifically in MCI, there is substantial evidence for increased levels of oxidative stress in the brains and in plasma of MCI subjects [11-23]. Our laboratory has reported an increase in the levels of protein carbonyls (PCO) [11, 16] and 3-nitrotyrosine (3NT)-modified proteins [21], both of which are markers of protein oxidation. Additionally, we have reported an increase in the levels of 4-hydroxynonenal-(HNE) bound proteins, indicating an increase in the levels of lipid peroxidation products [13]. Others have observed decreases in the levels of antioxidant enzymes and antioxidant enzymatic activity in brain and in plasma [22-24], increased levels of oxidative stress in nuclear and mitochondrial DNA [25, 26], increased levels of isoprostanes [27], and increased lipid peroxidation as measured by free HNE levels, thiobarbituic substances, and malondialdehyde [16, 20]. It is believed that oxidative stress also is related to several vascular factors, such as heart disease, hypertension, and diabetes mellitus that conceivably contribute to the conversion of MCI into AD.

It is important to understand more about the events that lead to the progression of AD from MCI in order to develop potential therapeutics that can delay or stop AD onset. Thus, proteomics can provide considerable insight into specific pathways that are influenced by MCI and which eventually aid in the progression of disease. To this end, we and others have investigated the changes associated with the proteomes of MCI subjects relative to normal age-matched healthy controls [11, 19, 28-33]. These studies include the search for candidate biomarkers of MCI which eventually lead to AD [29, 30, 33], changes in the expression levels of proteins [28], specific levels of protein oxidation as measured by PCO [11], 3NT-modified proteins [19], and lipid peroxidation as measured by HNE-bound proteins [32]. More recently, we have also investigated other post-translational modifications that change in subjects with MCI such as glycosylation [31]. This chapter summarizes the key findings from proteomics and redox proteomics studies in MCI and their implications in AD research.

2.0. Two-Dimensional (2D) Gel Electrophoresis (GE) Based Proteomics

The proteomics techniques used in the studies described herein follow the general approach outlined in Figure 1. Here proteins are extracted from brain, CSF, plasma, or other bodily tissues obtained from MCI subjects and normal age-matched controls. Extracted proteins are subjected to isoelectric focusing (IEF)/sodium dodecyl sulfate (SDS) polyacrylamide gel electrophoresis (PAGE), better known as 2DGE. In this approach, proteins are separated in a first dimension based on their isoelectric point and in a second dimension based on their migration rate through the gel, which often corresponds to molecular weight. Image analysis software is used to align spots across the gels obtained from all of the samples, and protein **spots that exhibit significant changes (based on Student's t-tests or analysis of variance)** in expression levels between MCI and controls are excised. Excised spots are subjected to in-gel trypsin digestion and analyzed using matrix assisted laser desorption ionization (MALDI) or electrospray ionization (ESI) mass spectrometry (MS). Data from MS experiments are then submitted to appropriate databases using search engines such as MASCOT [34] for protein identification.

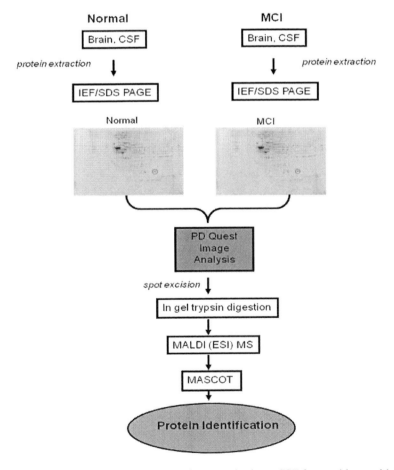

Figure 1. Schematic overview of the 2D GE experiment on brain or CSF from subjects with MCI and age-matched controls.

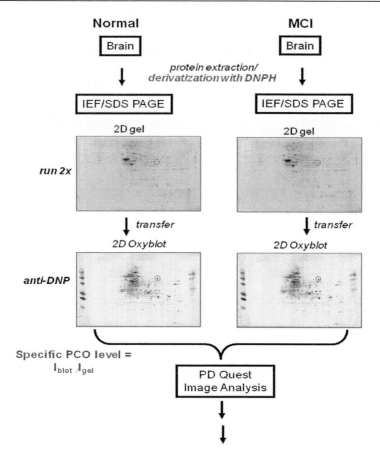

Figure 2. Schematic overview of the redox proteomics approach applied for the analysis of oxidatively modified proteins such as protein carbonyls (PCO), 3-nitrotyrosine (3NT) modified proteins and HNE-modified proteins. Extracted proteins are derivatized with 2,4-dinitrophenylhydrazine (DNPH) only for the analysis of PCO and separated with IEF/SDS PAGE. 2D gels are then transferred onto a nitrocellulose membrane and 2D Oxyblots are probed with anti-DNP (or anti-3NT, anti-HNE) antibodies and visualized using a secondary antibody linked with a colorimetric alkaline phosphatase assay. Specific oxidative levels of proteins are calculated by normalizing the intensity of spots in the 2D Oxyblot (I_{blot}) to the intensity of the corresponding spot in a 2D gel (I_{gel}). This calculation is carried out similarly for PCO, 3NT, and HNE. Protein spots exhibiting significant changes in oxidative modification are then excised, digested in-gel by trypsin, analyzed with MALDI or ESI-MS/MS, and identified with database searching as illustrated in Figure 1.

This general approach can also be adapted for the analysis of post-translational modifications. For example, changes in glycosylation of proteins can be analyzed by using affinity chromatography for purification of glycoproteins. Extracted proteins can be separated with Concanavalin-A lectin affinity columns which isolate proteins that contain asparagine (N)-linked carbohydrates. In some cases, Con-A may also have nonspecific interactions and isolate proteins based on its hydrophobic binding domain [35].

Oxidative modification of proteins can also be detected using the 2D GE approach with the incorporation of Western blotting analysis [36]. Figure 2 shows a schematic of the general approach used to detect PCO, 3NT-modified proteins, and HNE-bound proteins. As shown in Figure 2, for the detection of PCO, extracted proteins are derivatized with 2,4-dinitrophenylhydrazine (DNPH), which forms a Schiff base with carbonyl groups on proteins.

DNPH-derivatized proteins are then separated using 2D GE, and the spots on gels are transferred onto a nitrocellulose membrane forming a 2D Western blot or 2D Oxyblot. Immunochemical detection using a primary anti-DNP antibody that recognizes DNP hydrazone adducts is applied to the 2D Oxyblots, and oxidized spots are visualized with a secondary antibody linked to a colorimetric alkaline phosphatase assay. Similarly, for the detection of 3NT- and HNE-modified proteins non-derivatized extracted proteins are separated with 2D GE, transferred onto an 2D Oxyblot, and immunochemically detected with anti-3NT and anti-HNE, antibodies, respectively. Imaging analysis software and statistical approaches are applied as illustrated in Figure 1 in order to align 2D images and identify spots that have significant changes in oxidative modification. Specific carbonyl (or 3NT, HNE) levels of proteins are measured by normalizing the density of spots in the 2D Oxyblot, to the density of the same spot in a 2D gel analysis of the sample (separate experiment). Protein spots of interest (those exhibiting significant elevation or reduction in oxidative modification) are then excised from the gel, tryptically digested, analyzed by MS, and identified as described in Figure 1.

3.0. Candidate Biomarkers in CSF for the Progression of MCI to AD

CSF presents another biological fluid that can relay specific information about neurological molecular changes because the fluid encompasses the extracellular space surrounding the brain. Table 1 lists proteins that were identified as having significant changes in expression in CSF of MCI subjects relative to normal controls. Kim and coworkers have performed proteomics analysis on CSF from normal cognitive controls, MCI, and AD patients and identified three proteins which may be candidate markers for the diagnosis MCI and its progression into AD[29, 30]. The protein, fibrinogen γ-A chain, was detected as having a gradual elevation of expression in MCI and AD [30]. Fibrinogen γ -A chain is a part of the 340 kDa hexameric soluble glycoprotein (consisting of α, β, and γ chains) that is synthesized in the liver. This protein is involved in the polymerization of fibrin, blood coagulation, signal transduction, platelet activation and binding, and thrombin binding [37]. Fibrinogen has also been shown to have elevated expression levels during inflammation and in cardiovascular disease [37]; thus, its elevation in MCI may be reflective of early events of neuroinflammation. The other two proteins, plasma retinol-binding protein (RBP) and haptoglobin precursor allele 1, were detected as having a significant decrease in expression in CSF from MCI (and AD) patients relative to normal age-matched controls by 38% and 63%, respectively [29]. RBP is a 21 kDa carrier protein that tightly binds retinol allowing for it to freely circulate through plasma. Haptoglobin is a tetrameric protein that is a part of the acute phase response and binds free hemoglobin. Through binding of hemoglobin, haptoglobin inhibits oxidative activity of hemoglobin, prevents iron loss in the kidneys, and protects the kidneys against damage that could be caused by hemoglobin [38]. The effects of decreased expression of RBP and haptoglobin in MCI is not clearly understood [29].

Table 1. Candidate Biomarker Proteins in CSF of MCI

Protein	Change in MCI	Ref
C3a des-Arg	↑	Simonsen *et al.* 2007
C4a des-Arg	↑	Simonsen *et al.* 2007
Fibrinogen γ-chain A	↑	Lee *et al.* 2007
Haptoglobin precursor allele 1	↓	Jung *et al.* 2008
Phosphorylated osteopontin C-terminal fragment	↑	Simonsen *et al.* 2007
Plasma retinol-binding protein	↓	Jung *et al.* 2008
Ubiquitin	↑	Simonsen *et al.* 2007
β_2-Microglobulin	↑	Simonsen *et al.* 2007

Using 2D GE coupled to surface-enhanced laser desorption ionization (SELDI)-MS, Simonsen *et al.* identified a panel of 17 proteins that may be potential biomarkers of patients with MCI that convert to AD and of patients with MCI who do not progress to AD [33]. Of the 17 proteins, four proteins were down-regulated and 13 proteins were up-regulated in CSF of MCI patients that converted to AD relative to stable MCI patients and normal healthy controls [33]. Five proteins were identified with MS and have elevated expression in MCI patients that progress to AD: ubiquitin, C3a anaphylatoxin des-Arg, C4a anaphylatoxin des-Arg, β2-Microglobulin, and phosphorylated osteopontin C-terminal fragment. β-amyloid can bind to C1q and subsequently activate the complement cascade resulting in the production of C3a and C4a, as well as C5a peptides [39]. Osteopontin is a glycoprotein and proinflammatory cytokine involved in bone synthesis and various aspects of immunity such as chemotaxis [40], cell adhesion and wound healing [41], cell activation and cytokine production [41], and apoptosis [42, 43]. Elevation of the complement peptides and osteopontin in MCI patients that progress to AD suggests that innate immunity including inflammation in MCI patients may become activated and stay activated in the progression of disease. β2-Microglobulin is a part of the class I major histocompatibility complex and mediates amyloid fibril formation *in vitro* [44] and in the presence of transition metal cations [45]. Ubiquitin is used to target proteins for degradation by the 26S proteasome [46], and has been immunhistochemically shown to be present in neurofibrillary tangles (NFT) and senile plaques (SP) [47].

4.0. Proteomics Analyses of Brain from MCI Patients

An alternative approach to 2D GE that was recently used in proteomic comparisons of brain from MCI subjects relative to normal cognitive controls, is the PowerBlot proteomic approach (BD Transduction Laboratories). This approach uses a large-scale Western Blot approach to identify 750+ proteins simultaneously in a single experiment. Using the PowerBlot approach, Ho *et al.* detected 50 candidate proteins that had >2.0 fold-change in the EC region of MCI patients relative to normal cognitive controls [28]. Of the 50 proteins detected, 23 proteins were identified and could be functionally clustered into the following categories: neurotransmitter-related, cytoskeleton/cell adhesion, cell cycle/cell proliferation related, apopotosis related, transcription/translation related, and others. Neurotransmitter-related, apoptosis-related, and transcription/translation-related proteins were decreased in the EC of MCI patients, while cytoskeleton and cell cycle-related proteins included both increased and decreased proteins in MCI patients [28]. Several of these functional categories are similar to those observed in oxidatively modified proteins in MCI hippocampal brain regions and are discussed below.

5.0. Redox Proteomics Analyses of Brain from MCI Patients

Table 2 lists significantly elevated oxidatively modified proteins (i.e., PCO-, 3NT-, and HNE-modified) from the hippocampal and inferior parietal lobule (IPL) brain regions of MCI subjects relative to age-matched normal controls that were identified by our laboratory.

Table 2. Functional Categorization of Oxidatively Modified Proteins Identified in Brains of MCI Patients

Functional Categories	Protein	Oxidative Modifications
Energy/mitochondrial dysfunction	α-enolase	PCO, 3NT, HNE
	aldolase	3NT
	malate dehydrogenase	3NT
	glucose-regulated protein precursor	3NT
	lactate dehydrogenase	HNE
	phosphoglycerate kinase	HNE
	pyruvate kinase	PCO, HNE
	ATP synthase α-chain	HNE
Lipid abnormalities & cholinergic dysfunction	neuropolypeptide h3	HNE
Excitotoxicity	glutamine synthetase	PCO
Cell cycle, tau phosphorylation, A β production	PIN1	PCO
Neuritic abnormalities & structural dysfunction	DRP2	3NT
	fascin 1	3NT
	β actin	HNE
Antioxidant defense/Detoxification system dysfunction	GSTM3	3NT
	MRP3	3NT
	peroxiredoxin 6	3NT
	HSP70	3NT, HNE
	carbonyl reductase 1	HNE
Cell signaling dysfunction	14-3-3-γ	3NT
Protein synthesis alterations	Initiation factor α	HNE
	Elongation factor Tu	HNE

To-date these are the only reports of specific oxidatively modified proteins in MCI brain that may be relevant to the progression of AD [11, 19, 32]. These proteins can be grouped into several functional categories and were significantly oxidatively modified by one or more of the three oxidative parameters: PCO, 3NT, and HNE.

5.1. Energy or Mitochondrial Dysfunctions and Alterations

As listed in Table 2, several proteins involved in energy and/or mitochondrial-related pathways have significantly elevated levels of oxidative modification in the hippocampal or IPL regions of brains from subjects with MCI. These proteins are α-enolase, aldolase, pyruvate kinase (PK), malate dehydrogenase (MDH), lactate dehydrogenase (LDH), ATP synthase, phosphoglycerate kinase (PGK1), and glucose regulated protein precursor. Glycolysis plays an important role in supplying energy to the brain because glucose is the primary source of energy. Alterations in glucose metabolism and tolerance have been identified in brains of MCI and AD patients from positron emission tomography scanning, [48-50] and oxidatively modified glycolytic proteins have been identified in MCI and AD brain, and models thereof [12, 51]. α-Enolase, aldolase, PK, PGK1, and LDH are enzymes involved in or related to the glycolysis pathway. Increased oxidation of α-enolase, LDH, and PK has been shown to lead to loss of protein function measured by decreased enzymatic activity in MCI brain [11, 32]. Alterations in glycolysis could lead to less ATP production which is detrimental to cells requiring ATP to carry out normal functions, including signal transduction, maintenance of ion gradients, and protein synthesis, and detrimental to ATPases which are responsible for proper maintenance of ion pumps, lipid asymmetry, and intracellular communication. The observed impairments to glycolytic proteins in MCI brain suggest that energy metabolism is a key player in the progression of MCI to AD. This is further supported by the increased oxidation in MDH, ATP synthase α-chain, and glucose regulated protein precursor. ATP synthase α-chain is a component of complex V which plays a key role in energy production and undergoes a series of coordinated conformational changes in order to produce ATP. Oxidative modification to ATP synthase leads to reduced enzymatic activity [32]. Because ATP synthase is involved in the electron transport chain (ETC), alterations to its activity could result in an electron leakage from ETC carrier molecules, which would lead to an increase in reactive oxygen species (ROS). These ROS could then contribute to the observed increase in oxidative stress parameters in MCI brain [11, 13, 16, 20, 21, 25-27]. An overall decrease in ATP production due to dysfunction of glycolytic enzymes, ATP synthase, and glucose regulated protein precursor (from oxidative modification) could ultimately lead to Ca2+ dyshomeostasis and make neurons susceptible to excitotoxicity and cell death. From these studies it is apparent that potential preventative targets for AD could be targeted to restoring energy metabolism in earlier disease stages in MCI. In contrast to the usual observation of decreased enzymatic activity of oxidatively modified proteins, oxidative modification of MDH leads to increased activity [32]. The basis of this unusual observation remains unclear.

5.2. Neuritic Abnormalities/Structural Dysfunction

Oxidatively modified proteins in MCI related with neuritic and structural functions are dihydropyrimidinase like-2 (DRP2), β-actin, and fascin 1. DRP-2 is involved in axonal outgrowth and transmission and modulation of extracellular signals through the protein collapsin [52, 53]. In AD, DRP-2 also has increased oxidation [54, 55], which may be reflective of increasing neuritic degeneration, shortened dendritic length, and synapse loss as MCI progresses to AD. β-actin is crucial for proper maintenance of cellular and cytoskeletal integrity and morphology. High levels of actin can be found in growth cones, presynaptic terminals and dendritic spines, and thus its oxidation could lead to elongation of growth cones and synapse loss. Alterations in cellular integrity could be detrimental to cellular trafficking of key proteins involved in neurotransmission. Fascin 1, also known as p55, is a structural protein involved in cell adhesion and motility [56-58] and is used as a marker of normal dendritic function [59]. Overall, oxidation of structural proteins which could result in altered functionality ultimately can lead to impaired structural integrity, shortened dendritic lengths and faulty axonal growth, loss of interneuronal connections and poor neurotransmission. Neuritic abnormalities and structural dysfunction are documented in AD brain and thus may be key events in the loss of neurotransmission in MCI to AD.

5.3. Excitotoxicity

Overstimulation of neurons can result from an increase in the levels of extracellular glutamate. Glutamine synthetase, which converts glutamate to glutamine, was oxidatively modified in MCI brain and has been shown to have decreased activity [11]. Thus, decreased glutamine synthetase activity directly leads to a buildup in glutamate, which can lead to excitotoxicity. This phenomenon also affects Ca^{2+} homeostasis and eventually leads to neuronal death. Similar changes in glutamine synthetase oxidation and activity were observed in AD brain [60-62], and suggest that synapse loss observed in AD brain occurs early on in MCI with a contribution by excitotoxicity.

5.4. Lipid Abnormalities and Cholinergic Dysfunction

Neuropolypeptide h3 [(also known as phosphatidylethanolamine binding protein (PEBP)] is an enzyme involved in acetylcholine production and may play roles in phospholipid asymmetry. Oxidation of neuropolypeptide h3 in MCI brain and possible loss of function correlates well with the already known cholinergic loss observed in AD brain [63-66]. Also, loss of phospholipid asymmetry has been reported in MCI and AD brain [67-69], and thus oxidation of PEBP could play a role in lipid peroxidation events which lead to cellular apoptosis.

5.5. Antioxidant Defense/Detoxification System Dysfunction

Proteins involved in the antioxidant defense and detoxification system work to remove harmful species such as free radicals and toxic compounds from the cell. Peroxiredoxin 6 (PR6), multidrug resistance protein 3 (MRP3), glutathione-S-transferase Mu 3 (GSTM3), heat shock protein 70 (HSP70), and carbonyl reductase are oxidatively modified brain proteins in MCI. PR6 is an antioxidant enzyme that reduces the reactive nitrogen species, peroxynitrite, and was detected as having elevated nitration levels in MCI. PR6 also reduces reactive phospholipids and hydroperoxides [70] and has other roles which include cell differentiation, apoptosis, and detoxification [71]. Nitration of PR6, which could lead to loss of function, may result in increased levels of nitrated proteins, such as those detected in MCI brain [19, 21] . PR6 and GSTM3, a detoxification enzyme, form a complex that alters individual enzymatic activities [71] but which works to protect cells from toxic species such as HNE. GST catalyzes the conjugation of the low molecular weight intracellular thiol, glutathione, with toxins, and these toxins are transported out of the cell by MRP [72-74]. Oxidation and potential loss of function of PR6, MRP3, and GSTM3 could impair the cell's ability to remove toxicants leading to an increase in toxic species that subsequently attack cellular molecules (e.g., increased PCO, 3NT, or HNE) and lead to cell death. These observations in MCI brain are consistent with changes to MRP, GST, and PR6 which are observed in AD brain [73, 75], and demonstrate that proper antioxidant and detoxification defense systems may help to delay the progression of MCI to AD.

HSP70 is a molecular chaperone protein that repairs misfolded proteins and helps in the transportation of misfolded proteins to the proteasome. HSP70 belongs to the class of HSPs that also protect proteins from various stresses, such as oxidative damage [76]. Nitration of HSP70, leading to loss of function, could result in buildup of misfolded proteins and hence protein aggregates and "clogging" of the proteasome. Carbonyl reductase is an enzyme that reduces carbonyl compounds to their corresponding alcohols. HNE-modification of carbonyl reductase is rather interesting considering that it has been shown to reduce HNE levels [77], and thus its oxidative modification by HNE would lead to an increase in HNE available for attack on proteins such as those HNE-modified proteins identified in subjects with MCI [32].

5.6. Cell Signaling Dysfunction

14-3-3 γ belongs to a family of scaffolding proteins that normally bridges glycogen synthase kinase 3β (GSK-3β) and tau by forming a multiprotein tau phosphorylation complex [78]. Other functions include signal transduction, protein trafficking, and metabolism [79, 80]. 14-3-3 γ was observed as nitrated in MCI brain and has previously been observed to have elevated expression levels in AD brain [81, 82] and in AD CSF [83]. Nitration of 14-3-3 γ may contribute to tau hyperphosphorylation and thus NFT formation and dysfunction in cell signaling, events which are consistent with changes observed in AD brain.

5.7. Cell Cycle; Tau Phosphorylation; Aβ Production

Peptidyl-prolyl *cis/trans* isomerase 1 (Pin1) is a multifunctional protein involved in the cell cycle, tau phosphorylation, and Aβ production and regulates cellular processes such as protein folding, transcription, intracellular transport, and apoptosis [84-86]. Pin1 is oxidized in MCI brain [11] and has been previously reported as oxidized in AD brain [55]. Pin1, through its interactions with kinases and phosphatases such as GSK-3β, can directly regulate the phosphorylation of the tau protein [84, 87]. Thus, inactivation of Pin1 as a result of oxidative modification directly leads to hyperphosphorylation of tau and an increase in NFT. Pin1 has also been shown to bind to APP and influence the production of Aβ [84, 88]. Altered regulation of cell cycle processes by oxidized Pin1 may be related to elevation of cell cycle proteins in brain of MCI subjects [89]. Therefore, alterations to Pin1 activity also may lead to an increase in Aβ and SP. Oxidative impairment of Pin1 in early stages of AD, such as MCI, is consistent with and likely contributes to the major pathological hallmarks of AD: SP, NFT, and synapse loss.

5.8. Protein Synthesis Alterations

Initiation factor α (eIF-α) and elongation factor Tu (EF-Tu) are proteins involved in protein synthesis. eIF-α has been reported to have roles in cell proliferation and senescence [90], cytoskeletal organization [91], and apoptosis [92]. EF-Tu is a nuclear-encoded protein that assists in the translation of proteins in the mitochondria [93]. Specifically, EF-Tu binds aminoacylated tRNA in the cytoplasm and hydrolyzes GTP in order to allow the aminoacylated tRNA to enter the A site of the ribosome. Nitration of these proteins can directly influence protein synthesis. Decreased protein synthesis has been reported in MCI and AD [94-96], and thus these alterations are consistent with these reports. Alterations to protein synthesis in MCI may result in a reduction of key proteins necessary to combat many of the cellular insults observed in AD brain, which not only could result in compromised neuronal functions, but also contribute to progression of MCI to AD.

6.0. Concanavalin-A Associated Glycoproteins in Brain Regions from MCI Patients

DRP-2, glucose-regulated protein 78 (GRP78), protein phosphatase-related protein Sds-22, glial fibrillary acidic protein (GFAP), and β-synuclein were isolated in the ConA associated protein fraction using lectin affinity chromatography coupled to 2D GE and identified as having altered levels in the brains of subjects with MCI relative to age-matched controls [31]. DRP-2 and GRP78 were detected as significantly decreased and GFAP and protein phosphatase –related protein Sds22 as significantly increased in the hippocampal brain region of MCI patients, while β-synuclein was significantly decreased in the inferior parietal lobule region of MCI patients relative to age-matched controls [31]. DRP-2 as (discussed above) is a structural protein involved in axonal outgrowth and neuronal communication thus, its decreased expression may be indicative of neuritic dysfunction that occurs early in MCI and continues with disease progression into AD. GRP78 is an endoplasmic reticulum (ER) associated protein that belongs to the HSP70 protein family and is involved with the unfolfed protein response (UPR). HSP70 is significantly oxidatively modified in MCI brain (see Table 2). Because GRP78 normally reduces the levels of amyloid precursor protein (APP) and Aβ40 and 42 secretion [97], decreased expression of GRP78 in MCI brain could possibly play roles in the elevation of APP and Aβ levels found in AD brain. Also, alteration to GRP78 expression may disrupt Ca2+ homeostasis [98]. Conflicting reports of GRP78 expression in AD have been reported [99, 100], and thus its role in MCI progression to AD is not completely clear. Activation of the UPR in the ER might also mean that GRP78 is less available for refolding other damaged proteins or shuttling them to the 26S proteasome for degradation.

Protein phosphatase-related protein Sds22 is involved in the cell cycle and was detected as increased in MCI brain, the significance of this increase in the progression of MCI to AD is yet to be determined but as noted above, cell cycle proteins are elevated in brains of subjects with MCI [89]. β-synuclein is involved in synaptic functions, similar to the functions of α-synuclein in Parkinson's disease. β-synuclein also binds to Aβ [101] and has previously been shown by our laboratory to be oxidized *in vivo* following injection of Aβ(1-42) [102[. Decreased expression of β-synuclein in MCI brain could be related with altered synaptic functions which occur also due to the oxidatively modified proteins involved in synaptic functions mentioned above. GFAP, a glial specific intermediate filament protein is significantly increased in MCI brain. GFAP is involved in cytoskeletal integrity and maintenance of cellular shape and movement in astrocytes. Increased expression of GFAP in MCI brain is consistent with increased expression levels in AD [103] and with inflammation related to NFT and SP [104]. This finding provides further evidence supporting the notion that neuroinflammation is an event that occurs in the early stages of AD (MCI) and continues with disease progression.

7.0. Conclusion

This chapter has summarized some of the key findings from proteomics studies involving comparisons of brain and CSF in MCI subjects relative to normal age-matched controls. Candidate biomarkers of MCI that may help in early AD diagnosis were identified in CSF and may be useful as additional markers for AD diagnosis to the traditional tau (τ and p), and Aβ40 and 42 markers. Expression and redox proteomics analyses of various brain regions (e.g., EC, IPL, and hippocampus) revealed that a number of processes are altered in MCI including, neurotransmitter-related, apoptosis-related, energy/mitochondrial dysfunction, neuritic abnormalities/structural dysfunction, excitotoxicity, lipid abnormalities and cholinergic dysfunction, antioxidant defense/detoxification systems, cell signaling dysfunction, cell cycle/tau phosphorylation/Aβ production, and transcription/translation (protein synthesis) alterations. It is apparent that MCI to AD progression is a multifactorial process in which many pathways may be potential targets for intervening therapeutics. A large number of energy-related proteins were oxidatively modified in MCI further supporting the concept that normal energy maintenance is crucial and lacking in MCI and AD brain. In addition to oxidative modifications, concanavalin-A associated proteins have altered expression levels in IPL and hippocampal regions of MCI patients. These proteins are involved in structural integrity and molecular chaperoning, and altered levels of these proteins are congruent with the observed oxidative modifications and hence alterations of several structural and antioxidant defense/detoxification proteins. Proteomics has provided much insight into pathways that are related to MCI and with its progression to AD. Each of these pathways should be further investigated for their potential as therapeutic targets for early AD diagnosis, treatment, and/or prevention.

Acknowledgments

This work was supported by NIH grants to D.A.B. [AG-05119; AG-10836; AG-029839]. R.A.S. is a UNCF-Merck postdoctoral scholar.

References

[1] Morris, J. C., Mild cognitive impairment and preclinical Alzheimer's disease. *Geriatrics* 2005, *Suppl*, 9-14.

[2] Petersen, R. C., Mild cognitive impairment: transition between aging and Alzheimer's disease. *Neurologia* 2000, *15* (3), 93-101.

[3] Portet, F.; Ousset, P. J.; Touchon, J., [What is a mild cognitive impairment?]. *Rev. Prat.* 2005, *55* (17), 1891-4.

[4] Petersen, R. C.; Smith, G. E.; Waring, S. C.; Ivnik, R. J.; Tangalos, E. G.; Kokmen, E., Mild cognitive impairment: clinical characterization and outcome. *Arch Neurol.* 1999, *56* (3), 303-8.

[5] Petersen, R. C., Mild cognitive impairment clinical trials. *Nat. Rev. Drug Discov.* 2003, *2* (8), 646-53.

[6] Chertkow, H.; Bergman, H.; Schipper, H. M.; Gauthier, S.; Bouchard, R.; Fontaine, S.; Clarfield, A. M., Assessment of suspected dementia. *Can. J. Neurol. Sci.* 2001, *28 Suppl 1*, S28-41.

[7] Devanand, D. P.; Pradhaban, G.; Liu, X.; Khandji, A.; De Santi, S.; Segal, S.; Rusinek, H.; Pelton, G. H.; Honig, L. S.; Mayeux, R.; Stern, Y.; Tabert, M. H.; de Leon, M. J., Hippocampal and entorhinal atrophy in mild cognitive impairment: prediction of Alzheimer disease. *Neurology* 2007, *68* (11), 828-36.

[8] Du, A. T.; Schuff, N.; Amend, D.; Laakso, M. P.; Hsu, Y. Y.; Jagust, W. J.; Yaffe, K.; Kramer, J. H.; Reed, B.; Norman, D.; Chui, H. C.; Weiner, M. W., Magnetic resonance imaging of the entorhinal cortex and hippocampus in mild cognitive impairment and Alzheimer's disease. *J. Neurol. Neurosurg. Psychiatry* 2001, *71* (4), 441-7.

[9] Negash, S.; Petersen, L. E.; Geda, Y. E.; Knopman, D. S.; Boeve, B. F.; Smith, G. E.; Ivnik, R. J.; Howard, D. V.; Howard, J. H., Jr.; Petersen, R. C., Effects of ApoE genotype and mild cognitive impairment on implicit learning. *Neurobiol. Aging* 2007, *28* (6), 885-93.

[10] Ramakers, I. H.; Visser, P. J.; Aalten, P.; Bekers, O.; Sleegers, K.; van Broeckhoven, C. L.; Jolles, J.; Verhey, F. R., The association between APOE genotype and memory

dysfunction in subjects with mild cognitive impairment is related to age and Alzheimer pathology. *Dement Geriatr Cogn. Disord.*2008, *26* (2), 101-8.

[11] Butterfield, D. A.; Poon, H. F.; St Clair, D.; Keller, J. N.; Pierce, W. M.; Klein, J. B.; Markesbery, W. R., Redox proteomics identification of oxidatively modified hippocampal proteins in mild cognitive impairment: insights into the development of Alzheimer's disease. *Neurobiol. Dis.* 2006, *22* (2), 223-32.

[12] Butterfield, D. A.; Reed, T.; Newman, S. F.; Sultana, R., Roles of amyloid beta-peptide-associated oxidative stress and brain protein modifications in the pathogenesis of Alzheimer's disease and mild cognitive impairment. *Free Radic. Biol. Med.* 2007, *43* (5), 658-77.

[13] Butterfield, D. A.; Reed, T.; Perluigi, M.; De Marco, C.; Coccia, R.; Cini, C.; Sultana, R., Elevated protein-bound levels of the lipid peroxidation product, 4-hydroxy-2-nonenal, in brain from persons with mild cognitive impairment. *Neurosci. Lett.* 2006, *397* (3), 170-3.

[14] Cenini, G.; Sultana, R.; Memo, M.; Butterfield, D. A., Elevated levels of pro-apoptotic p53 and its oxidative modification by the lipid peroxidation product, HNE, in brain from subjects with amnestic mild cognitive impairment and Alzheimer's disease. *J. Cell Mol. Med.* 2008, *12* (3), 987-94.

[15] Ding, Q.; Markesbery, W. R.; Cecarini, V.; Keller, J. N., Decreased RNA, and increased RNA oxidation, in ribosomes from early Alzheimer's disease. *Neurochem. Res.* 2006, *31* (5), 705-10.

[16] Keller, J. N.; Schmitt, F. A.; Scheff, S. W.; Ding, Q.; Chen, Q.; Butterfield, D. A.; Markesbery, W. R., Evidence of increased oxidative damage in subjects with mild cognitive impairment. *Neurology* 2005, *64* (7), 1152-6.

[17] Lovell, M. A.; Markesbery, W. R., Oxidative damage in mild cognitive impairment and early Alzheimer's disease. *J. Neurosci. Res.* 2007, *85* (14), 3036-40.

[18] Murphy, M. P.; Beckett, T. L.; Ding, Q.; Patel, E.; Markesbery, W. R.; St Clair, D. K.; LeVine, H., 3rd; Keller, J. N., Abeta solubility and deposition during AD progression and in APPxPS-1 knock-in mice. *Neurobiol. Dis.* 2007, *27* (3), 301-11.

[19] Sultana, R.; Reed, T.; Perluigi, M.; Coccia, R.; Pierce, W. M.; Butterfield, D. A., Proteomic identification of nitrated brain proteins in amnestic mild cognitive impairment: a regional study. *J. Cell Mol. Med.* 2007, *11* (4), 839-51.

[20] Williams, T. I.; Lynn, B. C.; Markesbery, W. R.; Lovell, M. A., Increased levels of 4-hydroxynonenal and acrolein, neurotoxic markers of lipid peroxidation, in the brain in Mild Cognitive Impairment and early Alzheimer's disease. *Neurobiol. Aging* 2006, *27* (8), 1094-9.

[21] Butterfield, D. A.; Reed, T. T.; Perluigi, M.; De Marco, C.; Coccia, R.; Keller, J. N.; Markesbery, W. R.; Sultana, R., Elevated levels of 3-nitrotyrosine in brain from subjects with amnestic mild cognitive impairment: implications for the role of nitration in the progression of Alzheimer's disease. *Brain Res.* 2007, *1148*, 243-8.

[22] Guidi, I.; Galimberti, D.; Lonati, S.; Novembrino, C.; Bamonti, F.; Tiriticco, M.; Fenoglio, C.; Venturelli, E.; Baron, P.; Bresolin, N.; Scarpini, E., Oxidative imbalance in patients with mild cognitive impairment and Alzheimer's disease. *Neurobiol. Aging* 2006, *27* (2), 262-9.

[23] Rinaldi, P.; Polidori, M. C.; Metastasio, A.; Mariani, E.; Mattioli, P.; Cherubini, A.; Catani, M.; Cecchetti, R.; Senin, U.; Mecocci, P., Plasma antioxidants are similarly

depleted in mild cognitive impairment and in Alzheimer's disease. *Neurobiol. Aging* 2003, *24* (7), 915-9.

[24] Sultana, R.; Piroddi, M.; Galli, F.; Butterfield, D. A., Protein levels and activity of some antioxidant enzymes in hippocampus of subjects with amnestic mild cognitive impairment. *Neurochem. Res.* 2008, *33* (12), 2540-6.

[25] Migliore, L.; Fontana, I.; Trippi, F.; Colognato, R.; Coppede, F.; Tognoni, G.; Nucciarone, B.; Siciliano, G., Oxidative DNA damage in peripheral leukocytes of mild cognitive impairment and AD patients. *Neurobiol. Aging* 2005, *26* (5), 567-73.

[26] Wang, J.; Markesbery, W. R.; Lovell, M. A., Increased oxidative damage in nuclear and mitochondrial DNA in mild cognitive impairment. *J. Neurochem.* 2006, *96* (3), 825-32.

[27] Markesbery, W. R.; Kryscio, R. J.; Lovell, M. A.; Morrow, J. D., Lipid peroxidation is an early event in the brain in amnestic mild cognitive impairment. *Ann. Neurol.* 2005, *58* (5), 730-5.

[28] Ho, L.; Sharma, N.; Blackman, L.; Festa, E.; Reddy, G.; Pasinetti, G. M., From proteomics to biomarker discovery in Alzheimer's disease. *Brain Res. Brain Res. Rev.* 2005, *48* (2), 360-9.

[29] Jung, S. M.; Lee, K.; Lee, J. W.; Namkoong, H.; Kim, H. K.; Kim, S.; Na, H. R.; Ha, S. A.; Kim, J. R.; Ko, J.; Kim, J. W., Both plasma retinol-binding protein and haptoglobin precursor allele 1 in CSF: candidate biomarkers for the progression of normal to mild cognitive impairment to Alzheimer's disease. *Neurosci. Lett.* 2008, *436* (2), 153-7.

[30] Lee, J. W.; Namkoong, H.; Kim, H. K.; Kim, S.; Hwang, D. W.; Na, H. R.; Ha, S. A.; Kim, J. R.; Kim, J. W., Fibrinogen gamma-A chain precursor in CSF: a candidate biomarker for Alzheimer's disease. *BMC Neurol.* 2007, *7*, 14.

[31] Owen, J. B.; Di Domenico, F.; Sultana, R.; Perluigi, M.; Cini, C.; Pierce, W. M.; Butterfield, D. A., Proteomics-determined differences in the concanavalin-A-fractionated proteome of hippocampus and inferior parietal lobule in subjects with Alzheimer's disease and mild cognitive impairment: implications for progression of AD. *J. Proteome Res.* 2009, *8* (2), 471-82.

[32] Reed, T.; Perluigi, M.; Sultana, R.; Pierce, W. M.; Klein, J. B.; Turner, D. M.; Coccia, R.; Markesbery, W. R.; Butterfield, D. A., Redox proteomic identification of 4-hydroxy-2-nonenal-modified brain proteins in amnestic mild cognitive impairment: insight into the role of lipid peroxidation in the progression and pathogenesis of Alzheimer's disease. *Neurobiol. Dis.* 2008, *30* (1), 107-20.

[33] Simonsen, A. H.; McGuire, J.; Hansson, O.; Zetterberg, H.; Podust, V. N.; Davies, H. A.; Waldemar, G.; Minthon, L.; Blennow, K., Novel panel of cerebrospinal fluid biomarkers for the prediction of progression to Alzheimer dementia in patients with mild cognitive impairment. *Arch Neurol.* 2007, *64* (3), 366-70.

[34] Perkins, D. N.; Pappin, D. J.; Creasy, D. M.; Cottrell, J. S., Probability-based protein identification by searching sequence databases using mass spectrometry data. *Electrophoresis* 1999, *20* (18), 3551-67.

[35] Edelman, G. M.; Wang, J. L., Binding and functional properties of concanavalin A and its derivatives. III. Interactions with indoleacetic acid and other hydrophobic ligands. *J. Biol. Chem.* 1978, *253* (9), 3016-22.

[36] Sultana, R.; Newman, S. F.; Huang, Q.; Butterfield, D. A., Detection of carbonylated proteins in two-dimensional sodium dodecyl sulfate polyacrylamide gel electrophoresis separations. *Methods Mol. Biol.* 2008, *476*, 153-63.

[37] Mosesson, M. W., Fibrinogen gamma chain functions. *J. Thromb Haemost* 2003, *1* (2), 231-8.
[38] Wassell, J., Haptoglobin: function and polymorphism. *Clin. Lab.* 2000, *46* (11-12), 547-52.
[39] Akiyama, H.; Barger, S.; Barnum, S.; Bradt, B.; Bauer, J.; Cole, G. M.; Cooper, N. R.; Eikelenboom, P.; Emmerling, M.; Fiebich, B. L.; Finch, C. E.; Frautschy, S.; Griffin, W. S.; Hampel, H.; Hull, M.; Landreth, G.; Lue, L.; Mrak, R.; Mackenzie, I. R.; McGeer, P. L.; O'Banion, M. K.; Pachter, J.; Pasinetti, G.; Plata-Salaman, C.; Rogers, J.; Rydel, R.; Shen, Y.; Streit, W.; Strohmeyer, R.; Tooyoma, I.; Van Muiswinkel, F. L.; Veerhuis, R.; Walker, D.; Webster, S.; Wegrzyniak, B.; Wenk, G.; Wyss-Coray, T., Inflammation and Alzheimer's disease. *Neurobiol. Aging* 2000, *21* (3), 383-421.
[40] Koh, A.; da Silva, A. P.; Bansal, A. K.; Bansal, M.; Sun, C.; Lee, H.; Glogauer, M.; Sodek, J.; Zohar, R., Role of osteopontin in neutrophil function. *Immunology* 2007, *122* (4), 466-75.
[41] Wang, K. X.; Denhardt, D. T., Osteopontin: role in immune regulation and stress responses. *Cytokine Growth Factor Rev.* 2008, *19* (5-6), 333-45.
[42] Denhardt, D. T.; Noda, M.; O'Regan, A. W.; Pavlin, D.; Berman, J. S., Osteopontin as a means to cope with environmental insults: regulation of inflammation, tissue remodeling, and cell survival. *J. Clin. Invest* 2001, *107* (9), 1055-61.
[43] Standal, T.; Borset, M.; Sundan, A., Role of osteopontin in adhesion, migration, cell survival and bone remodeling. *Exp. Oncol.* 2004, *26* (3), 179-84.
[44] Hong, D. P.; Gozu, M.; Hasegawa, K.; Naiki, H.; Goto, Y., Conformation of beta 2-microglobulin amyloid fibrils analyzed by reduction of the disulfide bond. *J. Biol. Chem.* 2002, *277* (24), 21554-60.
[45] Eakin, C. M.; Miranker, A. D., From chance to frequent encounters: origins of beta2-microglobulin fibrillogenesis. *Biochim. Biophys. Acta* 2005, *1753* (1), 92-9.
[46] Hershko, A.; Leshinsky, E.; Ganoth, D.; Heller, H., ATP-dependent degradation of ubiquitin-protein conjugates. *Proc. Natl. Acad. Sci. USA* 1984, *81* (6), 1619-23.
[47] Perry, G.; Friedman, R.; Shaw, G.; Chau, V., Ubiquitin is detected in neurofibrillary tangles and senile plaque neurites of Alzheimer disease brains. *Proc. Natl. Acad. Sci. USA* 1987, *84* (9), 3033-6.
[48] Hoyer, S., Oxidative energy metabolism in Alzheimer brain. Studies in early-onset and late-onset cases. *Mol. Chem. Neuropathol.* 1992, *16* (3), 207-24.
[49] Messier, C.; Gagnon, M., Glucose regulation and brain aging. *J. Nutr. Health Aging* 2000, *4* (4), 208-13.
[50] Watson, G. S.; Craft, S., Modulation of memory by insulin and glucose: neuropsychological observations in Alzheimer's disease. *Eur. J. Pharmacol.* 2004, *490* (1-3), 97-113.
[51] Sultana, R.; Perluigi, M.; Butterfield, D. A., Oxidatively modified proteins in Alzheimer's disease (AD), mild cognitive impairment and animal models of AD: role of Abeta in pathogenesis. *Acta Neuropathol.* 2009, *in press*.
[52] Hamajima, N.; Matsuda, K.; Sakata, S.; Tamaki, N.; Sasaki, M.; Nonaka, M., A novel gene family defined by human dihydropyrimidinase and three related proteins with differential tissue distribution. *Gene* 1996, *180* (1-2), 157-63.

[53] Kato, Y.; Hamajima, N.; Inagaki, H.; Okamura, N.; Koji, T.; Sasaki, M.; Nonaka, M., Post-meiotic expression of the mouse dihydropyrimidinase-related protein 3 (DRP-3) gene during spermiogenesis. *Mol. Reprod. Dev.* 1998, *51* (1), 105-11.

[54] Castegna, A.; Aksenov, M.; Thongboonkerd, V.; Klein, J. B.; Pierce, W. M.; Booze, R.; Markesbery, W. R.; Butterfield, D. A., Proteomic identification of oxidatively modified proteins in Alzheimer's disease brain. Part II: dihydropyrimidinase-related protein 2, alpha-enolase and heat shock cognate 71. *J. Neurochem.* 2002, *82* (6), 1524-32.

[55] Sultana, R.; Boyd-Kimball, D.; Poon, H. F.; Cai, J.; Pierce, W. M.; Klein, J. B.; Merchant, M.; Markesbery, W. R.; Butterfield, D. A., Redox proteomics identification of oxidized proteins in Alzheimer's disease hippocampus and cerebellum: an approach to understand pathological and biochemical alterations in AD. *Neurobiol. Aging* 2006, *27* (11), 1564-76.

[56] Adams, J. C., Formation of stable microspikes containing actin and the 55 kDa actin bundling protein, fascin, is a consequence of cell adhesion to thrombospondin-1: implications for the anti-adhesive activities of thrombospondin-1. *J. Cell Sci.* 1995, *108 (Pt 5)*, 1977-90.

[57] Adams, J. C., Roles of fascin in cell adhesion and motility. *Curr. Opin. Cell Biol.* 2004, *16* (5), 590-6.

[58] Yamashiro, S.; Yamakita, Y.; Ono, S.; Matsumura, F., Fascin, an actin-bundling protein, induces membrane protrusions and increases cell motility of epithelial cells. *Mol. Biol. Cell* 1998, *9* (5), 993-1006.

[59] Pinkus, G. S.; Lones, M. A.; Matsumura, F.; Yamashiro, S.; Said, J. W.; Pinkus, J. L., Langerhans cell histiocytosis immunohistochemical expression of fascin, a dendritic cell marker. *Am. J. Clin. Pathol.* 2002, *118* (3), 335-43.

[60] Butterfield, D. A.; Hensley, K.; Cole, P.; Subramaniam, R.; Aksenov, M.; Aksenova, M.; Bummer, P. M.; Haley, B. E.; Carney, J. M., Oxidatively induced structural alteration of glutamine synthetase assessed by analysis of spin label incorporation kinetics: relevance to Alzheimer's disease. *J. Neurochem.* 1997, *68* (6), 2451-7.

[61] Castegna, A.; Aksenov, M.; Aksenova, M.; Thongboonkerd, V.; Klein, J. B.; Pierce, W. M.; Booze, R.; Markesbery, W. R.; Butterfield, D. A., Proteomic identification of oxidatively modified proteins in Alzheimer's disease brain. Part I: creatine kinase BB, glutamine synthase, and ubiquitin carboxy-terminal hydrolase L-1. *Free Radic. Biol. Med.* 2002, *33* (4), 562-71.

[62] Hensley, K.; Hall, N.; Subramaniam, R.; Cole, P.; Harris, M.; Aksenov, M.; Aksenova, M.; Gabbita, S. P.; Wu, J. F.; Carney, J. M.; et al., Brain regional correspondence between Alzheimer's disease histopathology and biomarkers of protein oxidation. *J. Neurochem.* 1995, *65* (5), 2146-56.

[63] Coyle, J. T.; Price, D. L.; DeLong, M. R., Alzheimer's disease: a disorder of cortical cholinergic innervation. *Science* 1983, *219* (4589), 1184-90.

[64] Davies, M. J.; Fu, S.; Wang, H.; Dean, R. T., Stable markers of oxidant damage to proteins and their application in the study of human disease. *Free Radic. Biol. Med.* 1999, *27* (11-12), 1151-63.

[65] Perry, E. K.; Curtis, M.; Dick, D. J.; Candy, J. M.; Atack, J. R.; Bloxham, C. A.; Blessed, G.; Fairbairn, A.; Tomlinson, B. E.; Perry, R. H., Cholinergic correlates of cognitive impairment in Parkinson's disease: comparisons with Alzheimer's disease. *J. Neurol. Neurosurg. Psychiatry* 1985, *48* (5), 413-21.

[66] Wevers, A.; Witter, B.; Moser, N.; Burghaus, L.; Banerjee, C.; Steinlein, O. K.; Schutz, U.; de Vos, R. A.; Steur, E. N.; Lindstrom, J.; Schroder, H., Classical Alzheimer features and cholinergic dysfunction: towards a unifying hypothesis? *Acta Neurol. Scand. Suppl.* 2000, *176*, 42-8.

[67] Bader Lange, M. L.; Cenini, G.; Piroddi, M.; Abdul, H. M.; Sultana, R.; Galli, F.; Memo, M.; Butterfield, D. A., Loss of phospholipid asymmetry and elevated brain apoptotic protein levels in subjects with amnestic mild cognitive impairment and Alzheimer disease. *Neurobiol. Dis.* 2008, *29* (3), 456-64.

[68] Cutler, R. G.; Kelly, J.; Storie, K.; Pedersen, W. A.; Tammara, A.; Hatanpaa, K.; Troncoso, J. C.; Mattson, M. P., Involvement of oxidative stress-induced abnormalities in ceramide and cholesterol metabolism in brain aging and Alzheimer's disease. *Proc. Natl. Acad. Sci. USA* 2004, *101* (7), 2070-5.

[69] Geula, C.; Nagykery, N.; Nicholas, A.; Wu, C. K., Cholinergic neuronal and axonal abnormalities are present early in aging and in Alzheimer disease. *J. Neuropathol. Exp. Neurol.* 2008, *67* (4), 309-18.

[70] Chowdhury, I.; Mo, Y.; Gao, L.; Kazi, A.; Fisher, A. B.; Feinstein, S. I., Oxidant stress stimulates expression of the human peroxiredoxin 6 gene by a transcriptional mechanism involving an antioxidant response element. *Free Radic. Biol. Med.* 2009, *46* (2), 146-53.

[71] Ralat, L. A.; Manevich, Y.; Fisher, A. B.; Colman, R. F., Direct evidence for the formation of a complex between 1-cysteine peroxiredoxin and glutathione S-transferase pi with activity changes in both enzymes. *Biochemistry* 2006, *45* (2), 360-72.

[72] Renes, J.; de Vries, E. E.; Hooiveld, G. J.; Krikken, I.; Jansen, P. L.; Muller, M., Multidrug resistance protein MRP1 protects against the toxicity of the major lipid peroxidation product 4-hydroxynonenal. *Biochem. J.* 2000, *350 Pt 2*, 555-61.

[73] Sultana, R.; Butterfield, D. A., Oxidatively modified GST and MRP1 in Alzheimer's disease brain: implications for accumulation of reactive lipid peroxidation products. *Neurochem. Res.* 2004, *29* (12), 2215-20.

[74] Tchaikovskaya, T.; Fraifeld, V.; Urphanishvili, T.; Andorfer, J. H.; Davies, P.; Listowsky, I., Glutathione S-transferase hGSTM3 and ageing-associated neurodegeneration: relationship to Alzheimer's disease. *Mech. Ageing Dev.* 2005, *126* (2), 309-15.

[75] Perluigi, M. S., R.; Cenini, G.; Di Domenico, F.; Memo, M.; Pierce, W.M.; Coccia, R.; Butterfield, D.A., Redox proteomics identification of HNE-modified brain proteins in Alzheimers disease: Role of lipid peroxidation in AD pathogenesis. *Proteomics Clin. Appli.* 2009, *Accepted*.

[76] Calabrese, V.; Scapagnini, G.; Colombrita, C.; Ravagna, A.; Pennisi, G.; Giuffrida Stella, A. M.; Galli, F.; Butterfield, D. A., Redox regulation of heat shock protein expression in aging and neurodegenerative disorders associated with oxidative stress: a nutritional approach. *Amino Acids* 2003, *25* (3-4), 437-44.

[77] Doorn, J. A.; Maser, E.; Blum, A.; Claffey, D. J.; Petersen, D. R., Human carbonyl reductase catalyzes reduction of 4-oxonon-2-enal. *Biochemistry* 2004, *43* (41), 13106-14.

[78] Agarwal-Mawal, A.; Qureshi, H. Y.; Cafferty, P. W.; Yuan, Z.; Han, D.; Lin, R.; Paudel, H. K., 14-3-3 connects glycogen synthase kinase-3 beta to tau within a brain

microtubule-associated tau phosphorylation complex. *J. Biol. Chem.* 2003, *278* (15), 12722-8.

[79] Dougherty, M. K.; Morrison, D. K., Unlocking the code of 14-3-3. *J. Cell Sci.* 2004, *117* (Pt 10), 1875-84.

[80] Takahashi, Y., The 14-3-3 proteins: gene, gene expression, and function. *Neurochem. Res.* 2003, *28* (8), 1265-73.

[81] Frautschy, S. A.; Baird, A.; Cole, G. M., Effects of injected Alzheimer beta-amyloid cores in rat brain. *Proc. Natl. Acad. Sci. USA* 1991, *88* (19), 8362-6.

[82] Layfield, R.; Fergusson, J.; Aitken, A.; Lowe, J.; Landon, M.; Mayer, R. J., Neurofibrillary tangles of Alzheimer's disease brains contain 14-3-3 proteins. *Neurosci. Lett.* 1996, *209* (1), 57-60.

[83] Burkhard, P. R.; Sanchez, J. C.; Landis, T.; Hochstrasser, D. F., CSF detection of the 14-3-3 protein in unselected patients with dementia. *Neurology* 2001, *56* (11), 1528-33.

[84] Butterfield, D. A.; Abdul, H. M.; Opii, W.; Newman, S. F.; Joshi, G.; Ansari, M. A.; Sultana, R., Pin1 in Alzheimer's disease. *J. Neurochem.* 2006, *98* (6), 1697-706.

[85] Gothel, S. F.; Marahiel, M. A., Peptidyl-prolyl cis-trans isomerases, a superfamily of ubiquitous folding catalysts. *Cell Mol. Life Sci.* 1999, *55* (3), 423-36.

[86] Lu, K. P.; Hanes, S. D.; Hunter, T., A human peptidyl-prolyl isomerase essential for regulation of mitosis. *Nature* 1996, *380* (6574), 544-7.

[87] Zhou, X. Z.; Kops, O.; Werner, A.; Lu, P. J.; Shen, M.; Stoller, G.; Kullertz, G.; Stark, M.; Fischer, G.; Lu, K. P., Pin1-dependent prolyl isomerization regulates dephosphorylation of Cdc25C and tau proteins. *Mol. Cell* 2000, *6* (4), 873-83.

[88] Pastorino, L.; Sun, A.; Lu, P. J.; Zhou, X. Z.; Balastik, M.; Finn, G.; Wulf, G.; Lim, J.; Li, S. H.; Li, X.; Xia, W.; Nicholson, L. K.; Lu, K. P., The prolyl isomerase Pin1 regulates amyloid precursor protein processing and amyloid-beta production. *Nature* 2006, *440* (7083), 528-34.

[89] Sultana, R.; Butterfield, D. A., Regional expression of key cell cycle proteins in brain from subjects with amnestic mild cognitive impairment. *Neurochem. Res.* 2007, *32* (4-5), 655-62.

[90] Thompson, J. E.; Hopkins, M. T.; Taylor, C.; Wang, T. W., Regulation of senescence by eukaryotic translation initiation factor 5A: implications for plant growth and development. *Trends Plant Sci.* 2004, *9* (4), 174-9.

[91] Condeelis, J., Elongation factor 1 alpha, translation and the cytoskeleton. *Trends Biochem. Sci.* 1995, *20* (5), 169-70.

[92] Tome, M. E.; Fiser, S. M.; Payne, C. M.; Gerner, E. W., Excess putrescine accumulation inhibits the formation of modified eukaryotic initiation factor 5A (eIF-5A) and induces apoptosis. *Biochem. J.* 1997, *328* (Pt 3), 847-54.

[93] Ling, M.; Merante, F.; Chen, H. S.; Duff, C.; Duncan, A. M.; Robinson, B. H., The human mitochondrial elongation factor tu (EF-Tu) gene: cDNA sequence, genomic localization, genomic structure, and identification of a pseudogene. *Gene* 1997, *197* (1-2), 325-36.

[94] Chang, R. C.; Wong, A. K.; Ng, H. K.; Hugon, J., Phosphorylation of eukaryotic initiation factor-2alpha (eIF2alpha) is associated with neuronal degeneration in Alzheimer's disease. *Neuroreport* 2002, *13* (18), 2429-32.

[95] Ding, Q.; Markesbery, W. R.; Chen, Q.; Li, F.; Keller, J. N., Ribosome dysfunction is an early event in Alzheimer's disease. *J. Neurosci.* 2005, *25* (40), 9171-5.

[96] Li, X.; An, W. L.; Alafuzoff, I.; Soininen, H.; Winblad, B.; Pei, J. J., Phosphorylated eukaryotic translation factor 4E is elevated in Alzheimer brain. *Neuroreport* 2004, *15* (14), 2237-40.

[97] Yang, Y.; Turner, R. S.; Gaut, J. R., The chaperone BiP/GRP78 binds to amyloid precursor protein and decreases Abeta40 and Abeta42 secretion. *J. Biol. Chem.* 1998, *273* (40), 25552-5.

[98] Falahatpisheh, H.; Nanez, A.; Montoya-Durango, D.; Qian, Y.; Tiffany-Castiglioni, E.; Ramos, K. S., Activation profiles of HSPA5 during the glomerular mesangial cell stress response to chemical injury. *Cell Stress Chaperones* 2007, *12* (3), 209-18.

[99] Hoozemans, J. J.; Veerhuis, R.; Van Haastert, E. S.; Rozemuller, J. M.; Baas, F.; Eikelenboom, P.; Scheper, W., The unfolded protein response is activated in Alzheimer's disease. *Acta Neuropathol.* 2005, *110* (2), 165-72.

[100] Katayama, T.; Imaizumi, K.; Sato, N.; Miyoshi, K.; Kudo, T.; Hitomi, J.; Morihara, T.; Yoneda, T.; Gomi, F.; Mori, Y.; Nakano, Y.; Takeda, J.; Tsuda, T.; Itoyama, Y.; Murayama, O.; Takashima, A.; St George-Hyslop, P.; Takeda, M.; Tohyama, M., Presenilin-1 mutations downregulate the signalling pathway of the unfolded-protein response. *Nat. Cell Biol.* 1999, *1* (8), 479-85.

[101] Jensen, P. H.; Hojrup, P.; Hager, H.; Nielsen, M. S.; Jacobsen, L.; Olesen, O. F.; Gliemann, J.; Jakes, R., Binding of Abeta to alpha- and beta-synucleins: identification of segments in alpha-synuclein/NAC precursor that bind Abeta and NAC. *Biochem. J.* 1997, *323 (Pt 2)*, 539-46.

[102] Boyd-Kimball, D.; Sultana, R.; Poon, H. F.; Lynn, B. C.; Casamenti, F.; Pepeu, G.; Klein, J. B.; Butterfield, D. A., Proteomic identification of proteins specifically oxidized by intracerebral injection of amyloid beta-peptide (1-42) into rat brain: implications for Alzheimer's disease. *Neuroscience* 2005, *132* (2), 313-24.

[103] Beach, T. G.; Walker, R.; McGeer, E. G., Patterns of gliosis in Alzheimer's disease and aging cerebrum. *Glia* 1989, *2* (6), 420-36.

[104] Korolainen, M. A.; Auriola, S.; Nyman, T. A.; Alafuzoff, I.; Pirttila, T., Proteomic analysis of glial fibrillary acidic protein in Alzheimer's disease and aging brain. *Neurobiol. Dis.* 2005, *20* (3), 858-70.

In: Encyclopedia of Neuroscience Research
Editors: Eileen J. Sampson and Donald R. Glevins
ISBN 978-1-61324-861-4
© 2012 Nova Science Publishers, Inc.

Chapter XXVI

Animal Models for Cerebrovascular Impairment and its Relevance in Vascular Dementia

Veronica Lifshitz and Dan Frenkel *
Department of Neurobiology, The George S Wise Faculty of Life Sciences,
Tel-Aviv University, Tel Aviv, Israel

Abstract

Dementia is one of the most predominant neurological disorders in the elderly. The prevalence and incidence of degenerative diseases, such as Alzheimer's disease (AD) and vascular dementia, strongly correlate with age. Vascular dementia (VaD) is recognized as the second most prevalent type of dementia and described as a multifaceted cognitive decline resulting from cerebrovascular injury to brain regions associated with memory, cognition, and behavior. Vascular risk factors including diabetes, insulin resistance, hypertension, heart disease, smoking and obesity are each independently associated with an increased risk of cognitive impairment and dementia. This review will summarize relevant animal models for studying cognitive deficits as well as diagnostic assessment of VaD in humans with the intention of developing both an early diagnostic capability and future therapeutic interventions.

Introduction

Dementia, derived from the Latin *demens* (without mind), is a clinical syndrome characterized by a cluster of symptoms and indications manifested by difficulties in memory, disturbances in language, psychological and psychiatric changes, and a general impairment in successfully completing everyday tasks [1, 2]. Dementia is a common and growing problem,

---

* Corresponding author: Dan Frenkel, Sherman building, Room 424 Tel Aviv, Israel 69978, Telephone: 972-3-6409484, Fax: 972-3-6409028. e-mail: dfrenkel@post.tau.ac.il.

affecting 5% of the population over the age of 65 and 20% over the age of 80. The number of people living with dementia will almost double every 20 years, to 42 million by 2020 and 81 million by 2040, assuming no changes in mortality and no effective prevention strategies or curative treatments [3].

The Importance of the Vascular System in Neuronal Function

A normal neuronal-vascular relationship is critical for normal brain functioning. A healthy brain relies on all of the cells of the neurovascular unit - neurons, astrocytes, brain endothelium, pericytes, vascular smooth muscle cells (VSMC), microglia and perivascular macrophages - to function properly and communicate with each other in order for neuronal synapses and circuitries to maintain normal cognitive functions [4].

Some brain disorders may have a vascular origin. Vascular cells, i.e., endothelium and pericytes, can directly affect neuronal and synaptic functioning through changes in the blood flow, BBB permeability, nutrient supply, faulty clearance of toxic molecules, failure of enzymatic functions, altered secretion of trophic factors and matrix molecules, abnormal expression of vascular receptors and induction of ectoenzymes [5] (Figure 1). In response to a vascular insult, signals from neurons and astrocytes recruit microglia, which, when activated, secrete several proinflammatory cytokines [6]. This further aggravates the neuronal injury and synaptic dysfunction. In the case of a primary neuronal disorder, vasculo-glial activation follows and may critically modify progression of the disease [5, 7].

Vascular Dysfunction Can Lead to Cognitive Impairment

Vascular dementia (VaD) is the second-most-common cause of dementia in the elderly, after Alzheimer's disease (AD) [8]. VaD is a highly heterogenous disorder and is defined as a loss of cognitive function resulting from ischemic, hypoperfusive, or hemorrhagic brain lesions due to cerebrovascular disease or cardiovascular pathology [9]. As the brain lacks a reserve of glucose and oxygen, it is wholly dependent on a constant blood supply. A threat to cerebral perfusion is likely to have dramatic consequences on neuronal functions. Vascular diseases may lead to activated astrocytes and microglial cells resulting in elevated expressions of inducible nitric oxide synthase (iNOS) and release of neurotoxic reactive oxygen species and nitric oxide (Figure 2). The inflammatory mechanisms may be aggravated by continuous release of chemokines such as CCL2 and CCL3 by astrocytes. Furthermore, oxidative stress and glucose starvation may lead to the impairment of astrocyte glutamate uptake which in turn results in glutamate neurotoxicity.

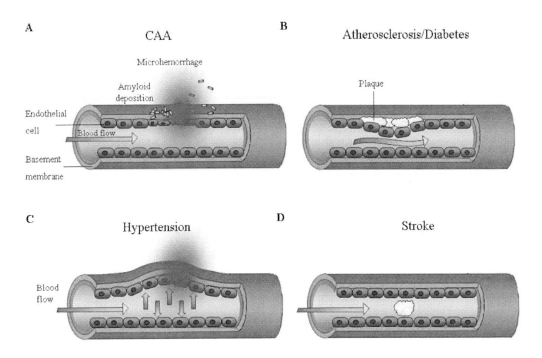

Figure 1. Different types of diseases with vascular pathology. (A) CAA is characterized in the deposition of amyloid proteins around the blood vessels in the brain which may disrupt vascular endothelium and induce microhemorrhage. (B) Diabetes and Atherosclerosis are characterized by intimal plaques containing lipids resulting in reduce blood flow. (C) Hypertension promotes cerebral artery disruption and fibrinoid necrosis of the vascular wall. (D) Ischemic stroke blocks blood flow, as a result embolus forms, resulting in small-vessel occlusions with multiple artery and brain lesions.

Neurodegenerative Diseases with Cognitive Impairment and Vascular Pathology

Disorders of dementia are multi-factorial and are caused by an interaction of genetic and environmental factors acting throughout the life of an individual. Different types of dementia can have different courses, with different patterns of symptoms and can respond differently to treatment. The main types of dementia are: AD (about 50% of cases), VaD (about 25%), Mixed Alzheimer's disease and vascular dementia (included in the above, 25%), Lewy body dementia (15%) and all others (about 5% combined) including frontotemporal dementia, focal dementias (such as progressive aphasia), subcortical dementias (such as Parkinson's disease dementia) and secondary causes of dementia syndrome (such as intracranial lesions) [2].

Although the underlying cause for most VaD cases have not clearly been established [10], several reports suggest that vascular risk factors could affect the neurodegenerative processes leading to dementia, in general, as well as Alzheimer's disease and vascular dementia, in particular [11]. Thus, for example, smoking [12], diabetes [11, 13], high blood pressure [11], hypercholesterolemia [14], poor respiratory function [15] and/or previous history of cardiovascular disease [16] have been associated with both the incidence of dementia and a higher risk of dementia related mortality [10].

Alzheimer's Disease

Alzheimer's disease is the most common type of dementia. In its early stages, AD is characterized by subtle impairments in memory associated with other cognitive deficits. In patients aged 65 years or older, who have some kind of cognitive decline, it accounts for over 50% of cases. Progression to full dementia may take several years following the signs of mild cognitive impairment (MCI) at the early stage of AD [17]. The main cause of AD is generally attributed to the increased production and accumulation of amyloid-β (Aβ), as well as neurofibrillary tangle (NFT) formation. The main role of Aβ, as a mediator in AD, is inferred from its accumulation in the brain several decades before the disease becomes evident. Furthermore, tri-chromosome 21 Down syndrome patients present with a significantly higher incidence of AD symptoms by their late 30's than the average population. As a result of this observation and since the gene for the amyloid precursor protein is located on chromosome 21, strength the important role of Aβ in the pathogenesis of AD. While early research indicated that the fibril form of Aβ was toxic, in more recent papers, the toxicity to the neurons seems also to be derived from soluble Aβ oligomers [18, 19]. The classical view is that Aβ is deposited extracellularly, however, emerging evidence from transgenic mice and human patients indicate that this peptide can also accumulate intra-neuronally, which may contribute to disease progression [20]. Aβ can adversely affect distinct molecular and cellular pathways, thereby facilitating tau phosphorylation, aggregation, mis-localization and accumulation. Intracellular abnormally phosphorylated tau and Aβ exhibit synergistic effects that ultimately lead to an acceleration of the neurodegenerative mechanisms related to metabolism, cellular detoxification, mitochondrial dysfunction and energy deficiency. These mechanisms, in turn, result in the formation of neuritis plaques [21].

The fact that cerebrovascular diseases and AD share common risk factors may indicate on common pathogenic mechanisms. In AD, the deposition of Aβ is found both in brain tissue (neuritic plaques) and on the wall of cerebral blood vessels (cerebral amyloid angiopathy or CAA). Moreover, the vascular pathology in AD is not limited to the accumulation of Aβ on the vessel wall, but is also characterized by atherosclerosis, vascular fibrosis, and structural and inflammatory changes of blood vessels [22]. Evidence suggests that cerebrovascular pathologies, such as atherosclerotic lesions, structural and inflammatory alterations as well as impaired hemodynamic responses, are early features of AD [23, 24]. Indeed, although Aβ is the major component of the amyloid deposits, other molecules are also associated with these deposits (e.g., ferritin, components of the complement pathway, α1-antichymotrypsin, α2-macroglobulin, LDL-receptor related protein, APP, acetylcholinesterase, laminin, glycosaminoglycans and the apolipoproteins, E and J [25-27].

In AD, there is evidence of denervated cortical microvessels. Perivascular cholinergic (ACh) nerve terminals from the basal forebrain, which are important regulators of cortical cerebral blood flow (CBF) [28, 29], are also denervated. These alterations impair the ability of the brain to maintain CBF. In addition, there is evidence of degenerative changes in endothelial and smooth muscle cells in AD [30]. Indeed, perfusion deficits in AD often precede the neurodegenerative changes [5].

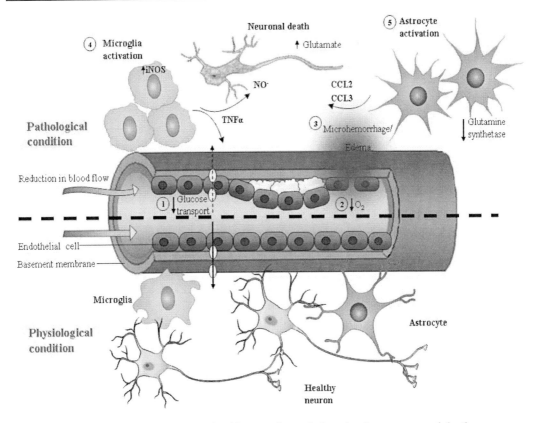

Figure 2. Hypothetical mechanisms involved in vascular pathology leading to neuronal death. Mechanisms resulting in vascular pathology lead to alteration of the blood-brain barrier and decreases in glucose concentration (1) and oxygen transport (2). Stress to vessels will lead to micro-hemorrhage and edema (3) and to activated astrocytes and microglial cells resulting in elevated expressions of inducible nitric oxide synthase (iNOS), release of neurotoxic reactive oxygen species and nitric oxide (4). The inflammatory mechanisms may be aggravated by continued release of chemokines such as CCL2 and CCL3 by astrocytes as well as impairment in glutamate uptake (5). Vascular pathology leads to neuronal stress and may eventually lead to neuronal death with cognitive consequences.

Cerebral Amyloid Angiopathy

Cerebral amyloid angiopathy (CAA) is defined as deposition of vascular amyloid in the walls of the meningeal and parenchymal arteries, arterioles, capillaries and, albeit rarely, veins. Veins can also present with thickening walls and smooth muscle cells (SMC) degeneration. These symptoms can further spread into the surrounding neuropil, deposit in the glia limitans or in the capillary wall [31, 32]. The vascular amyloid deposits in CAA, like senile plaques in AD, are composed primarily of the β-amyloid peptide (Aβ), a 39–43 amino acid fragment of the β-amyloid precursor protein (APP) [32]. Furthermore, the short form of Aβ (Aβ1–40) is the predominant form in cerebral blood vessels, whereas the long form (Aβ 1–42) is more abundant in brain parenchyma. However, a number of other molecules, arising from both local and circulating sources, can be associated with Aβ deposits; these include extracellular matrix proteins [33], α1-antichymotrypsin [34], apolipoprotein E [25], ubiquitin [35], serum amyloid P [36], acetylcholinesterase [37], cystatin C [38] and complement components [39].

Although numerous conditions can promote cerebrovascular amyloidogenesis, the most salient risk factors for CAA are advancing age, Alzheimer's disease, and genetic influences [40]. While most cases of CAA or AD are sporadic, various mutations have been found to associate with cases of familial disease, including several mutations of APP. CAA incidence increases with age to almost 100% after 80 years and AD ranges from 70 to 97.6% [41]. The most widely recognized clinical manifestation of CAA is intracerebral hemorrhage (ICH) [32]. ICH related to CAA typically occurs in the cortical or cortico-subcortical brain regions most heavily involved in CAA. Brains with ICH caused by CAA demonstrate extensive replacement of blood vessel walls with amyloid as well as evidence of breakdown of amyloid-laden vessels such as concentric cracking, microaneurysms and fibrinoid necrosis [32, 42].

Cognitive impairment has been observed in both familial [43] and sporadic [44] instances of severe CAA. These observations suggest that CAA can cause clinically important vascular dysfunction [45, 46] . The type of vessel involved and distribution of CAA may differently affect cognition. Capillary CAA has been shown to be more closely associated with impaired cognition than CAA of larger blood vessels [47]. Although CAA is most prevalent in the posterior regions of the cerebral hemispheres, the association between CAA and clinical dementia has been reported to be strongest for frontal CAA [48, 49].

Diabetes

Diabetes is another risk factor for stroke and VaD. Diabetic patients have twice the risk of developing cognitive impairment [50, 51]. A possible mechanism is the impaired brain perfusion due to endothelial oxidative damage and alteration of the blood-brain barrier resulting from excessive glycosylation [52] (Figure 1). Diabetic patients also present high blood viscosity, which is a blood flow-reducing factor. Although these vascular factors may increase dementia risk through cerebral small- or large-vessel disease, they may also act through noncerebrovascular mechanisms (e.g. directly affect amyloid accumulation in the brain [53] and disrupt brain amyloid-β clearance by competing for the insulin degrading enzyme.)

Hypertension

High blood pressure (HBP) or hypertension is defined as high pressure (tension) in the arteries. Hypertension is also associated with increased vascular permeability and protein extravasation. This permeability leads, in turn, to chronic edema and tissue necrosis [54] (Figure 1). Hypertension is a risk factor for stroke, ischemic white-matter lesions, cardiovascular disorders, and vascular dementia. Several studies have suggested a positive correlation between hypertension and cognitive impairment, such as AD. Possible mechanisms may correspond to incidents of micro- hemorrhaging (Figure 2) that link to a reduction in glutamate and oxygen absorbance of neurons. Furthermore, local inflammation can increase production of reactive oxygen species (ROS) and neurodegeneration. High blood pressure precedes AD by decades. In the studies that demonstrated a significant positive correlation between cognitive impairment and hypertension, the neuropsychologic domains predominantly affected were learning, memory, attention, and mental flexibility [55].

Ischemic Stroke

Dementia may occur in 25% to 33% of ischemic stroke cases at age 65 and older. Post stroke VaD may be caused by large-vessel disease with multiple strokes (multi infarct dementia) or by a single stroke (strategic stroke VaD). A common form is subcortical ischemic VaD caused by small-vessel occlusions with multiple lacunas and by hypoperfusive lesions resulting from stenosis of medullary arterioles. The evidence also suggests that a neuronal energy crisis, brought on by a relentless chronic brain hypoperfusion (CBH), is responsible for protein synthesis defects that later result in the classic AD neurodegenerative lesions. Subcortical ischemic vascular dementia (SIVD) has been proposed as a subtype of vascular cognitive impairment [56]. A rare disorder known as cerebral autosomal dominant arteriopathy with subcortical infarctions and leukoencephalopathy (CADASIL) is the most common hereditary subcortical vascular dementia and may lead to SIVD in the absence of Alzheimer's disease. It is caused by a defective *NOTCH3* gene, which encodes a transmembrane receptor. Smooth muscle cells in small arteries throughout the body degenerate and vessel walls become fibrotic. In the brain, this results in circulatory disturbances and lacunar infarcts, mainly in cerebral white matter and deep gray matter, resulting in VaD.

Diagnostic of Vascular Dementia

Vascular dementia is diagnosed based on history, symptoms and cognitive tests as well as ruling out other causes of dementia. History may include a history of stroke or hypertension. Diagnosis of vascular dementia requires two main criteria: 1) cognitive impairment and 2) vascular brain lesions that can be imaged. Criteria for the diagnosis of VaD was described by the National Institute of Neurological Disorder and Stroke—Association Internationale pour Recherche et l'Enseignment en Neuroscience (NINDS-AIREN) [8, 57]. These criteria recognized the identity of VaD as a separate etiological condition and led to a growing number of international epidemiological studies for early diagnosis and controlled clinical trials for the symptomatic treatment of VaD.

Cognitive Impairment

The most commonly used cognitive assessment tool is the mini-mental state examination (MMSE). The test is based on a brief 30-point questionnaire in a number of areas including arithmetic, language use and comprehension and basic motor skills [2, 58]. A score of <24 suggests dementia. Nevertheless, the test results may be affected by mental disorders, a physical disability (inability to hear or read instructions properly) or a motor deficit (that affects writing and drawing skills, for example). Therefore, the MMSE test is usually performed with other cognition screening tests such as the Mini-Cog assessment instrument and the memory impairment screen [2]. The Mini-Cog assessment test combines two simple cognitive tasks (three-item word memory and clock drawing) with an empirical algorithm for scoring and was developed for screening people of low education and language skills. This test appears to satisfy most standards for brief cognitive screens and is currently being used by doctors even when dementia is in a mild or subclinical stage [59]. The Memory Impairment Screen (MIS) test is a brief, 4-minute four-item delayed free- and cued-recall memory impairment test. The MIS uses controlled learning to ensure attention, induce specific semantic processing and optimize encoding specificity to improve detection of dementia. The MIS was reported to be valid screening for AD and other dementias [60].

Dementia Imaging Methods

i) Magnetic Resonance Imaging

Magnetic resonance imaging (MRI) commonly demonstrates three types of abnormalities in patients with vascular dementia disease [61]: i) Cortical infarctions provide evidence of cerebrovascular abnormality and are more common in VaD rather than other types of dementia [62]. ii) Areas of abnormally high signal on T2 weighted MRI are commonly seen in patients with dementia [63, 64]. iii) Accelerated atrophy compared with normal elderly individuals [65]. There are evidences that both the distribution and rate of atrophy differ depending on the cerebrovascular disease [66]. Recently, global MRI diffusion changes have been shown to be sensitive to clinically relevant microstructural alterations and may be useful markers of CAA-related tissue damage [67].

ii) Positron Emission Tomography

Positron emission tomography (PET) studies with $^{(18)}$F-2-fluoro-2-deoxy-D-glucose (FDG) have demonstrated that individuals with mild cognitive impairment (MCI), linked to vascular dementia, have reduced glucose transport across the BBB prior to neurodegeneration and brain atrophy [68]. The significantly reduced glucose uptake is specifically located in the right precuneus, posterior cingulate, right angular gyrus, and bilateral middle temporal cortices [68]. A longitudinal study using FDG-PET imaging with follow-up PET exams has suggested that reductions of glucose utilization by the hippocampus during normal aging can predict cognitive decline years in advance of the clinical diagnosis [69].

Animal Models for Vascular Dementia

Animal models are essential for understanding mechanisms of disease and to development of therapeutic intervention. Nevertheless, animal models rarely mimic human disease progression. In this review, we aim to summarize different animal models that present with some of the affects of cerebrovascular dysfunction and may be used to investigate VaD.

Several tests have been developed to study cognitive functions in animals. Different tests have been designed to examine different cognitive domains. Studies on cognitive functions, for example in rodents, have so far preferentially explored learning and reference memory using tests such as the Water Maze test and the Radial Arm Maze test [70, 71]. These tests include a training phase and are frequently associated with reward or punishment. The results of these studies nearly always demonstrate alteration in learning and reference or visuospatial memory. Other studies for cognitive tests explore functions associated with the frontal-subcortical circuits [72], as opposed to the hippocampus [73], such as the Y Maze Test [74] and the Object Recognition Test [75]. These tests focus on spatial and non-spatial working memory using no-reward, no-punishment procedures. Nevertheless, the relevance of cognition tests in animals to VaD aspects in humans remains an open question.

Animal Models of Cerebral Amyloid Angiopathy

There are two naturally occurring models of CAA that are currently being studied. Both non-human primates and canines develop vascular β-amyloidosis naturally as they grow old [40] and are thus good models for CAA. The main obstacles to studying these animals are their relatively long life spans. Different transgenic mice models were established that experience accelerated cerebrovascular pathology. Nevertheless, in most animal models, cerebrovascular pathology is usually part of a global cerebral pathology, which makes it difficult to isolate vascular impairment from the effectors of the observed pathology.

i) Non-human Primate Model of CAA

Nonhuman primates probably represent the best model because their vascular architecture and their gyrencephalic brain with extensive white matter more closely resembles that of humans [76]. CAA is a common finding in aged non-human primates, occurring in species ranging from lemurs (Microcebus murinus [77]) to chimpanzees (Pan troglodytes [78]). Intriguingly, whether amyloid is deposited mainly in the parenchyma of the brain or in the vascular wall is species-specific to a certain degree, presenting unique opportunities for probing the cellular and molecular correlates of CAA. For example, CAA abounds specifically in squirrel monkeys starting at age 15 years[40]. Unfortunately, in squirrel monkeys, CAA also correlates with cerebral plaque so the direct relevance of CAA to cognition is unclear.

ii) Canine Model for CAA

Multiple investigators have confirmed CAA in aged dogs [79, 80]. As in humans, the presence of CAA in dogs is positively correlated with the incidence of intracerebral hemorrhage [81]. Within the vessel wall, canine amyloid deposits are usually associated most closely with the intercellular spaces of the tunica media, similar to deposits in humans [82] and non-human primates [83], and supporting the potential role of vascular smooth muscle cells in the generation of CAA [84].

In arteries, smooth muscle cell atrophy is evident in advanced CAA in canine [40]. Amyloid angiopathy appears to begin in larger brain-supplying arteries, and progresses into smaller vessels [85]. Canine CAA particularly afflicts the leptomeningeal arteries [85] and the parenchymal vessels, including capillaries [81]. Cognitive damage correlates with the extent of Aβ deposits in the cerebral cortex and CAA [86]. However, in spite of the dog's relevance as a wild-type model for the study of AD, there is no clear evidence regarding the effect of cerebrovascular dysfunction on cognition in this model.

iii) Transgenic Mice Model for CAA

Since, naturally occurring animal models of CAA are not well suited for the laboratory, a variety of transgenic murine models have been developed. These models demonstrate cerebral amyloidosis similar to that of humans and are mostly based on the involvement of Alzheimer's Aβ in CAA as part of parenchymal amyloid, or primarily parenchymal amyloid [87]. Two molecules have been implicated in the cerebrovascular pathology of AD: (i) Aβ, which in soluble form directly alters cerebrovascular functions; and (ii) Transforming Growth Factor-β1(TGF-β1), which promotes the synthesis of extracellular matrix proteins and contributes to vascular remodeling following injury or lesion [88].

Transgenic mice that over-express mutated forms of the human amyloid precursor protein (APP; APP mice) exhibit mostly cerebral amyloid deposition of Aβ1-42 along with cerebrovascular deposition of Aβ1-40. The Aβ-induced oxidative stress impairs cerebrovascular dilatory responses to acetylcholine (ACh), calcitonin gene-related peptide (CGRP) and other endothelium-dependent dilators [24, 89], in addition to altering autoregulation [90] and functional hyperaemia [89]. Transgenic mice overexpressing the Swedish mutation of APP develop cognitive impairment at 6 months of age and amyloid plaques at 9–12 months [91]. However, the earliest abnormality observed in these mice is a profound alteration in the regulation of the cerebral circulation at 2–3 months of age. In mouse models of familial AD harboring the Dutch and Iowa vasculotropic APP mutations, the dense plaques may develop initially on blood vessels or as classical CAA because of inefficient Aβ clearance across the blood-brain barrier (BBB) and/or perivascular Virchow-Robin arterial spaces, respectively [5, 92]. The E693Q mutation in the amyloid beta precursor protein (APP) leads to cerebral CAA, with recurrent cerebral hemorrhagic strokes and dementia [93]. In contrast to AD, the brains of those affected by hereditary cerebral hemorrhage with amyloidosis–Dutch type (HCHWA-D) show few parenchymal amyloid plaques. Neuronal overexpression of human E693Q APP in mice (APPDutch mice) caused extensive CAA, smooth muscle cell degeneration, hemorrhages and neuroinflammation. This HCHWA-D mouse model is the first to develop robust CAA in the absence of parenchymal amyloid, highlighting the key role of neural Aβ production to vascular amyloid pathology and emphasizing the differing roles of Aβ 1-40 and Aβ 1-42 in vascular and parenchymal amyloid pathology. Dementia in humans with the HCHWA-D mutation is associated with CAA, independent of hemorrhagic strokes [94]. Nevertheless, cognition impairment in HCHWA-D has yet to be reported.

Transforming growth factor-β1 (TGF-β1) is a multifunctional cytokine that is a major regulator of the immune response; it has profound effects on vasculogenesis, angiogenesis and the maintenance of vessel wall integrity. Furthermore, cortical TGF-β1 messenger RNA (mRNA) levels correlate positively with the degree of cerebrovascular amyloidosis in AD cases, and TGF-β1 immunoreactivity in such cases is elevated along the cerebral blood vessels [95]. There are also positive correlations found between TGF-β1 polymorphisms and CAA [96, 97]. Co-expression of TGF-β1 with hAPP/Aβ in bigenic mice result in cerebrovascular and meningeal amyloid deposition at 2–3 months of age [98]. APP/TGF-β1 mice show a prominent perivascular astrocytosis and age-related deposition of amyloid around cerebral blood vessels [98]. These structural alterations translate into regional decreases in CBF [99], compatible with a reported decrease in glucose metabolism in several brain regions [100]. They also exhibit deficits in cerebrovascular dilatory responses to ACh and CGRP and, with aging, in contractile responses to endothelin-1 (ET-1; [24]), in addition to cerebral hypoperfusion [99] and hypometabolism [100]. Recent results from our lab [101] suggest a correlation between cognitive impairment and CAA levels in TGF-β1 mice.

Animal Models of Hypertension

Spontaneously hypertensive rats (SHR) and SHR-stroke-prone (SHRSP) are considered among the best animal models for assessing hypertension and stroke, respectively and might serve as a suitable animal model for vascular dementia in humans caused by cerebrovascular lesions. SHR were developed in Japan in 1963 by pairing Wistar rats (defined Wistar Kyoto rats, WKY) with high blood pressure. Successive breeding of SHR produced a substrain with higher blood pressure than SHR [102]. Males of this substrain, in which stroke occurs spontaneously in 100% of cases, were labeled SHRSP [102, 103].

Significantly decreased acetylcholine levels in both the cerebral cortex and the hippocampus in SHRSP compared to age-matched Wistar Kyoto (WKY) rats (normotensive animals of the same SHR strain) were observed. This decrease may be related to impairment of learning-memory function and abnormal behavior. In SHRSP, increases in blood viscosity, hematocrit and fibrinogen might produce the formation of thrombus and induce cerebral infarction.

In SHR, hyperactivity, hyper-reactivity, lower levels of anxiety, and a decrement in habituation capability as well as a severe deficit in attention and decreased learning ability in several behavioral tasks were described by various groups [104, 105]. However, it has been suggested that behavioral changes observed in SHR reflect a decreased ability to deal with threatening situations [106]. Therefore, these animal models cannot be considered to have been conclusively investigated in terms of learning and memory skills. Although SHR and SHRSP may represent animal models of VaD, they were characterized primarily from a morphological point of view [107], further work is necessary to characterize their behavioral and neuro-chemical attributes.

Animal Models of Diabetics

Hyperglycaemia is strongly related to brain and vascular disorders [108, 109]. Animal models of 'induced diabetes' suggest that diabetes has direct neurodegenerative affects. The majority of studies show results in the hippocampus – the area associated with learning and memory and the first structure to be affected by the neurodegeneration in Alzheimer's disease. Risk-enhancing mechanisms of vascular dysfunction include the affects of increased advanced glycation end products, oxidative stress, inflammation, macrovascular and microvascular injury. Both large and small blood vessels are susceptible to alterations from diabetes. Endothelial cell dysfunction associated with small vessel (known as microangiopathy) is a primary factor in the development and progression of diabetes-related disabilities. Studies of effects of streptozotocin-induced diabetes have shown that diabetes-induced perturbations to cerebral microvessels may disrupt homeostasis and contribute to long-term cognitive and functional deficits of the central nervous system [110]. Obese animals with insulin resistance mimicked the human form of the early stages of Type 2 Diabetes. These animals have also shown promise in modeling VaD through protein –lipid vascular deposition.

Animal Model for Ischemic Stroke

Subcortical vascular dementia is characterized by small vessel disease involving white matter changes and lacunar infarctions. White matter lesions are observed in rodent models of chronic cerebral hypoperfusion in which the bilateral carotid arteries are stenosed or ligated [111] as well as in stroke-prone spontaneously hypertensive rats which have small vessel pathology [112]. Such WM lesions seem contribute to frontal hypometabolism and executive dysfunction [113, 114]. Recently, a model of vascular dementia rats was established by inducing a short middle cerebral artery occlusion (MCAO) without concomitant sensorimotor deficits. These animals have shown significant cognitive deficiency in different cognition tasks [115].

Conclusions

The cerebrovascular system plays a critical role in maintaining normal brain function. The development of dementia in several neurodegenerative diseases has been shown to correlate well to impairment of the cerebrovascular system and in these cases has been defined as vascular dementia. Therefore, therapeutic approaches that aim to reduce vascular toxicity and re-establish normal function may prevent or reverse cognitive impairment. At the moment, once vascular dementia is clinically established, treatment aims primarily at slowing the progression of the disease by prevention of stroke recurrence. Although the efficacy of antithrombotic therapy in stroke prevention is well documented, the role of this approach in preventing cognitive decline is still to be proven. Animal models are essential for understanding the mechanisms of vascular disease and for developing therapeutic interventions. Nevertheless, whether animal models accurately mimic all different aspects of VaD progression in human remains an open question. Elucidating the molecular pathways linking the individual vascular risk factors and cerebrovascular changes leading to cognitive impairment is the crucial next step in the study of vascular cognitive impairment. Therefore, studies that aim to explicate the aspects of vascular disease that correlate to cognitive decline in animal models, are crucial for identifying future treatment strategies.

Acknowledgments

This work is supported by grants from the HFSP organization and Dana Foundation (to D.F.).

References

[1] Fratiglioni L, Paillard-Borg S, Winblad B. An active and socially integrated lifestyle in late life might protect against dementia. *Lancet Neurol.* 3(6): 343-53, 2004.
[2] Burns A, Iliffe S. Dementia. *Bmj* 338(b75, 2009.
[3] Ferri CP, Prince M, Brayne C, Brodaty H, Fratiglioni L, Ganguli M, Hall K, Hasegawa K, Hendrie H, Huang Y, Jorm A, Mathers C, Menezes PR, Rimmer E, Scazufca M. Global prevalence of dementia: a Delphi consensus study. *Lancet* 366(9503): 2112-7, 2005.
[4] Bell RD, Zlokovic BV. Neurovascular mechanisms and blood-brain barrier disorder in Alzheimer's disease. *Acta Neuropathol.* 2009.
[5] Zlokovic BV. The blood-brain barrier in health and chronic neurodegenerative disorders. *Neuron* 57(2): 178-201, 2008.
[6] Man S, Ubogu EE, Ransohoff RM. Inflammatory cell migration into the central nervous system: a few new twists on an old tale. *Brain Pathol.* 17(2): 243-50, 2007.
[7] Lok J, Gupta P, Guo S, Kim WJ, Whalen MJ, van Leyen K, Lo EH. Cell-cell signaling in the neurovascular unit. *Neurochem. Res.* 32(12): 2032-45, 2007.
[8] Roman GC. Vascular dementia: distinguishing characteristics, treatment, and prevention. *J. Am. Geriatr. Soc.* 51(5 Suppl Dementia): S296-304, 2003.

[9] Bigler ED, Kerr B, Victoroff J, Tate DF, Breitner JC. White matter lesions, quantitative magnetic resonance imaging, and dementia. *Alzheimer Dis. Assoc. Disord.* 16(3): 161-70, 2002.

[10] Alonso A, Jacobs DR, Jr., Menotti A, Nissinen A, Dontas A, Kafatos A, Kromhout D. Cardiovascular risk factors and dementia mortality: 40 years of follow-up in the Seven Countries Study. *J. Neurol. Sci.* 280(1-2): 79-83, 2009.

[11] Hayden KM, Zandi PP, Lyketsos CG, Khachaturian AS, Bastian LA, Charoonruk G, Tschanz JT, Norton MC, Pieper CF, Munger RG, Breitner JC, Welsh-Bohmer KA. Vascular risk factors for incident Alzheimer disease and vascular dementia: the Cache County study. *Alzheimer Dis. Assoc. Disord.* 20(2): 93-100, 2006.

[12] Anstey KJ, von Sanden C, Salim A, O'Kearney R. Smoking as a risk factor for dementia and cognitive decline: a meta-analysis of prospective studies. *Am. J. Epidemiol.* 166(4): 367-78, 2007.

[13] Luchsinger JA, Tang MX, Stern Y, Shea S, Mayeux R. Diabetes mellitus and risk of Alzheimer's disease and dementia with stroke in a multiethnic cohort. *Am. J. Epidemiol.* 154(7): 635-41, 2001.

[14] Anstey KJ, Lipnicki DM, Low LF. Cholesterol as a risk factor for dementia and cognitive decline: a systematic review of prospective studies with meta-analysis. *Am. J. Geriatr. Psychiatry* 16(5): 343-54, 2008.

[15] Guo X, Waern M, Sjogren K, Lissner L, Bengtsson C, Bjorkelund C, Ostling S, Gustafson D, Skoog I. Midlife respiratory function and Incidence of Alzheimer's disease: a 29-year longitudinal study in women. *Neurobiol. Aging* 28(3): 343-50, 2007.

[16] Newman AB, Fitzpatrick AL, Lopez O, Jackson S, Lyketsos C, Jagust W, Ives D, Dekosky ST, Kuller LH. Dementia and Alzheimer's disease incidence in relationship to cardiovascular disease in the Cardiovascular Health Study cohort. *J. Am. Geriatr. Soc.* 53(7): 1101-7, 2005.

[17] Petersen RC, Smith GE, Waring SC, Ivnik RJ, Tangalos EG, Kokmen E. Mild cognitive impairment: clinical characterization and outcome. *Arch Neurol.* 56(3): 303-8, 1999.

[18] Kayed R, Head E, Thompson JL, McIntire TM, Milton SC, Cotman CW, Glabe CG. Common structure of soluble amyloid oligomers implies common mechanism of pathogenesis. *Science* 300(5618): 486-9, 2003.

[19] Klein WL. Abeta toxicity in Alzheimer's disease: globular oligomers (ADDLs) as new vaccine and drug targets. *Neurochem. Int.* 41(5): 345-52, 2002.

[20] Li M, Chen L, Lee DH, Yu LC, Zhang Y. The role of intracellular amyloid beta in Alzheimer's disease. *Prog. Neurobiol.* 83(3): 131-9, 2007.

[21] Braak H, Braak E, Bohl J. Staging of Alzheimer-related cortical destruction. *Eur. Neurol.* 33(6): 403-8, 1993.

[22] Hamel E, Nicolakakis N, Aboulkassim T, Ongali B, Tong XK. Oxidative stress and cerebrovascular dysfunction in mouse models of Alzheimer's disease. *Exp. Physiol.* 93(1): 116-20, 2008.

[23] Iadecola C. Neurovascular regulation in the normal brain and in Alzheimer's disease. *Nat. Rev. Neurosci.* 5(5): 347-60, 2004.

[24] Tong XK, Nicolakakis N, Kocharyan A, Hamel E. Vascular remodeling versus amyloid beta-induced oxidative stress in the cerebrovascular dysfunctions associated with Alzheimer's disease. *J. Neurosci.* 25(48): 11165-74, 2005.

[25] Strittmatter WJ, Roses AD. Apolipoprotein E and Alzheimer disease. *Proc. Natl. Acad. Sci. USA* 92(11): 4725-7, 1995.
[26] Bronfman FC, Garrido J, Alvarez A, Morgan C, Inestrosa NC. Laminin inhibits amyloid-beta-peptide fibrillation. *Neurosci. Lett.* 218(3): 201-3, 1996.
[27] Selkoe DJ. Alzheimer's disease: genes, proteins, and therapy. *Physiol. Rev.* 81(2): 741-66, 2001.
[28] Hamel E. Cholinergic modulation of the cortical microvascular bed. *Prog. Brain Res.* 145(171-8, 2004.
[29] Tong XK, Hamel E. Regional cholinergic denervation of cortical microvessels and nitric oxide synthase-containing neurons in Alzheimer's disease. *Neuroscience* 92(1): 163-75, 1999.
[30] Kalaria RN. Arteriosclerosis, apolipoprotein E, and Alzheimer's disease. *Lancet* 349 (9059): 1174, 1997.
[31] Revesz T, Ghiso J, Lashley T, Plant G, Rostagno A, Frangione B, Holton JL. Cerebral amyloid angiopathies: a pathologic, biochemical, and genetic view. *J. Neuropathol. Exp. Neurol.* 62(9): 885-98, 2003.
[32] Greenberg SM. Cerebral amyloid angiopathy and vessel dysfunction. *Cerebrovasc. Dis.* 13 Suppl 2(42-7, 2002.
[33] van Duinen SG, Maat-Schieman ML, Bruijn JA, Haan J, Roos RA. Cortical tissue of patients with hereditary cerebral hemorrhage with amyloidosis (Dutch) contains various extracellular matrix deposits. *Lab. Invest* 73(2): 183-9, 1995.
[34] Abraham CR, Selkoe DJ, Potter H. Immunochemical identification of the serine protease inhibitor alpha 1-antichymotrypsin in the brain amyloid deposits of Alzheimer's disease. *Cell* 52(4): 487-501, 1988.
[35] Cruz-Sanchez FF, Marin C, Rossi ML, Cardozo A, Ferrer I, Tolosa E. Ubiquitin in cerebral amyloid angiopathy. *J. Neurol. Sci.* 112(1-2): 46-50, 1992.
[36] Kalaria RN, Grahovac I. Serum amyloid P immunoreactivity in hippocampal tangles, plaques and vessels: implications for leakage across the blood-brain barrier in Alzheimer's disease. *Brain Res.* 516(2): 349-53, 1990.
[37] Mesulam M, Carson K, Price B, Geula C. Cholinesterases in the amyloid angiopathy of Alzheimer's disease. *Ann. Neurol.* 31(5): 565-9, 1992.
[38] Levy E, Jaskolski M, Grubb A. The role of cystatin C in cerebral amyloid angiopathy and stroke: cell biology and animal models. *Brain Pathol.* 16(1): 60-70, 2006.
[39] Fan R, DeFilippis K, Van Nostrand WE. Induction of complement proteins in a mouse model for cerebral microvascular A beta deposition. *J. Neuroinflammation* 4(22, 2007.
[40] Walker LC. Animal models of cerebral beta-amyloid angiopathy. *Brain Res. Brain Res. Rev.* 25(1): 70-84, 1997.
[41] Attems J. Sporadic cerebral amyloid angiopathy: pathology, clinical implications, and possible pathomechanisms. *Acta Neuropathol.* 110(4): 345-59, 2005.
[42] Vonsattel JP, Myers RH, Hedley-Whyte ET, Ropper AH, Bird ED, Richardson EP, Jr. Cerebral amyloid angiopathy without and with cerebral hemorrhages: a comparative histological study. *Ann. Neurol.* 30(5): 637-49, 1991.
[43] Grabowski TJ, Cho HS, Vonsattel JP, Rebeck GW, Greenberg SM. Novel amyloid precursor protein mutation in an Iowa family with dementia and severe cerebral amyloid angiopathy. *Ann. Neurol.* 49(6): 697-705, 2001.

[44] Greenberg SM, Vonsattel JP, Stakes JW, Gruber M, Finklestein SP. The clinical spectrum of cerebral amyloid angiopathy: presentations without lobar hemorrhage. *Neurology* 43(10): 2073-9, 1993.
[45] Thomas T, Thomas G, McLendon C, Sutton T, Mullan M. beta-Amyloid-mediated vasoactivity and vascular endothelial damage. *Nature* 380(6570): 168-71, 1996.
[46] Greenberg SM, Gurol ME, Rosand J, Smith EE. Amyloid angiopathy-related vascular cognitive impairment. *Stroke* 35(11 Suppl 1): 2616-9, 2004.
[47] Jellinger KA. Alzheimer disease and cerebrovascular pathology: an update. *J. Neural Transm.* 109(5-6): 813-36, 2002.
[48] Keage HA, Carare RO, Friedland RP, Ince PG, Love S, Nicoll JA, Wharton SB, Weller RO, Brayne C. Population studies of sporadic cerebral amyloid angiopathy and dementia: a systematic review. *BMC Neurol.* 9(3, 2009.
[49] Attems J, Quass M, Jellinger KA, Lintner F. Topographical distribution of cerebral amyloid angiopathy and its effect on cognitive decline are influenced by Alzheimer disease pathology. *J. Neurol. Sci.* 257(1-2): 49-55, 2007.
[50] Pasquier F, Boulogne A, Leys D, Fontaine P. Diabetes mellitus and dementia. *Diabetes Metab.* 32(5 Pt 1): 403-14, 2006.
[51] Sorrentino G, Migliaccio R, Bonavita V. Treatment of vascular dementia: the route of prevention. *Eur. Neurol.* 60(5): 217-23, 2008.
[52] Mooradian AD. Central nervous system complications of diabetes mellitus--a perspective from the blood-brain barrier. *Brain Res. Brain Res. Rev.* 23(3): 210-8, 1997.
[53] Selkoe DJ. The origins of Alzheimer disease: a is for amyloid. *JAMA* 283(12): 1615-7, 2000.
[54] Nag S. Cerebral changes in chronic hypertension: combined permeability and immunohistochemical studies. *Acta Neuropathol.* 62(3): 178-84, 1984.
[55] Rigaud AS, Hanon O, Seux ML, Forette F. Hypertension and dementia. *Curr. Hypertens Rep.* 3(6): 454-7, 2001.
[56] Chui HC. Subcortical ischemic vascular dementia. *Neurol. Clin.* 25(3): 717-40, vi, 2007.
[57] Roman GC, Tatemichi TK, Erkinjuntti T, Cummings JL, Masdeu JC, Garcia JH, Amaducci L, Orgogozo JM, Brun A, Hofman A, et al. Vascular dementia: diagnostic criteria for research studies. Report of the NINDS-AIREN International Workshop. *Neurology* 43(2): 250-60, 1993.
[58] Kalmijn S, Foley D, White L, Burchfiel CM, Curb JD, Petrovitch H, Ross GW, Havlik RJ, Launer LJ. Metabolic cardiovascular syndrome and risk of dementia in Japanese-American elderly men. The Honolulu-Asia aging study. *Arterioscler. Thromb Vasc. Biol.* 20(10): 2255-60, 2000.
[59] Borson S, Scanlan JM, Chen P, Ganguli M. The Mini-Cog as a screen for dementia: validation in a population-based sample. *J. Am. Geriatr. Soc.* 51(10): 1451-4, 2003.
[60] Buschke H, Kuslansky G, Katz M, Stewart WF, Sliwinski MJ, Eckholdt HM, Lipton RB. Screening for dementia with the memory impairment screen. *Neurology* 52(2): 231-8, 1999.
[61] Varma AR, Adams W, Lloyd JJ, Carson KJ, Snowden JS, Testa HJ, Jackson A, Neary D. Diagnostic patterns of regional atrophy on MRI and regional cerebral blood flow change on SPECT in young onset patients with Alzheimer's disease, frontotemporal dementia and vascular dementia. *Acta Neurol. Scand.* 105(4): 261-9, 2002.

[62] Charletta D, Gorelick PB, Dollear TJ, Freels S, Harris Y. CT and MRI findings among African-Americans with Alzheimer's disease, vascular dementia, and stroke without dementia. *Neurology* 45(8): 1456-61, 1995.

[63] Barber R, Scheltens P, Gholkar A, Ballard C, McKeith I, Ince P, Perry R, O'Brien J. White matter lesions on magnetic resonance imaging in dementia with Lewy bodies, Alzheimer's disease, vascular dementia, and normal aging. *J. Neurol. Neurosurg. Psychiatry* 67(1): 66-72, 1999.

[64] Doddy RS, Massman PJ, Mawad M, Nance M. Cognitive consequences of subcortical magnetic resonance imaging changes in Alzheimer's disease: comparison to small vessel ischemic vascular dementia. *Neuropsychiatry Neuropsychol. Behav. Neurol.* 11(4): 191-9, 1998.

[65] Fox NC, Freeborough PA. Brain atrophy progression measured from registered serial MRI: validation and application to Alzheimer's disease. *J. Magn. Reson. Imaging* 7(6): 1069-75, 1997.

[66] O'Brien JT, Desmond P, Ames D, Schweitzer I, Chiu E, Tress B. Temporal lobe magnetic resonance imaging can differentiate Alzheimer's disease from normal ageing, depression, vascular dementia and other causes of cognitive impairment. *Psychol. Med.* 27(6): 1267-75, 1997.

[67] Viswanathan A, Patel P, Rahman R, Nandigam RN, Kinnecom C, Bracoud L, Rosand J, Chabriat H, Greenberg SM, Smith EE. Tissue microstructural changes are independently associated with cognitive impairment in cerebral amyloid angiopathy. *Stroke* 39(7): 1988-92, 2008.

[68] Hunt A, Schonknecht P, Henze M, Seidl U, Haberkorn U, Schroder J. Reduced cerebral glucose metabolism in patients at risk for Alzheimer's disease. *Psychiatry Res.* 155(2): 147-54, 2007.

[69] Mosconi L, De Santi S, Li J, Tsui WH, Li Y, Boppana M, Laska E, Rusinek H, de Leon MJ. Hippocampal hypometabolism predicts cognitive decline from normal aging. *Neurobiol. Aging* 29(5): 676-92, 2008.

[70] Sekhon LH, Morgan MK, Spence I, Weber NC. Chronic cerebral hypoperfusion: pathological and behavioral consequences. *Neurosurgery* 40(3): 548-56, 1997.

[71] Sarti C, Pantoni L, Bartolini L, Inzitari D. Cognitive impairment and chronic cerebral hypoperfusion: what can be learned from experimental models. *J. Neurol. Sci.* 203-204(263-6, 2002.

[72] Bartolini L, Risaliti R, Pepeu G. Effect of scopolamine and nootropic drugs on rewarded alternation in a T-maze. *Pharmacol. Biochem. Behav.* 43(4): 1161-4, 1992.

[73] Mumby DG. Perspectives on object-recognition memory following hippocampal damage: lessons from studies in rats. *Behav. Brain Res.* 127(1-2): 159-81, 2001.

[74] Sarti C, Pantoni L, Bartolini L, Inzitari D. Persistent impairment of gait performances and working memory after bilateral common carotid artery occlusion in the adult Wistar rat. *Behav. Brain Res.* 136(1): 13-20, 2002.

[75] Ennaceur A, Delacour J. A new one-trial test for neurobiological studies of memory in rats. 1: Behavioral data. *Behav. Brain Res.* 31(1): 47-59, 1988.

[76] Durukan A, Tatlisumak T. Acute ischemic stroke: overview of major experimental rodent models, pathophysiology, and therapy of focal cerebral ischemia. *Pharmacol. Biochem. Behav.* 87(1): 179-97, 2007.

[77] Bons N, Mestre N, Ritchie K, Petter A, Podlisny M, Selkoe D. Identification of amyloid beta protein in the brain of the small, short-lived lemurian primate Microcebus murinus. *Neurobiol. Aging* 15(2): 215-20, 1994.

[78] Gearing M, Rebeck GW, Hyman BT, Tigges J, Mirra SS. Neuropathology and apolipoprotein E profile of aged chimpanzees: implications for Alzheimer disease. *Proc. Natl. Acad. Sci. USA* 91(20): 9382-6, 1994.

[79] Cummings BJ, Su JH, Cotman CW, White R, Russell MJ. Beta-amyloid accumulation in aged canine brain: a model of early plaque formation in Alzheimer's disease. *Neurobiol. Aging* 14(6): 547-60, 1993.

[80] Giaccone G, Verga L, Finazzi M, Pollo B, Tagliavini F, Frangione B, Bugiani O. Cerebral preamyloid deposits and congophilic angiopathy in aged dogs. *Neurosci. Lett.* 114(2): 178-83, 1990.

[81] Uchida K, Nakayama H, Goto N. Pathological studies on cerebral amyloid angiopathy, senile plaques and amyloid deposition in visceral organs in aged dogs. *J. Vet. Med. Sci.* 53(6): 1037-42, 1991.

[82] Yamaguchi H, Yamazaki T, Lemere CA, Frosch MP, Selkoe DJ. Beta amyloid is focally deposited within the outer basement membrane in the amyloid angiopathy of Alzheimer's disease. An immunoelectron microscopic study. *Am. J. Pathol.* 141(1): 249-59, 1992.

[83] Uno H, Alsum PB, Dong S, Richardson R, Zimbric ML, Thieme CS, Houser WD. Cerebral amyloid angiopathy and plaques, and visceral amyloidosis in aged macaques. *Neurobiol. Aging* 17(2): 275-81, 1996.

[84] Kawai M, Kalaria RN, Cras P, Siedlak SL, Velasco ME, Shelton ER, Chan HW, Greenberg BD, Perry G. Degeneration of vascular muscle cells in cerebral amyloid angiopathy of Alzheimer disease. *Brain Res.* 623(1): 142-6, 1993.

[85] Wegiel J, Wisniewski HM, Dziewiatkowski J, Tarnawski M, Nowakowski J, Dziewiatkowska A, Soltysiak Z. The origin of amyloid in cerebral vessels of aged dogs. *Brain Res.* 705(1-2): 225-34, 1995.

[86] Head E, McCleary R, Hahn FF, Milgram NW, Cotman CW. Region-specific age at onset of beta-amyloid in dogs. *Neurobiol. Aging* 21(1): 89-96, 2000.

[87] Herzig MC, Van Nostrand WE, Jucker M. Mechanism of cerebral beta-amyloid angiopathy: murine and cellular models. *Brain Pathol.* 16(1): 40-54, 2006.

[88] Chronic overproduction of transforming growth factor-beta1 by astrocytes promotes Alzheimer's disease-like microvascular degeneration in transgenic mice. 156(1): 139-50, 2000.

[89] Park L, Anrather J, Forster C, Kazama K, Carlson GA, Iadecola C. Abeta-induced vascular oxidative stress and attenuation of functional hyperemia in mouse somatosensory cortex. *J. Cereb. Blood Flow Metab.* 24(3): 334-42, 2004.

[90] Niwa K, Kazama K, Younkin L, Younkin SG, Carlson GA, Iadecola C. Cerebrovascular autoregulation is profoundly impaired in mice overexpressing amyloid precursor protein. *Am. J. Physiol. Heart Circ. Physiol.* 283(1): H315-23, 2002.

[91] Kawarabayashi T, Younkin LH, Saido TC, Shoji M, Ashe KH, Younkin SG. Age-dependent changes in brain, CSF, and plasma amyloid (beta) protein in the Tg2576 transgenic mouse model of Alzheimer's disease. *J. Neurosci.* 21(2): 372-81, 2001.

[92] Bell RD, Deane R, Chow N, Long X, Sagare A, Singh I, Streb JW, Guo H, Rubio A, Van Nostrand W, Miano JM, Zlokovic BV. SRF and myocardin regulate LRP-mediated amyloid-beta clearance in brain vascular cells. *Nat. Cell Biol.* 11(2): 143-53, 2009.

[93] Herzig MC, Winkler DT, Burgermeister P, Pfeifer M, Kohler E, Schmidt SD, Danner S, Abramowski D, Sturchler-Pierrat C, Burki K, van Duinen SG, Maat-Schieman ML, Staufenbiel M, Mathews PM, Jucker M. Abeta is targeted to the vasculature in a mouse model of hereditary cerebral hemorrhage with amyloidosis. *Nat. Neurosci.* 7(9): 954-60, 2004.

[94] Natte R, Maat-Schieman ML, Haan J, Bornebroek M, Roos RA, van Duinen SG. Dementia in hereditary cerebral hemorrhage with amyloidosis-Dutch type is associated with cerebral amyloid angiopathy but is independent of plaques and neurofibrillary tangles. *Ann. Neurol.* 50(6): 765-72, 2001.

[95] Wyss-Coray T, Lin C, Sanan DA, Mucke L, Masliah E. Chronic overproduction of transforming growth factor-beta1 by astrocytes promotes Alzheimer's disease-like microvascular degeneration in transgenic mice. *Am. J. Pathol.* 156(1): 139-50, 2000.

[96] Peila R, Yucesoy B, White LR, Johnson V, Kashon ML, Wu K, Petrovitch H, Luster M, Launer LJ. A TGF-beta1 polymorphism association with dementia and neuropathologies: the HAAS. *Neurobiol. Aging* 28(9): 1367-73, 2007.

[97] Hamaguchi T, Okino S, Sodeyama N, Itoh Y, Takahashi A, Otomo E, Matsushita M, Mizusawa H, Yamada M. Association of a polymorphism of the transforming growth factor-beta1 gene with cerebral amyloid angiopathy. *J. Neurol. Neurosurg. Psychiatry* 76 (5): 696-9, 2005.

[98] Wyss-Coray T, Lin C, Yan F, Yu GQ, Rohde M, McConlogue L, Masliah E, Mucke L. TGF-beta1 promotes microglial amyloid-beta clearance and reduces plaque burden in transgenic mice. *Nat. Med.* 7(5): 612-8, 2001.

[99] Gaertner RF, Wyss-Coray T, Von Euw D, Lesne S, Vivien D, Lacombe P. Reduced brain tissue perfusion in TGF-beta 1 transgenic mice showing Alzheimer's disease-like cerebrovascular abnormalities. *Neurobiol. Dis.* 19(1-2): 38-46, 2005.

[100] Galea E, Feinstein DL, Lacombe P. Pioglitazone does not increase cerebral glucose utilisation in a murine model of Alzheimer's disease and decreases it in wild-type mice. *Diabetologia* 49(9): 2153-61, 2006.

[101] Lifshitz V, Benromano T, Blumenfeld-Katzir T, Yaniv Assaf Y, Xia W, Weiner H, Frenkel D. Novel therapeutic approach in cerebral amyloid angiopathy. Blood-Brain Barrier Physiology meeting Cold Spring Harbor Laboratory, 2008.

[102] Tayebati SK. Animal models of cognitive dysfunction. *Mech. Ageing Dev.* 127(2): 100-8, 2006.

[103] Kimura S, Saito H, Minami M, Togashi H, Nakamura N, Nemoto M, Parvez HS. Pathogenesis of vascular dementia in stroke-prone spontaneously hypertensive rats. *Toxicology* 153(1-3): 167-78, 2000.

[104] Togashi H, Kimura S, Matsumoto M, Yoshioka M, Minami M, Saito H. Cholinergic changes in the hippocampus of stroke-prone spontaneously hypertensive rats. *Stroke* 27(3): 520-5; discussion 525-6, 1996.

[105] Meneses A, Castillo C, Ibarra M, Hong E. Effects of aging and hypertension on learning, memory, and activity in rats. *Physiol. Behav.* 60(2): 341-5, 1996.

[106] Sagvolden T, Hendley ED, Knardahl S. Behavior of hypertensive and hyperactive rat strains: hyperactivity is not unitarily determined. *Physiol. Behav.* 52(1): 49-57, 1992.

[107] Tomassoni D, Avola R, Di Tullio MA, Sabbatini M, Vitaioli L, Amenta F. Increased expression of glial fibrillary acidic protein in the brain of spontaneously hypertensive rats. *Clin. Exp. Hypertens* 26(4): 335-50, 2004.

[108] Rosenbloom AL. Hyperglycemic crises and their complications in children. *J. Pediatr. Endocrinol. Metab.* 20(1): 5-18, 2007.

[109] Alvarez EO, Beauquis J, Revsin Y, Banzan AM, Roig P, De Nicola AF, Saravia F. Cognitive dysfunction and hippocampal changes in experimental type 1 diabetes. *Behav. Brain Res.* 198(1): 224-30, 2009.

[110] Huber JD, VanGilder RL, Houser KA. Streptozotocin-induced diabetes progressively increases blood-brain barrier permeability in specific brain regions in rats. *Am. J. Physiol. Heart Circ. Physiol.* 291(6): H2660-8, 2006.

[111] Kurumatani T, Kudo T, Ikura Y, Takeda M. White matter changes in the gerbil brain under chronic cerebral hypoperfusion. *Stroke* 29(5): 1058-62, 1998.

[112] Lin JX, Tomimoto H, Akiguchi I, Wakita H, Shibasaki H, Horie R. White matter lesions and alteration of vascular cell composition in the brain of spontaneously hypertensive rats. *Neuroreport* 12(9): 1835-9, 2001.

[113] Shibata M, Yamasaki N, Miyakawa T, Kalaria RN, Fujita Y, Ohtani R, Ihara M, Takahashi R, Tomimoto H. Selective impairment of working memory in a mouse model of chronic cerebral hypoperfusion. *Stroke* 38(10): 2826-32, 2007.

[114] de Groot JC, de Leeuw FE, Oudkerk M, Hofman A, Jolles J, Breteler MM. Cerebral white matter lesions and subjective cognitive dysfunction: the Rotterdam Scan Study. *Neurology* 56(11): 1539-45, 2001.

[115] Hattori K, Lee H, Hurn PD, Crain BJ, Traystman RJ, DeVries AC. Cognitive deficits after focal cerebral ischemia in mice. *Stroke* 31(8): 1939-44, 2000.

In: Encyclopedia of Neuroscience Research
Editors: Eileen J. Sampson and Donald R. Glevins

ISBN 978-1-61324-861-4
© 2012 Nova Science Publishers, Inc.

Chapter XXVII

The Critical Role of Cognitive Function in the Effective Self-administration of Inhaler Therapy

S. C. Allen*
The Royal Bournemouth Hospital, Bournemouth,
Dorset, United Kingdom

Abstract

The morbidity and mortality from asthma remain high in patients over the age of 75 years in developed countries, in contrast to the improving mortality from that condition in children and younger adults. One of the reasons for this might be the relatively poor performance with inhaler devices confirmed by surveys of technique in elderly patients. Also, it has been shown that the most prominent determinants of adequate inhaler technique in old age are global cognitive function, frontal executive function and ideomotor praxis. This review explores the pattern of observed and measured errors in the use of various inhalers in old age. The cognitive barriers to a successful grasp of inhaler technique are explained. The review then proceeds to describe a systematic research sequence by the author and others to identify factors that determine the likelihood of an aged person being able reliably to self-administer medication by the inhaled route, also drawing on parallel evidence from studies of spirometry technique. Relatively simple tests of cognition, praxis and executive function are described which have been shown to have predictive value in assessing frail elderly subjects for potential inhaler therapy or spirometry. The evidence presented can enable clinicians, particularly geriatricians, internists and pulmonologists to approach inhaled therapy for asthma and chronic obstructive pulmonary disease in a logical and effective way for their frail elderly patients. Further, it will be argued that the relation between congitive function and inhaler technique can be seen as a clinical metaphor for other self-administered tests or treatments, such as insulin therapy. There is scope for research to determine the predictive role of cognitive function tests in a wide range of self-treatment settings, particularly for patients with relatively mild unrecognized cognitive impairment.

---

* e-mail: Stephen.allen@rbch.nhs.uk.

Keywords: cognitive function, executive function, praxis, inhaler technique, asthma, COPD, elderly

Introduction

Patients with severe cognitive impairment, for example those with advanced Alzheimer's disease, are clearly not capable of coping with complex aspects of self-care, so arrangements are made to compensate for this disability. Overtly demented patients, if they need for example inhaled drug treatment, have alternative therapies such as nebulisers and direct supervision by attendants. At the other end of the spectrum, elderly patients with intact cognition can almost always manage their own medications, including sequenced devices such as inhalers. Probably the most vulnerable group are those with subtle non-apparent cognitive impairment who are often prescribed devices that they are unable to learn to use effectively. It is that group that will be the main object of this review. Since much of the work in this domain has been conducted on patients' use of inhaler devices and spirometry, the review will largely set out the problem in the context of inhaled respiratory drugs. To set the scene; many elderly patients require maintenance respiratory therapy through delivery of drugs to the airways by means of inhaler devices to achieve optimal management of their obstructive airways disease. This applies to asthmatic patients of all ages, and also confers benefit in an important proportion of people with chronic obstructive pulmonary disease (COPD). A large proportion of elderly subjects with these conditions have a degree of cognitive impairment, including problems with executive function and praxis. However, to achieve an adequate therapeutic effect, there must be sufficient deposition of drugs in the medium and small airways [1]. If the delivery device is patient-activated, this requires a competent inhaler technique, irrespective of the design and relative complexity of the specific inhaler. Even an optimal inhaler technique results in the deposition of only between 15 and 30% of the inhaled dose in the medium and small airways, depending on the design of the inhaler [1,2]. More importantly, that proportion is reduced significantly by minor inadequacies of technique and drastically reduced to sub-therapeutic proportions for patients with an overtly inadequate technique [1-7]. Consequently, it can be seen that issues of inhaler technique over ride all other considerations when trying to achieve satisfactory maintenance inhaler therapy in patients with asthma and COPD. In this review it will be explained how cognitive function is the main determining factor for inhaler technique, and it can be argued that the relationship between cognitive function and the effective use of inhalers can be seen as a metaphor for a wide range of self-applied treatments in frail old age. Indeed, in the case of respiratory therapy, while a substantial amount of research has been carried out to compare the relative efficacies of different types of bronchodilator and inhaled corticosteroid medications, the real issue for many patients is the suitability, patient preference and usability of the inhaler itself [8-11], and this trend is exaggerated in the presence of cognitive impairment. This review explores the use of inhaler devices in frail elderly patients. Both asthma and COPD have a high prevalence in old age so it is commonplace for family physicians, specialists in respiratory medicine, geriatricians, and specialists in internal medicine to face the dilemma of how best to assess an elderly patient for inhaled maintenance therapy and choose an appropriate means of delivery. This clinical challenge will be used to illustrate the need to account for a patient's cognitive state, including subtle degrees of impairment, when deciding how best to apply patient-managed treatments in old age.

The Scale of the Problem

In most countries the prevalence of asthma and COPD are not known accurately, though both are clearly common illnesses. In the United Kingdom the prevalence of asthma has been demonstrated at between 3 and 7% of the population over the age of 65, depending on geographical region and the definitions used [12-14]. Similar limitations apply to prevalence studies carried out in the United States of America, though figures in the region of 3% of elderly adults are indicated [15]. It is not known whether asthma is becoming commoner in older adults, as has been shown to be the case in children in some developed northern countries, though there is evidence of underdiagnosis of the condition in old age [14,16,17]. The figures for COPD are also unreliable. As mean life expectancy rises it is to be expected that the number of people surviving with obstructive airways disease who also have impaired cognition will grow in proportion to the prevalence of dementia. It has been demonstrated, using overall mortality data that the death rate from asthma in the United Kingdom has fallen progressively over the last three decades in children, young adults and those of middle age [18]. However, there has been no similar fall in the mortality from asthma in old age [18]. Reasons for this will clearly be complex, and are likely to be related to co-morbidity patterns and improvements in the diagnosis of asthma in old age, which will proportionally inflate the mortality statistics with time. Factors relating to the reduced ability of older people subjectively to detect deterioration in their airways resistance might also contribute to delays in seeking appropriate medical help. In addition to this, an important consideration is the fact that a proportion of elderly patients develop physical and cognitive problems that prevent the acquisition and retention of a useful inhaler technique, and that proportion rises with increasing age [19].

Errors in Inhaler Technique Observed in Elderly Patients

Several surveys, conducted in community and hospital settings, have shown that in old age a significant proportion of patients, ostensibly established on inhaler therapy, have an inadequate technique [20-22]. Interest in this field was stimulated by a study carried out by the author in 1986 that showed that 40 percent of community dwelling patients with a mean age of 80 years had a completely inadequate technique with a metered dose inhaler [20]. A number of subsequent surveys demonstrated a similar finding, and the use of less technically demanding devices has been shown to reduce this percentage only moderately, unless a substantial programme of screening and training was undertaken [23-27]. In most studies, the commonest errors in technique have been related to poor coordination between actuation of devices and inhalation; sometimes referred to as hand-lung in-coordination [20,28,29]. These include, for metered dose inhalers: failure to actuate the device in the early phase of inhalation, actuation without inhalation and actuation at the end of inhalation or during exhalation [20,]. In extreme cases, patients were completely unable to use a device, for example by attempting to inhale the dose through the nose, or were non-compliant [20,30]. Lesser errors included failure to breath-hold at the end of inhalation and slightly late inhalation or actuation [20,31]. For dry powder devices the main errors relate to problems with capsule loading and priming, inadequate inspiratory effort with some models, and uncertainty about remaining dose content. The majority of such errors inevitably lead to a reduction in therapeutic efficacy. Certain other commonplace errors are related to the specific technology of devices, for example, failure to turn the scraper ring on the Turbohaler^{TM}.

Non-cognitive Barriers to an Effective Inhaler Technique

The most important factor in this domain centres around manual dexterity, and to a lesser extent vision [8,10,19,25,28]. There is ample evidence that patients with weak or painful hands are unable to use inhalers that require actuation by compression of a valve, such as a pressurized metered dose inhaler [19]. Patients with chronic inflammatory arthropathies, such as rheumatoid arthritis, are a typical example. Adapter aids, based on the fulcrum principle, are available for such devices and they can be of help, but for patients with conditions such as rheumatoid disease, pain and weakness often remain insurmountable barriers. The same also applies if patients have painful elbow or shoulder joints and are therefore unable easily to raise an inhaler to their mouth. Finger weakness and poor dexterity, often due to neurological conditions, is an important reason for failure to use correctly inhaler devices with complex loading sequences, as is the case with some dry powder inhalers. Of course, there are some aspects of dexterity that are determined by cognitive processes such as ideo-motor and ideational praxis, so there is a clear overlap between physical and cognitive disabilities in many individuals. Reduced vision can make it impossible for some patients to manipulate complex devices, particularly the loading sequences, and also makes it impossible for some patients to read dose meters or instructions. Visual problems are also a barrier to adequate interpretation of colour coded inhaler canisters. Impaired hearing can also cause some problems, particularly during the training phase of inhaler usage, though this barrier is relatively easy to surmount, providing patients have good vision and adequate cognitive function.

The Importance of Cognitive and Executive Function

This part of the review will examine a sequence of pieces of research that has explored the relationship between inhaler technique and cognitive function in old age. It will be shown that these factors are overwhelmingly the most important in determining whether or not an elderly person can acquire and retain adequate inhaler technique. Research by other workers will be referred to throughout, and for clarity, the work will be presented in approximate chronological order.

Overall Cognitive Function and the Role of Screening Tests

In the early 1980s there was a strong clinical impression that many elderly patients were not using their inhaler devices correctly. A systematic scrutiny of the literature at that time showed that there were no properly collected data to demonstrate the scale of the problem or to look into factors associated with a poor technique in older adults. Consequently, we embarked upon a preliminary community-based survey in Manchester, UK, which demonstrated that only 60% of the patients surveyed had a technique sufficiently competent to deliver an adequate dose from a metered dose inhaler [20]. Indeed, only 10% of those surveyed had an ideal technique and 40% had a completely incompetent technique (certain to provide no delivery of the drug to the target airways), despite being on prescribed inhaled therapy for a mean of more than four years. Further exploration of associated characteristics revealed that an incompetent technique within this age group was not related to age, the underlying respiratory diagnosis, the duration of inhaled therapy or the place of initial inhaler training. However, a very strong association was demonstrated between an incompetent technique and the abbreviated mental test score (AMTS) of <7/10. In that study no patients with a score of <7/10 had a competent technique. In some ways this small study was a turning point in our understanding of the determinants of inhaler technique in old age. A clear-cut association with cognitive function had been demonstrated. Several larger studies followed in the subsequent few years that showed a similar proportion of community dwelling patients with an incompetent inhaler technique in the elderly age group, and also confirmed the strong association with an abnormal score on cognitive screening tests. This was shown to be the case, not only for the abbreviated mental test score, but also for the Mini Mental State Examination (MMSE), in which case a score of <24/30 was strongly associated with an inadequate technique with a metered dose inhaler. This finding was reinforced by a study of cognitively intact elderly subjects that showed a high (around 80 percent) proportion with an adequate technique [22]. This body of evidence led to recommendations that cognitive function should be taken into account when assessing elderly patients for inhaled respiratory therapy and was embedded in national guidelines for asthma, nebuliser therapy and COPD [32-34].

Despite the strong association described above, a number of unanswered questions remained. For example, in our original study published in 1986, we observed that while patients with a low abbreviated mental test score almost invariably could not use a metered dose inhaler, there were other patients with a normal abbreviated mental test score who were also unable to use an inhaler correctly. This led to the question as to whether other higher functions, if impaired, might also be a barrier to a good inhaler technique. Furthermore, the late 1980s saw the appearance of a wide range of alternative inhaler devices, many of which were clearly easier to use than standard metered dose inhalers in that they required fewer operational steps and less coordination. We therefore felt it was important to establish whether cognitive screening tests could be as useful for the new inhaler devices as they had proved to be for metered dose inhalers. We conducted a study of the ability of patients (50 patients age 70 years or more), 36 female, mean age 81 years with normal cognition (abbreviated mental test score 8-10/10), borderline cognition (abbreviated mental test score 7/10) and three groups of patients with impairment, each with abbreviated mental test scores of 6/10, 5/10 and 4/10 respectively. These patients were inhaler-naïve and their ability to acquire and retain a competent technique with three different inhaler devices was studied. We included a relatively complex device (standard pressurized metered dose inhaler) that requires a 5-stage closely coordinated technique, a 4-stage device (metered dose inhaler with large-volume spacer) and a 3-stage device (breath-actuated inhaler). Patients in each of the cognitive groups received training and testing on all of the inhalers in sequence. In essence, we showed that cognitive scores remained highly predictive of the likelihood of a patient acquiring and retaining an adequate inhaler technique. Furthermore, a threshold analysis enabled us to demonstrate that some patients with cognitive impairment could learn to use the simpler devices, and again we demonstrated that an abbreviated mental test score of <7/10 predicted the likelihood that some patients would fail to learn the technique. In the case of the most complex devices, all failed [35]. Around the time that the study was published, other researchers were conducting comparative technique-acquisition studies with various inhaler devices, often taking into account cognitive screening scores in older patients. Broadly speaking, those studies confirmed our original finding that patients with impaired cognition on screening tests are not able to learn to use complex inhaler devices and only a small proportion can manage the relatively simple devices [25,29,30].

The Importance of Praxis and Ideo-motor Function

Having established the importance of global cognitive function in this context, we turned our attention to the exploration of other cognitive states with demonstrable functional deficits, to try to explain why some patients with apparently normal cognition on screening tests are still unable to master the use of an inhaler. Using an inhaler, particularly a relatively complex device requiring close hand-lung coordination, would be expected to be impaired by dyspraxia. Patients with overt dyspraxic conditions due, for example, to hemispheric stroke are often not able to perform complex sequenced movements. However, it is known that a number of conditions common in old age can result in a more global form of dyspraxia that may not be clinically obvious or easily demonstrated by superficial examination. Such conditions include Alzheimer's Disease, fronto-temporal dementias and multi-infarct brain disease. To explore this, we performed a study of patients above the age of 75 with a normal abbreviated mental test score (8-10/10) [37]. Patients were inhaler-naïve and received standardized training in the use of the metered dose inhaler. Retained technique was assessed against a scoring scale of 0 – 10 where 10 represented the perfect technique and 0 represented the worst possible technique. The scale also had a threshold embedded in it at the 5-6 level, such that a score of 5 or less indicated an incompetent technique and a score of 6 or more indicated a competent technique. Patients also had the Mini Mental State Examination [38] performed as well as the Ideational Dyspraxia Test [39], Ideo-motor Dyspraxia Test (IMDT) [40], Geriatric Depression Scale [41] and Barthel ADL (Activities of Daily Living) Index [42]. We found that the inhaler score correlated positively and significantly with the Ideo-motor Dyspraxia Test ($r = 0.45$, $P=0.039$) and the Mini Mental State Examination ($r = 0.48$, $P=0.032$). There was no relationship with the Barthel ADL Index, Geriatric Depression Scale or Ideational Dyspraxia Test. This established a broad relationship between inhaler competence, global cognition and praxis. More importantly, we demonstrated a threshold effect in that no subject with a Mini Mental State Examination score of <23/30 had an adequate inhaler score. Furthermore, about 90 percent of patients with an adequate inhaler score had an Ideo-motor Dyspraxia Test score of 14/20 or more (14 is usually accepted as the threshold of normal) [43](P<0.01). This study gave us two further important insights. Firstly, some patients failed to learn to use inhalers despite a normal abbreviated mental test score simply because that test had not detected their globally reduced cognition. The more sophisticated Mini Mental State Examination is therefore probably more accurate in that context. Secondly, very clear correlative and threshold relationships have been demonstrated between the ability to acquire inhaler technique and a normal Ideo-motor Dyspraxia Test score. Of course, it is necessary to acknowledge the tautology embedded in this observation in that acquiring an inhaler technique is in itself a test of ideo-motor praxis. Nevertheless, this was an important step in our understanding of the factors that determine whether or not an elderly subject can learn to use inhalers.

Executive Function

Learning complex purposeful movement sequences is dependent on executive function for which intact frontal lobe processing is required [44]. This is subtly different from the function of praxis, described above, which is largely a parietal lobe or fronto-parietal function. Therefore, it might be expected that impairment of executive function would be an even more powerful predictor of poor inhaler technique acquisition than global cognitive function or praxis. To determine this, we conducted a study in which 30 inhaler-naïve elderly subjects, 21 female, with a mean age of 85, received standardized training in metered dose inhaler and Turbohaler^{TM} techniques, the retention of which was scored on an analogue scale (for metered dose inhalers) or as a competence threshold for the Turbohaler^{TM}. The patients also had their Mini Mental State Examination performed as well as an established test of executive function, suitable for elderly subjects, known as EXIT25 [45]. All the subjects had normal abbreviated mental test screening scores. We confirmed a significant positive correlation between the metered dose inhaler analogue score and the Mini Mental State Examination ($r = 0.54$, $P < 0.002$) and demonstrated a highly significant negative correlation with EXIT25 (in which higher scores indicate greater impairment)($r = -0.702$, $P < 0.0001$). A similar threshold to that in the above study was found for the Mini Mental State Examination, in this case <24/30, for predicting inability to acquire a competent Turbohaler^{TM} technique. Furthermore, an EXIT25 score of >14/25 was highly predictive of being unable to learn to use, and retain, a Turbohaler^{TM}. Similar thresholds applied to the standard metered dose inhaler [46].

Thus, in this sequence of studies we established and confirmed the importance of cognitive and executive function in inhaler self-administration, and demonstrated the practical utility of quick and simple screening tests as predictors of successful acquisition and retention of adequate inhaler technique.

Another Example – Cognitive Function as a Determinant of Acceptable Spirometry Technique

The importance of these higher brain functions in establishing inhaler technique is reinforced by similar findings in research conducted to find the factors that determine an adequate forced spirometry technique in older subjects. Such research has confirmed that global cognitive function is also the most important factor in that context, the only other significant factor being an individual's level of educational attainment [47-49]. As with inhaler technique, age in itself is not a bar to the acquisition of the technique. One important factor, highlighted by the work on spirometry technique and inhaler use, is the importance of a good training programme, sympathetic trainers and technicians and reinforcement. In inhaler research, several studies have demonstrated the substantially higher success rates in inhaler training programmes conducted by trained and dedicated staff [24,27,30]. This applies at all ages and there is some evidence that it is effective in elderly subjects with normal cognitive scores or very slight cognitive impairment. However, further research is required to confirm this for a range of inhalers, and to establish the forms of training most helpful in older subjects. It is also possible that those with overt impairment of cognition, praxis or executive function might respond to enhanced training, though the evidence base in that domain is very thin and deserves systematic study.

The most recent phase in the research sequence described above has been a study to determine whether even simpler and quicker tests can be used to identify older patients who are unlikely to be able to self-administer an inhaler. The preliminary work has shown that ability to copy a diagram consisting of two intersecting pentagons to the standard required by the mini-mental state examination [38] has a crude predictive power comparable with the whole mini-mental test [50, 51]. This is perhaps not surprising given that the act of complying with an instruction to copy two overlapping pentagons is in itself an example of integrated cognitive, executive and ideo-motor function. Another candidate test is CLOX [52], a clock face drawing test validated for assessing executive function in elderly people. In a recent study comparing CLOX 1 and 2 with the MMSE it was found that the MMSE had better overall predictive power, with an MMSE below 24/30 predicting inability to perform spirometry, or learn inhaler technique, with a higher specificity than CLOX [53, 54]

Implications for the Design of Devices for the Self-administration of Drugs

There is an obvious need to take into account cognitive, executive and ideo-motor factors when choosing a delivery device for a particular inhaled therapy, otherwise therapeutic failure can occur despite using an inherently efficacious drug. Perhaps the best example of this applies to inhaled post-exposure influenza prophylaxis. The patients most likely to benefit from such treatment are those with chronic cardio-pulmonary conditions, many of whom are frail and elderly. If the delivery device is technically complex, and there is a need to begin treatment immediately, there is little time for inhaler technique training. In a study conducted in the United Kingdom [36], it was shown that approximately 60% of the intended target population for post-exposure influenza prophylaxis were unable to master a sufficiently good technique to use the treatment effectively. On the other hand, younger cognitively intact patients could be expected to acquire the technique without difficulty, though they were the least likely to benefit from the treatment. Parallel considerations need to be taken into account for other devices, such as insulin pens, topical applicators, multi-dose drug regimens etc. Further, with the development of a wide range of nano-engineered drug preparations for systemic drug delivery through the pulmonary route, it can be expected that patients with poor inhalation technique due to cognitive impairment will not be able to benefit reliably from the technology [55].

Summary

Research has shown that relatively quick and simple cognitive screening tests can be used to determine the likelihood of an elderly patient being able to learn to use a self-administered inhaler. The evidence, as it stands, supports the following points:

- patients with an Abbreviated Mental Test Score of < 7/10 are almost never able to learn to use the more complex inhalers such as pressurized metered dose inhalers, and can not usually master even the simplest of devices
- this is also the case for a Mini-Mental State Examination score of < 24/30, and/or an EXIT25 score of > 14/25
- patients who are overtly dyspraxic or who have an abnormal score on screening for sub-clinical dyspraxia are also usually not able to learn to use an inhaler
- weak and/or painful hands and poor vision are a substantial barrier to the successful use of an inhaler
- clinicians should prescribe the most operationally simple device as clinical considerations allow to elderly patients in anticipation of the high likelihood of some degree of cognitive/executive impairment at the time of prescription or in the future.
- Training and reinforcement improve the acquisition and retention of a satisfactory technique in some older patients with normal cognitive screening test scores and might be useful for patients with slight cognitive impairment.
- Patients with definite cognitive impairment will not be able to use an inhaler independently and will require assisted administration with, for example, a large volume spacer or nebuliser

References

[1] Newman SP, Moren F, Pavia D, Sheahan NF, Clarke SW. Deposition of pressurized aerosols in the human respiratory tract. *Thorax* 1981;63:52-5.

[2] Newman SP, Pavia D, Clarke SW. How should a pressurized beta-adrenergic bronchodilator be inhaled? *Eur. J. Resp. Dis.* 1981;62:3-20.

[3] Usmani OS, Biddiscombe MF, Nightingale JA, Underwood SR, Barnes P. Effects of bronchodilator particle size in asthmatic patients using monodisperse aerosols. *J. Appl. Physiol.* 2003;95:2106-12.

[4] Lawford P, McKenzie D. Pressurized aerosol technique; influence of breath-holding time and relationship of inhaler to mouth. *Br. J. Dis. Chest* 1982;76:229-33.

[5] Newman SP, Pavia D, Garland N, Clarke SW. Effects of various inhalation modes on the deposition of radio-active pressurized aerosols. *Eur. J. Resp. Dis.* 1982;63(supp 119): 57-65.

[6] Newman SP, Clarke SW. Therapeutic aerosols 1 – physical and practical considerations. *Thorax* 1983;38:881-6.

[7] Laube BL, Norman PS, Adams GK. The effect of aerosol distribution on airway responsiveness to inhaled methacholine in patients with asthma. *J. Allergy Clin. Immunol.* 1992;89:510-18.

[8] Molimard M, Raherison C, Lignot S, Depont F, Abouelfath A, Moore N. Assessment of handling of inhaler devices in real ilfe: an observational study in 3811 patients in primary care. *J. Aerosol. Med.* 2003;16:249-54.

[9] Lenney J, Innes JA, Crompton GK. Inappropriate inhaler use: assessment of use and patient preference of seven inhalation devices. EDICI. *Respir. Med.* 2000;94:496-500.

[10] Todd MA, Baskett JJ, Richmond DE. Inhaler devices and the elderly. *N. Z. Med. J.* 1990; 103:43-6.

[11] Welch MJ, Nelson HS, Shapiro G et al. Comparison of patient preference and ease of teaching inhaler technique for Pulmicort Turbohaler versus pressurized metered dose inhalers. *J. Aerosol. Med.* 2004;17:129-39.

[12] Soriano JB, Kiri VA, Maier WC, Strachan D. Increasing prevalence of asthma in UK primary care during the 1990s. *Int. J. Tuberc. Lung Dis.* 2003;7:415-21.

[13] Bannerjee DK, Lee GS, Malik SK, Daly S. Underdiagnosis of asthma in the elderly. *Br. J. Dis. Chest* 1987;81:23-29.

[14] Dow L, Coggon D, Osmond C, Holgate ST. A population survey of respiratory symptoms in the elderly. *Eur. Respir. J.* 1991;4:267-72.

[15] Gwynn RC. Risk factors for asthma in US adults: survey from the 2000 Behavioural Risk Factor Surveillance System. *J. Asthma* 2004;41:91-8.

[16] Dow L, Fowler L, Phelps L et al. Prevalence of untreated asthma in a population sample of 6000 older adults in Bristol, UK. *Thorax* 2001;56:472-6.

[17] Enright PL. The diagnosis and management of asthma is much tougher in older patients. *Curr. Opin. Allergy Clin. Immunol.* 2002;2:175-81.

[18] Campbell MJ, Cogma GR, Holgate ST, Johnston SL. Age specific trends in asthma mortality in England and Wales 1983-95. The results of an observational study. *Brit. Med. J.* 1997;314:1439-41.

[19] Gray SL, Williams DM, Pulliam CC et al. Characteristics predicting incorrect metered-dose inhaler technique in older subjects. *Arch Intern. Med.* 1996;156:984-8.

[20] Allen SC, Prior A. What determines whether an elderly patient can use a metered dose inhaler correctly? *Br. J. Dis. Chest* 1986;80:45-9.

[21] Armitage JM, Williams SJ. Inhaler technique in the elderly. *Age Ageing* 1988;17:275-8.

[22] Ho SF, O'Mahony MS, Steward JA, Breay P, Burr ML. Inhaler technique in older people in the community. *Age Ageing* 2004;33:185-8.

[23] Jones V, Fernandez C, Diggory P. A comparison of large volume spacer, breath-activated and dry powder inhalers in older people. *Age Ageing* 1999;28:481-4.
[24] Connolly M. Inhaler technique of elderly patients: comparison of metered dose inhalers and large volume spacer devices. *Age Ageing* 1995;24:190-2.
[25] Diggory P, Bailey R, Vallon A. Effectiveness of inhaled bronchodilator delivery systems for older patients. *Age Ageing* 1991;20:379-82.
[26] Renwick DS, Connolly MJ. Improving outcomes in elderly patients with asthma. *Drugs Aging* 1999;14:1-9.
[27] Dow L. Asthma in older people. *Clin. Exp. Allergy* 1998;28(supp 5):195-202.
[28] Goodman DE, Israel E, Rosenberg M et al. The influence of age, diagnosis and gender on proper use of metered dose inhalers. *Am. J. Respir. Crit. Care Med.* 1994;150:1256-61.
[29] Chapman KR, Love L, Brubaker H. A comparison of breath-actuated and conventional metered-dose inhaler inhalation techniques in elderly subjects. *Chest* 1993;104:1332-7.
[30] Johnson DH, Robart P. Inhaler technique of outpatients in the home. *Respir. Care* 2000;45:1182-7.
[31] van Beerendonk I, Mesters I, Mudde AN, Tan TD. Assessment of the inhalation technique in outpatients with asthma or chronic obstructive pulmonary disease using a metered-dose inhaler or dry powder device. *J. Asthma* 1998;35:273-9.
[32] The British Thoracic Society. The British guidelines on asthma management; 1995 review and position statement. *Thorax* 1997;52;S1.
[33] The British Thoracic Society. Guidelines on the use of nebulisers. *Thorax* 1997;52:S2.
[34] The British Thoracic Society. Guideline for the management of COPD in primary and secondary care. *Thorax* 2004;59:S1.
[35] Allen SC. Competence thresholds for the use of inhalers in people with dementia. *Age Ageing* 1997;26:83-6.
[36] Diggory P, Fernandez C, Humphrey A, Jones V, Murphy M. Comparison of elderly people's technique in using two dry powder inhalers to deliver zanamivir: randomized controlled trial. *Brit. Med. J.* 2001;322:577-9.
[37] Hodkinson HM. Evaluation of a mental test score for assessment of mental impairment in the elderly. *Age Ageing* 1972;1:233-8.
[38] Folstein FM, Folstein SE, McHugh PR. Mini-mental state: a practical method for grading the cognitive state of patients for the clinician. *J. Psychiatr Res.* 1975;12:189-198.
[39] Nelson HE. A modified card sorting test sensitive to frontal cortex defects. *Cortex* 1976; 12:313-24.
[40] Delis DC, Squire LR, Bihrle A et al. Componential analysis of problem-solving ability: performance of patients with frontal lobe damage and amnesic patients on a new sorting test. *Neuropsychologia* 1992;30:683-97.
[41] Yesavage JA. Development and validation of a depression screening scale: a preliminary report. *J. Psychiatr Res.* 1983;17:37-49.
[42] Mahoney F, Barthel D. Functional evaluation. The Barthel Index. *Maryland State Med. J.* 1965;14:61-5.
[43] Allen SC, Ragab S. Ability to learn inhaler technique in relation to cognitive scores and tests of praxis in old age. *Postgrad. Med. J.* 2002;78:37-9.

[44] Royall DR, Chiobo LK, Polk MJ. Correlates of disability among elderly retirees with 'subclinical' cognitive impairment. *J. Gerontol. A Biol. Sci. Med. Sci.* 2000;55:M541-6.

[45] Royall DR, Mahurin RK, Gray KF. Bedside assessment of executive cognitive impairment: the executive interview. *J. Am. Geriatr. Soc.* 1992;40:1221-6.

[46] Allen SC, Jain M, Ragab S, Malik N. Acquisition and short term retention of inhaler techniques require intact executive function in elderly subjects. *Age Ageing* 2003; 32:299-302.

[47] Bellia V, Pistelli R, Catalano F et al. Quality control of spirometry in the elderly – the SARA study. *Am. J. Respir. Crit. Care Med.* 2000;161:1094-1100.

[48] Pezzoli L, Giardini G, Consonni S et al. Quality of spirometric performance in older people. *Age Ageing* 2003;32:43-6.

[49] Carvalhaes-Neto N, Lorino H, Gallinari C et al. Cognitive function and assessment of lung function in the elderly. *Am. J. Respir. Crit. Care Med.* 1995;152:1611-5.

[50] Allen SC, Yeung P. Inability to draw intersecting pentagons as a predictor of unsatisfactory spirometry technique in elderly hospital inpatients. *Age Ageing* 2006; 35:204-5.

[51] Allen SC, Yeung P, Janczewski M, Siddique N. Predicting inadequate spirometry technique and the use of FEV1/FEV3 as an alternative to FEV1/FVC for patients with mild cognitive impairment.

[52] Royall DR, Cordes JA, Polk M. CLOX: an executive clock drawing task. *J. Nerol. Neurosurg. Psychiatry* 1998;64:588-94.

[53] Baxter M, Warwick-Sanders M, Allen SC. Comparison of four tests of cognition as predictors of ability to acquire metered dose inhaler technique in old age *Age Ageing* 2009: in press (abstract)

[54] Baxter M, Warwick-Sanders M, Allen SC. Comparison of CLOX, MMSE and pentagon copying for predicting spirometry performance. *Age Ageing* 2009: in press (abstract)

[55] Allen SC. Are inhaled systemic therapies a viable option for the treatment of the elderly patient? *Drugs Aging* 2008;25:89-94.

In: Encyclopedia of Neuroscience Research
Editors: Eileen J. Sampson and Donald R. Glevins
ISBN 978-1-61324-861-4
© 2012 Nova Science Publishers, Inc.

Chapter XXVIII

Foetal Alcohol Spectrum Disorders: The 21st Century Intellectual Disability

Teresa Whitehurst[*]
Sunfield Research Institute, West Midlands,
United Kingdom

Abstract

The pattern of childhood disability now facing practitioners and clinicians across the world is an ever changing landscape. Both educators and medical professionals are challenged to develop new skills and new practices in order to address the complex array of symptoms and behaviours presented to them by the children and families seeking their support. In stark contrast to the 'traditional' range of intellectual disabilities, whose causes are known and researched, such as Fragile X Syndrome, Down's syndrome and autism spectrum disorders, practitioners are now faced with the challenge of 'man-made' intellectual disabilities such as Foetal Alcohol Spectrum Disorders (FASD). Despite this disorder being regarded as the leading known cause of non-genetic intellectual disability in the Western World (Abel and Sokel 1987) with prevalence rates estimated at 1 in 100 (Autti-ramo 2002) there is scant public awareness of the dangers of drinking whilst pregnant, insufficient or contradictory guidance on 'safe' limits, little or no professional knowledge base and a paucity of strategies to support those affected or their families. In a culture which sees binge drinking on the increase, the numbers of children with FASD are set to escalate. To what extent are our professionals ready for the challenges of this 21st Century intellectual disability? This chapter considers the neurological deficits of those affected by FASD, the implications for education both as a proactive and reactive strategy, the importance of early diagnosis and intervention, set within a context of need.

---

[*] Woodman Lane, Clent, Stourbridge, West Midlands, DY9 9PB, United Kingdom. Telephone: 01562 881320. Email: Teresaw@sunfield.org.uk.

Background

Care, education and support for children with disabilities and their families is high on the agenda of developing countries. A wide range of government policies and initiatives in the United Kingdom seek to protect and include such children to ensure they are treated as valued and equal members of society with much done to change historic conceptualisations of disability (DfES, 2003; DfES, 2007; DoH, 2004; Whitehurst, 2007; United Nations, 1989). Ideally, children diagnosed with a disability at birth are surrounded by plethora of services which aim to maximise the potential of early intervention, seeking the best possible quality of life for the child and their family. These opportunities, however, are dependent upon early diagnosis, the provision of local services and a sound body of professional knowledge.

Much is now known about the aetiology, diagnosis and assessment of intellectual and neurodevelopmental disabilities such as Autistic Spectrum Disorder, Down's Syndrome and Fragile X Syndrome with subsequent interventions now well in place. However, whilst medical advances have greatly improved the detection of these conditions, either at birth or as a consequence of later neurodevelopmental assessment, many children affected by maternal prenatal alcohol consumption remain undetected. Even where a full diagnosis can be made knowledge and understanding of the disorder in the United Kingdom, both from a medical and educational perspective, is poor. Families remain therefore bereft of the services and support they need.

Britain's Binge Drinking Culture

British government concern regarding drinking habits, especially amongst teenagers, has been the focus of much debate in recent years. Alcohol misuse, in the form of binge drinking, is a distinctive characteristic of the British drinking culture (Parliamentary Office of Science and Technology, 2005), with over 26% of teenagers reporting 3 or more binge drinking episodes within the previous month (Plant and Plant, 2006). A binge is defined in the UK as the consumption of 6 or more units for a woman and 8 or more for a man, drinking on a single occasion. This definition is based guidelines issued by the Department of Health (1995) which recommend a maximum daily alcohol intake of 3-4 units for men and 2-3 units for women (Parliamentary Office of Science and Technology, 2005). Numbers of female teenagers (aged 15 and 16) admitting to binge drinking 3 or more times in the past month now exceed the number of males of the same age (Plant and Plant, 2006).

Research by Plant (2001) revealed that 8% of British women aged 18-24 women were drinking at high risk consuming over 35 units of alcohol per week. Coincidentally this is also the age group of women most at risk of unplanned pregnancy. Dex and Joshi (2005) reported that 84% of mothers under 20 years of age had unplanned pregnancies. More generally, mean alcohol consumption by young teenagers aged 11-15 has almost doubled since 1990 rising from an average of 5 units per week to almost 10 (DCSF 2008; DoH, 2008). With alcohol now more affordable consumption has increased in the general population (DoH, 2008) leading unsurprisingly to an increase in hospital admissions for alcohol related harm (Jones et al, 2008). The number of female hospital admissions in the UK for alcohol related harm was approximately 250,000 for the period 2006/07. With trends towards an increase in drinking amongst women it is inevitable that numbers of children born damaged as a result of maternal prenatal alcohol consumption will continue to increase.

The Effects of Alcohol on the Developing Foetus

First officially recognised by Jones and Smith in 1973, Foetal Alcohol Syndrome (FAS) has received considerable worldwide attention in recent years. Prenatal alcohol exposure has now been recognised as a major public health issue in all industrialized nations as well as in many developing countries where alcohol abuse by women of childbearing age is common. Indeed, foetal alcohol exposure is the leading known preventable cause of intellectual disabilities in the Western world (Abel and Sokel, 1987, Stratton, Howe and Battaglia, 1996).

Alcohol is a teratogenic compound (ie: a substance which interferes with the normal development of the embryo or foetus) that readily crosses the placenta (British Medical Association, 2007). The immature blood filtration system and liver function in the foetus cannot process and eliminate alcohol as it can in the adult thus resulting in damage to the unborn child. The following diagram outlines the impact of alcohol throughout the duration of pregnancy:

Development of Embryo

Coles, 1994.

Diagram One: The impact of Alcohol on the Foetus throughout pregnancy.

The darker areas in the diagram represent times when alcohol damage to the foetus is at its greatest, whilst lighter areas represent times when less significant harm, but harm nonetheless, may be inflicted.

The essential principles of the teratogenic effects of prenatal alcohol involve:

a) The dosage of alcohol – even a low dosage may be teratogenic
b) The timing (trimester) of the exposure during pregnancy. Each trimester of pregnancy has specific teratogenic effects; drinking in the first trimester results in facial dysmorphology and growth deficiencies whilst second and third trimester exposure has the most effect on central nervous system development, especially the neurotransmitter development.
c) Medical factors particular to the mother. These include chronicity of alcoholism in the mother and her general nutritional status, as well as the protective factor of her specific genetic endowment.

(adapted from O'Malley, 2007)

The term Foetal Alcohol Spectrum Disorders (FASD) has been developed in more recent years, operating as an umbrella term for a set of disorders caused by the consumption of alcohol by a mother whilst pregnant (Mukherjee, Hollins and Turk, 2006). These disorders exist on a spectrum ranging from the full presentation of Foetal Alcohol Syndrome (FAS) to a set of conditions affecting the neuro-behavioural presentations of the condition without encompassing all of the features.

Full FAS is defined by evidence of pre and post natal growth retardation, central nervous system dysfunction (eg: neurocognitive deficits, learning disability, mental retardation) specific facial dysmorphology (eg; short palpebral fissures, flattened philtrum and thin upper lip) and a history of prenatal alcohol exposure (Hoyme et al, 2005; Bishop, Gahagan and Lord, 2007). The diagram below illustrates these features:

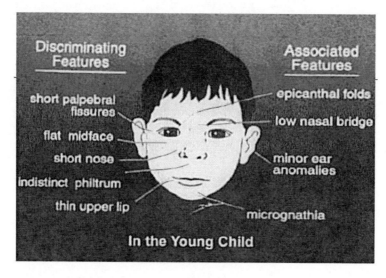

Diagram Two: Facial Dysmorphology associated with FAS.

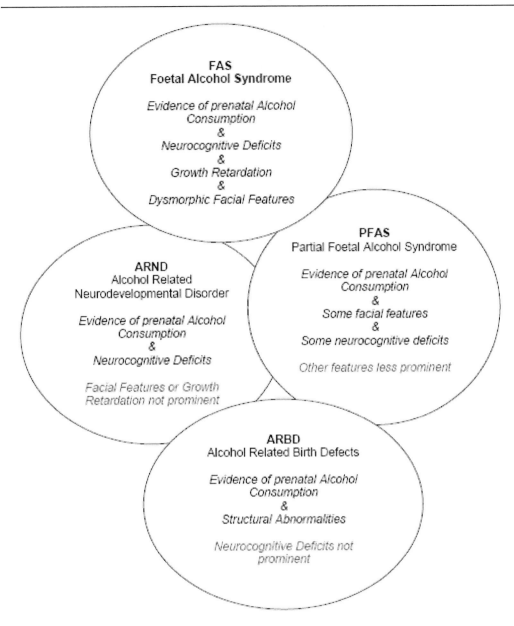

Figure Three: Differentiating the presentation of FASD.

Facial dysmorphia occurs only at a discrete stage of the pregnancy, within a two week period, and thus not all children damaged by alcohol present with the full range of criteria necessary for a diagnosis of FAS. Estimates of the number of diagnosable cases of FAS among heavily drinking women have been placed at between 10-40% (Jones and Smith, 1975), although it is accepted that this figure will vary internationally (Abel, 1995) and be subject to constant revision as diagnostic criteria are refined. However, escaping facial dysmorphia does not protect children from being at serious risk of developing behavioural, cognitive and physical problems associated with alcohol damage which may occur at other times throughout the pregnancy. In reviewing the literature Evenson (1991) identifies that only the most severely affected children are discussed while the other 80%, many undiagnosed, are at risk for developing significant problems. Where prenatal exposure to alcohol can be confirmed two 'labels' exist under the umbrella term Foetal Alcohol Spectrum Disorders (O'Malley 2007); Foetal Alcohol Syndrome (FAS) and Alcohol Related Neurodevelopmental Disorder (ARND). A diagnosis of ARND is made when maternal alcohol consumption can be confirmed alongside prominent neurocognitive deficits but in the absence of facial features or growth retardation. Two other labels are also commonly used; Alcohol Related Birth Defects (ARBD) and Partial Foetal Alcohol Syndrome (PFAS). A diagnosis of ARBD is made when maternal alcohol consumption can be confirmed alongside behaviour or structural abnormalities but where facial features may be absent whilst PFAS requires evidence of maternal drinking combined with some facial features and neurocognitive deficits but other features may be absent. It is important to note that these disorders are experienced on a continuum with mild to severe presentation. However, whilst FAS is a clinical diagnosis, FASD is not (Astley and Clarren 2000).

An overview of the characteristic features of these disorders can be seen in the diagram above.

Epidemiology of FASD

Worldwide incidence of FAS is estimated at 0.97 per 1000 but it is important to note that this data is based almost exclusively on research conducted within the USA (British Medical Association, 2007). It has been estimated that in Western countries as many as 9 per 1,000 live births involve children affected by FAS, Partial Foetal Alcohol Syndrome or Alcohol Related Neurodevelopmental Disorder (Autti-Rämö, 2002). The incidence of FASD within the UK is currently unknown. Figures are collected in England and Scotland for FAS (but not FASD) whilst there is no data at all collected in Northern Ireland and Wales. It is, however, generally accepted that the incidence of children presenting with the effects of prenatal alcohol exposure in its variety of forms is around the 1 in 100 figure reported by Autti-Rämö, (2002). It is important also to note that as more complete diagnoses and broader definitions of FASD become available, prevalence and incidence may increase.

It should also be noted that these epidemiological figures represent only the clinical picture of those cases which come to the attention of the medical profession. Irrespective of difficulties with differential diagnosis there may be personal and moral factors which prevent the true extent of the problem being known.

The Impact of Alcohol upon the Developing Brain

Alcohol induced damage to the brain of a developing child has been replicated in animal studies and in nerve cell cultures (Michaelis and Michaelis, 1994) whilst techniques such as MRI scans have facilitated the study of the living brain enabling researchers to determine the extent and location of the structural damage. Ikonomidou et al (2000) report that an apoptotic degenerative reaction which deletes neurones from developing brain sites can be induced through exposure to alcohol in a rat brain at specific developmental times resulting in generalised loss of brain mass with specific loss of cerebellar and hippocampal neurons. This is mediated via a dual mechanism; blocking NMDA Glutamate Receptors and through excessive activation of GABA receptors. Synaptogenesis, neuronal migration and fasciculation may all be compromised even with low alcohol consumption (Charness, Safran and Perides, 1994). Neuroimaging studies have revealed specific abnormalities in the brains of children exposed to maternal prenatal consumption of alcohol. A reduction in the size of the basal ganglia is noted (Mattson et al, 1996a) which has been associated with spatial memory deficits and problems with set shifting in animals (Mattson et al 1996b; Mattson and Riley, 1999). The basal ganglia/caudate pathway is particularly important for relaying information to the frontal lobes whilst also acting as an emotional regulator. Deficits in this region may therefore also extend to perseveration tendencies, attentional problems and impaired executive functioning. Frontal lobes have been noted to be thin and small (Riley, 2008). This area of the brain has a major role in higher intellectual functioning communicating with other brain regions to facilitate executive functioning (organising, prioritising, controlling impulses, self-awareness, initiating and ending tasks, problem solving), motor function, spontaneity, memory, risk taking, interpreting feedback from the environment, rule compliance, judgment, language, social and sexual behaviour. The complex interplay between the frontal lobe and other brain regions is vulnerable to a range of neurodevelopmental frontostriatal disorders such as **Tourette's Syndrome, Attention Deficit** Hyperactivity Disorder, Obsessive Compulsive Disorder and Autism, with many of these disorders sharing common phenotypical presentation with FASD. Abnormalities in the parietal lobe have been noted giving rise to problems with spatial representations, writing, mathematics and self care skills. Another feature typical of this disorder is a reduction in the size of the cerebellum, a structure important for balance, gait, cognition and co-ordination. Additional tasks of the cerebellum include measuring time, directing attention and implications for classical conditioning. Hypoplasia or agenesis of the corpus callosum, an important bundle of nerve fibres connecting the left and right hemispheres and implicated in communication, is also evident. Seven percent of children with FAS may lack a corpus callosum which is at a rate of twenty percent higher than that found in the general population (Riley et al 1995). Diffusion Tensor Imaging has shown that a brain affected by prenatal alcohol has abnormalities in grey matter structures, such as the basal ganglia and the thalamus which act as relay station to integrate incoming sensory and monitor input before passing this to the cortex , alongside a reduction in volume of white matter with the corpus callosum and tracts connecting to the temporal lobe being the most affected (Rasmussen, 2008). This would indicate that there is a deficit of myelin which is essential to ensure the brains 'communication system' works effectively. **Brain imaging** data has been shown to be concordant with both physical and behavioural manifestations of this disorder.

Children affected by prenatal maternal alcohol consumption suffer from both primary and secondary disabilities in a range of domains, the effects of which are experienced throughout the life span. Although intellectual disability is not a feature of the diagnostic criteria of FAS it is well reported that children prenatally exposed to alcohol have a compromised level of intellectual function. Studies by Streissguth, Barr and Sampson (1990) suggest that exposure to as little as one ounce of alcohol per day has been associated with decrements of six to seven points in an IQ score. These children have an inability to link cause and effect and often experience serious problems with maths due to the effect of alcohol on the development of the parietal lobe (Goswami and Bryant, 2007). Affected children experience communication difficulties presenting as speech delays or impediments, receptive and expressive difficulties

(Shaywitz, Caparulo and Hodgson, 1981; Tenbrinck and Buchin, 1975) and problems with word comprehension. Verbal learning has been shown to be impaired with affected children experiencing problems at the encoding level rather than retention and recollection (Mattson et al, 1996b). Behaviourally these children present with attentional problems, poor impulse control, working memory problems and poor adaptive functioning (O'Malley, 2007) and are often misdiagnosed with Attention Deficit Hyperactivity Disorder (ADHD) (Coles et al, 1997). Qualitative differences in attention were noted by Coles et al (1997) in a comparison between children with and without FAS. Children in the former group displayed evidence of difficulty in focusing and sustaining attention whilst children in the latter group were able to maintain and focus attention but displayed difficulties in the subsequent shifting of their attention. Children with FASD experience problems in the domain of social interactions. Whilst they may be eager to make friends, they simply don't understand the nuances required to formulate friendships. Similarities in the difficulties experienced by children with FASD have been compared to those of children with autistic spectrum disorders. Phenotypical presentation is startlingly similar but has been found to differentiate at the level of social interaction (Bishop, Gahagan and Lord, 2007). However, in the absence of full birth mother history and distinctive facial features, differential diagnosis at a phenomenological level would be difficult. Children and young people affected by prenatal maternal alcohol consumption frequently meet the diagnostic criteria for autistic spectrum disorder and Asperger's Syndrome. Whilst their presenting features may be similar, the underlying causal factors may be different. However, until diagnostic systems are developed which allow for discrimination between disorders on the autistic spectrum and those which emanate from prenatal maternal alcohol consumption, interpretation is open to misdiagnosis.

The large variations seen in the presentation of the spectrum of foetal alcohol disorders may be explained as a consequence of the timing and varying amount of prenatal alcohol consumed by the mother impacting to varying degrees upon the circuits which comprise the frontrostriatal system. These systems are crucial to willed action and are intimately involved in planning the what, how and when of our actions (Bradshaw, 2001).

Table 1. Percentage of those with FASD suffering Secondary Disabilities

Problem	Percentage
Mental Health/Psychiatric Problems	94%
Problems with Employment	80%
Patient in Dependent Living	80%
Inappropriate Sexual Behaviour	45%
Disrupted School Experience	43%
Trouble with the law	42%
Confinement for a Crime	35%
Mental Hospital Admission	23%
Drug/Alcohol in-patient treatment Admission	15%

Reported in O'Malley, 2007 p10.

Subsequent secondary disabilities have been noted including psychiatric and mental health problems, drug and alcohol dependency and encounters with the criminal justice system. Foetal Alcohol Syndrome is estimated to cost the USA $2.9 million per individual over the lifespan with much of this cost attributable to the secondary disabilities including disrupted education, contact with the law, mental health problems, alcohol and drug misuse, inappropriate sexual behaviour and inability to obtain and maintain employment and independent living (Peadon, Fremantle, Bower and Elliott, 2008). The table below shows the relationship between psychiatric/social dysfunction and FASD:

Children affected by prenatal alcohol consumption also suffer from a range of physical symptoms. Eyes are a continuum of the central nervous system and are a sensitive indicator of heavy prenatal alcohol exposure (Autti-Rämö, 2002). Consequent problems with their eyes include drooping eyelids, myopia, blindness, underdeveloped optic nerve. Children also experience problems with their ears (hearing loss, recurrent ear infections, central auditory processing disorder secondary to brain damage), teeth (faulty enamel, improper alignment or misshapen teeth) musculo-skeletal (minor problems with hands, fingers, arms, toes; cervical spine abnormalities; some problems with joint movement; thoracic abnormalities), internal organs (septal defects of heart, underdeveloped or misplaced kidneys) and genitourinary problems (abnormal genital development) (Mukherjee, 2007).

Professional Knowledge of FASD

With the majority of research having been conducted in the United States knowledge of FASD in the UK is limited both at public and professional levels. Whilst information regarding the dangers of drinking is a point of political and media concern, its relationship to the potential harm to the unborn child is rarely mentioned. Within the educational arena there has been almost no systematic research on the needs of students with FASD or on the best educational strategies (Ryan and Ferguson, 2006a) or any systematic training for teachers to educate young people on the consequences of maternal alcohol consumption. FAS has been identified as the leading cause of birth defects in the United States (Evensen 1991) and yet still these children are not recognised and included in the Individuals with Disabilities Education Improvement Act (IDEIA) thus rending them ineligible for services. With FASD barely on the UK radar the situation is even more disconcerting. Ryan and Ferguson (2006a, 2006b) point out that most children with FASD are not placed in special schools and therefore it is important for all teachers to have at their disposal a sound knowledge of the learning needs of this group of students and a range of interventions and strategies at their disposal in order to modify and adapt their instructional techniques. Within the medical profession there is little evidence of the knowledge or understanding of the disorder either at a General Practitioner level or even paediatric level (Nanson et al, 1995). Even when children are born displaying facial features and were a result of high risk pregnancies routine paediatric screening failed to identify the disorder (Stoler and Holmes 1999). Furthermore, despite the mothers in the latter study being within a high risk group, 73% of the case notes contained no record of maternal alcohol consumption. Expectant mothers are rarely advised at point of contact with the midwife of the potential dangers of continuing drinking during pregnancy. Kesmodel et al (2002) found that 65% of pregnant women received little or no information from their midwife with a majority of 74% believing it was acceptable to continue drinking. Peadon et al, (2008) suggest that 'training needs to be accessible to all health professionals who are in the position to identify children who need assessment (p8).

Implications for Education

It would appear that the role of the educational system with regards to FASD is two fold. Firstly, educationalists have the potential to influence the likelihood of children being born with FASD by providing sound education on the consequences of drinking during pregnancy. Young people cannot be expected to take action and responsibility for drinking during pregnancy if they remain unaware of the dangers of alcohol to their unborn child. Secondly, educationalists must be prepared for the challenge of FASD in the classroom. Many children are currently slipping through the net and have no diagnosis, yet remain seated in classrooms unable to engage in the learning environment as a consequence of life long damage to their neurological systems caused by their birth mother's prenatal alcohol consumption.

These challenges very clearly demonstrate the need for neuroscience and education to set aside their inherent difficulties and differences and work together to produce pedagogies underpinned by sound evidence and a wider understanding of the implications and impact of the neurological limitations of certain groups of learners. Geake & Cooper (2003) argue that 'cognitive neuroscience advances our understanding of the very basics of learning' (p11). This venture can surely only succeed if our educationalists are equipped with the foundations of knowledge necessary to understand the challenges that lie before them. Teachers are in the best position to judge how to engage young people. Empowering them to design pedagogies which are synthesised with an understanding of the contribution that neuroscience can make to educational research (Goswami, 2004) is surely the way forward.

Conclusion

Whist alcohol has featured in all cultures and societies for centuries, the impact it has on the developing foetus has only recently emerged. Education to support sensible drinking patterns amongst women is vital to reduce the numbers of children born damaged by foetal alcohol spectrum disorders. Developing a sound professional understanding of the difficulties faced by this group of children and their families amongst our medical and educational practitioners is imperative. Both proactive and reactive strategies are required to ensure that professionals work together, from the earliest opportunity, to ensure that the needs of those affected by this 21st Century intellectual disability are addressed and met.

References

Abel, E.L. and Sokel, R.J. (1987). 'Incidence of fetal alcohol syndrome and economic impact of FAS-related anomalies. *Drug and Alcohol Dependence* 19, 51-70

Abel, E.L. (1995) 'An update on incidence of FAS: FAS is not an equal opportunity birth defect. *Neurotoxicology & Teratology,* 17 (4), 437-443

Astley, S.J. and Clarren, S.K. (2000) Diagnosing the full spectrum of fetal alcohol-exposed individuals: introducing the 4-digit diagnostic code. *Alcohol and Alcoholism* 35 (4), 400-410

Autti-Rämö, I. (2002) Foetal alcohol syndrome – a multifaceted condition. *Developmental Medicine and Child Neurology* **44**: 141-4.

Bishop, S., Gahagan, S. and Lord, C. (2007). Re-examining the core features of autism; a comparison of autistic spectrum disorder and fetal alcohol spectrum disorder. *Journal of Child Psychology and Psychiatry,* 48, 1111-1121.

Bradshaw, J.L. (2001) *Developmental Disorders of the Frontostriatal System: Neuropsychological, Neuropsychiatric and Evolutionary Perspectives.* Hove : Psychology Press Limited

British Medical Association (2007) 'Fetal Alcohol Spectrum Disorders : A Guide for Healthcare Professionals. *BMA* : London

Charness, M.E., Safran R.M. and Perides, G. (1994) Ethanol inhibits neural cell-cell adhesion *Journal of Biological Chemistry* Vol. 269 (12) 9304-9309

Coles, C.D. (1994) Critical periods for prenatal alcohol exposure: Evidence from animal and human studies. *Alcohol Health Research World.* 18 (1), 22-29

Coles, C.D.; Platzman, K.A.; Raskind-Hood, C.L., Brown, R.T., Falexk, A. and Smith, I.E. (1997) A comparison of children affected by prenatal alcohol exposure and attention deficit, hyperactivity disorder. *Alcohol Clinical and Experimental Research* 21(1):150-161

Department for Children, Schools and Families, The Home Office, Department of Health, (2008) *The Youth Alcohol Action Plan (cm. 7387).* London : TSO

Department for Education and Skills (2003) *Every Child Matters*: *Summary.* London: DfES

Department for Education and Skills (2007) *Aiming High for Disabled Children: better support for families.* London : DfES

Department of Health. (2004*). National Service Framework for Children,Young People and Maternity Services.* London : DoH

Department of Health (2008). Safe, Sensible, Social – Consultation on Further Action. London : DoH

Dex, D., and Joshi, H., (eds) (2005) Children of the 21st Century: From Birth to Nine Months, Policy Press, Bristol

Evenson, D (1991*) Bridging the Gap for Fetal Alcohol Effect Children.* www.acbr.com/fas/j.htm Accessed 21.3.2008.

Geake, J and Cooper, P. (2003) *'Cognitive Neuroscience: Implications for Education?'* Westminster Studies in Education, vol 6 (1), 7-20

Goswami, U. (2004) Neuroscience, education and special education. *British Journal of Special Education,* Vol 31 (4) 175-183

Goswami, U and Bryant, P. (2007). *Children's cognitive development and learning* (Primary Review Research Survey 2/1a). Cambridge : University of Cambridge Faculty of Education

Hoyme, H.E., May, P.A., Kalberg, W.O., Kodutuwakku, P., Gossage, J.P., Trujillo, P.M. (2005). A practical clinical approach to the diagnosis of fetal alcohol spectrum disorder: clarification of the 1996 institute of medicine criteria. *Paediatrics;* 115 (1) 39-47

Ikonomidou, C., Bittigau, P., Ishimaru, M.J., Wozniak, D.F., Koch, C., Genz, K., M.T. Price., Stefovska, V., Hörster, F.,Tenkova, T., Dikranian, K. and Olney, J.W. (2000) Ethanol-Induced Apoptotic Neurodegeneration and Fetal Alcohol Syndrome. *Science,* vol 287 (5455) p1056-1060

Jones, K.L. and Smith, D.W. (1973) Recognition of the fetal alcohol syndrome in early infancy. *Lancet* 2 (7836) 999-1001

Jones, K.L. and Smith, D.W. (1975). The fetal alcohol syndrome. *Teratology, 12,* 1-10

Jones, L., Bellis, M.A., Dedman, D., Sumnal, H and Tocque, K. (2008). *Alcohol Attributable Fractions for England : Alcohol attributable mortality and hospital admissions.* www.nwph.net/nwpho/publications/AlcoholAttributableFractions. Accessed 6.12.2008

Kesmodel, U. and Kesmodel, P.S. (2002) Drinkind during pregnancy: attitudes and knowledge among pregnant Danish women in 1998. *Alcoholism Clinical and Experimental Research,* 26 (10), 1553-60

Mattson, S.N.; Riley, E.P.; Sowell, E.R., Jernigan, T.L., Sobel, D.F. and Jones, K.L. (1996a). A decrease in the size of the basal ganglia in children with fetal alcohol syndrome. *Alcohol Clinical and Experimental Research* 20(6):1088-1093

Mattson, S.N.; Riley, E.P.; Delis, D.C.; Stern, C.; and Jones, K.L. (1996b) Verbal learning and memory in children with fetal alcohol syndrome. *Alcohol Clin Exp Res* 20(5):810-816

Mattson, S.N., and Riley, E.P. (1999) Implicit and explicit memory functioning in children with heavy prenatal alcohol exposure. *J. Int. Neuropsychol. Soc.* 5(5):462-471.

Michaelis, E.K., and Michaelis, M.L. (1994) Cellular and molecular bases of alcohol's teratogenic effects. *Alcohol Health Res World* 18(1):17-21.

Mukherjee, R (2007) Foetal Alcohol Syndrome : NOFAS-UK Conference, London.

Mukherjee, R., Hollins, S and Turk, J. (2006) Fetal Alcohol Spectrum Disorder: An Overview. *Journal of the Royal Society of Medicine,* 99 (6) 298-302

Nanson, J.L., Bolaria, R., Snyder, R.E., Morse, B.A., Weiner, L. (1995) Physician awareness of FAS; survey of paediatricians and GP. *Canadian Medical Association Journal* 152 (7), 1071-1076

O'Malley, K. (2007) Fetal Alcohol Spectrum Disorders: An Overview. *ADHD and Fetal Alcohol Spectrum Disorders (FASD).* New York : Nova Science Publishers 1-23

Parliamentary Office of Science and Technology (2005) *Binge Drinking and Public Health.* http://www.parliament.uk/parliamentary_offices/post/pubs2005.cfm Accessed 26.4.09

Peadon, E., Fremantle, E., Bower, C and Elliott, EJ (2008) International survey of diagnostic services for children with Fetal Alcohol Spectrum Disorders. *BioMed Central Pediatrics* published online April 15 2008 http://www.pubmedcentral.nih.gov/articlerender.fcgi?artid=2377245 accessed 17th April 2009.

Plant, M.L. and Plant, M.A. (2001) Heavy drinking by young British women gives cause for concern. *British Medical Journal* 323 1183

Plant, M.L. and Plant, M.A. (2006). *Binge Britain: Alcohol and the National Response.* Oxford: Oxford University Press

Rasmussen, C. (2008) Water-diffusion Technology Identifies Brain Regions Damaged By Prenatal Alcohol Exposure. *ScienceDaily.* http://www.sciencedaily.com /releases/2008/07/080718180726.htm accessed 6.12.2008

Riley, E.P. (2008) FASD – Not just another pretty face; the effects of prenatal alcohol on brain and behaviour. Binge Britain Conference, Ormskirk.

Riley, E.P.; Mattson, S.N.; Sowell, E.R., Jernigan, T.L., Sobel, D.F. and Jones, K.L. (1995) Abnormalities of the corpus callosum in children prenatally exposed to alcohol. *Alcohol Clinical and Experimental Research* 19(5):1198-1202.

Ryan, S and Ferguson, D.L. (2006a) On, yet under, the radar: Students with Fetal Alcohol Syndrome Disorder. *Exceptional Children, 72 (3) 363-5*

Ryan, S and Ferguson, D.L. (2006b) The person behind the face of fetal alcohol spectrum disorder: student experiences and family and professionals' perspectives on FASD. *Rural Special Education Quarterly*, 25 (1) 32-40.

Shaywitz, S.E., Caparulo, B.K. and Hodgson, E.S. (1981) Developmental Language Disability as a consequence of Prenatal Exposure to Ethanol. *Pediatrics* Vol 68 (6) 850-855

Stoler, J.M. and Holmes, L.B. (1999) Under recognition of prenatal alcohol effects in infants of known alcohol abusing women. *Journal of Paediatrics,* 135 (4) 430-436

Stratton, K., Howe, C., and Battaglia, F. (1996) *Fetal Alcohol Syndrome: Diagnosis, Epidemiology, Prevention, and Treatment.* Washington, D.C.: National Academy Press.

Streissguth, A.P., Barr, H.M., and Sampson, P.D. (1990). Moderate prenatal alcohol exposure: Effects on child IQ and learning problems at age 71/2 years. *Alcoholism: Clinical and Experimental Research,* 14 (5), 662-669

Tenbrinck, M.S. and Buchin, S.Y. (1975) Fetal Alcohol Syndrome : Report of a case. *The Journal of the American Medical Association* Vol 232 (11) 1144-1147

United Nations (1989) *Convention on the rights of the child.* Geneva : United Nations

Whitehurst, T (2007) 'Changing Perspectives on Inclusion and Disability : The Monkey King Arts Project' in Barry Carpenter and Jo Egerton (Eds) *New Horizons in Special Education.* Clent : Sunfield Publications 155-171

In: Encyclopedia of Neuroscience Research
Editors: Eileen J. Sampson and Donald R. Glevins
ISBN 978-1-61324-861-4
© 2012 Nova Science Publishers, Inc.

Chapter XXIX

Where There are no Tests: A Systematic Approach to Test Adaptation

Penny Holding[1], Amina Abubakar[2] and Patricia Kitsao Wekulo[3]

[1]. Africa Mental Health Foundation, Kenya/
Case Western Reserve University, USA
[2]. Tilburg University, Netherlands/ Utrecht University, Netherlands/
Kenya Medical Research Institute-Centre for
Geographic Medicine Research Coast
[3]. Africa Mental Health Foundation, Kenya

In this chapter we outline a systematic approach to test adaptation developed through more than a decade of research in child development in both urban and rural settings in East Africa (Abubakar, Holding, Van Baar, Newton, and van de Vijver, 2008; Alcock, Holding, Mung'ala Odera, and Newton, 2008a; Carter et al., 2005; Holding et al., 2004).

Rationale: The literature on the use of psychological tests in diverse cultures is rich in examples that illustrate the necessity of making modifications to test content and administration techniques (Holding et al., 2004; Kathuria and Serpell, 1998; Sigman et al., 1989). The need to make modifications stems from fundamental differences in cultural experience between test takers in rural Africa and test takers in North America and Europe, for whom the majority of assessments have been designed. Without these modifications the psychometric qualities of the test data may be questionable and the distribution of scores elicited show a lack of sensitivity to within-population differences.

Based upon the premise that no instrument can claim to be culture-free (Nell, 2000) the *appropriateness* of an instrument to a specific target population needs to be determined by evaluating its psychometric properties in context, as well as its cultural, developmental and health or educational relevance.

The Special Case of Test Administrators: Beyond the characteristics of the test taker, those developing tests need also to consider the expertise of the potential test administrators. Few countries in Africa posses a psychological service, and therefore also lack training opportunities in test administration and interpretation. It is preferable to use test administrators who are more familiar to the client group and are fluent in the local languages to expert strangers who are unable to understand the child's responses. Research studies in resource-limited settings are therefore likely to be dependent upon technicians with limited experience to administer their test batteries. This lack of experience needs to be taken into account in both test selection and the design of test instructions.

Approaches to Test Development – Relative Merits

Three approaches to test development, Adoption, Adaptation, and Assembly, have been proposed to address the shortage of measures of childhood outcome that are appropriately standardized, validated and with documented reliability (Van de Vijver and Tanzer, 2004).

The approach known as *Adoption* involves taking in its entirety a test already in use in another population. The language of the test is then translated into that of the new target population, with test content and procedures administered as per the original standardization. The purported advantage of this approach is the availability of pre-existing information on the psychometric properties of the instrument. However, any change made to content or procedure undermines the assumptions upon which the original standardization was made. It is therefore incorrect to presume that the psychometric properties will remain intact in a new population.

Examples of direct adoption of measures from other cultural settings have been shown to constrain within-population variance, remove expected sensitivity to changes with age, and even mask true group differences (Baddeley, Gardener, and Grantham-McGregor, 1995; Connolly and Grantham-McGregor, 1993; Oluyomi and Houser, 2002). The inadequacy of the adoption approach largely results from the fact that activities used to measure psychological concepts reflect values, knowledge and communication strategies of their culture of origin (Ardila, 2005). When a test is applied to members of minority groups residing within the majority culture of the standardization sample, systematic selection bias can be observed (Reynolds, 1983). Significant differences even between similar cultures is acknowledged in the revisions of major test batteries, such as the British Ability Scales, modified as the Differential Ability Scales for use in the USA (Elliott, Smith, and McCullouch, 1990).

Test adoption, however, does allow for the direct comparison of two different population groups. Adoption may therefore be a valid approach in cross-cultural studies that are designed to describe the influence of cultural differences on performance, or where the test itself is the unit of analysis. In the latter instance the researcher may be interested in determining which cognitive skills are applied to solving the problem proposed by the test. In both these examples potential differences between population groups are anticipated, and the interpretation of test performance is therefore unlikely to be marred by issues of test bias.

Two other approaches, *Adaptation* and *Assembly,* better acknowledge the specific influence of different cultural contexts (Malda and Van de Vijver, 2005). Assembly is the production of a totally novel test, based upon the cultural practices of the target population, and makes no assumptions about conceptual or performance comparability. Assembly is most appropriate where no test already exists to measure the concept being assessed, or where an underlying psychological concept is most readily observed through an activity that is culturally specific (Kearins, 1976; Sternberg et al., 2001; WHO, 2007). For instance, Western tests of intelligence emphasize skills such as reasoning, memory and acquired knowledge, but lack the social component of African conceptualizations of intelligence (Mpofu, 2001). Therefore, while sub-tests of the KABC (Kaufman and Kaufman, 1983) or WPPIS-R (Weschsler, 1989) may provide an adequate appraisal of specific cognitive skills, they do not provide an appropriate definition of intelligence in the African context.

Adaptation, in contrast, can be followed when an existing instrument provides a proven measure of an underlying psychological concept, but where the specific methodology used in one context, test language, materials, and/or administration procedures, requires modifications to make it suitable to the new context (Foxcroft, 2002; Holding et al., 2004). *Adaptation* therefore acknowledges the existence of underlying psychological universals, and attempts to enable the measurement of cognitive or behavioral skills in a universally comparable manner.

The first step to enabling comparisons between test versions is the establishment of equivalence between forms. Then, in accordance with sound psychometric practice, each test version should include its own standardization and normative population. While being time-consuming, establishing equivalence ensures that the adaptation maintains acceptable reliability and validity that allows for a meaningful interpretation of test scores.

Selecting an Approach. Choosing an approach to test development will depend upon the target construct and the ultimate purpose of the test application. A common application of psychological tests in the African context has been seeking to understand the influence of specific health exposures on development. For this purpose the test needs to be sensitive to true group differences. The ability to make comparisons between different studies of the same condition also makes it valuable to have information on common constructs that can be used to summarize disease effects. The advantage of *Adaptation* over *Assembly* is the degree of comparability that it affords between study sites, and over *Adoption* is the potential for increased sensitivity of the test materials.

Through extensive experience in the field we have developed the following systematic procedure to maximize the reliability, validity and sensitivity of instruments developed, whether through *Adaptation* or *Assembly*. The procedure of test development can be divided into four main stages, with the equivalence of the instrument to other tests or test versions being evaluated by at least one of each of Herdman et al's forms equivalence (Herdman, 1998). Herdman et al.'s (1998) universalist model for translation and cultural adaptation of instruments suggests at aiming for the following forms of equivalence: conceptual, item, semantic, operational, measurement and functional. Only if the last two of these, measurement and functional equivalence, are achieved will it be possible to compare performance scores between one test version and another. .

The Four-Stage Approach to Test Adaptation

STEP 1- *Construct definition:* In this first stage the aim is to achieve conceptual clarity, and for adapted tests, conceptual equivalence, through defining the construct to be measured, as well as the cultural parameters that will capture the definition of the concepts involved.

Multiple sources should be used to generate a description of constructs in ways that are both culture-specific (emic) as well as in culturally neutral terms (etic), that will facilitate universal comparisons across cultures (Pike, 1967).

Some psychological constructs show greater functional universality than others. Tests of these constructs can be more readily adapted. One example is psychomotor development, where motor control and co-ordination can be seen to develop in a universal sequence. Despite this, assessment items and normative tables will vary from one setting to another to take into account not only the different rates of development seen across settings, (de Vries, 1999; Leiderman, Babu, Kagia, Kraemer, and Leiderman, 1973; Lynn, 1998; Neil, 1972) but also the different activities with which children are familiar. Universality therefore does not extend to the definition of appropriate item selection. For example, in most rural settings in Africa children do not have stairs in their homes; and therefore assessing psychomotor development based on the ability to climb stairs may be inappropriate.

Specific activities undertaken by test developers will define skill levels in the function of interest, a conceptual vocabulary through which the concept can be described in the language of interest, and finally the training needs of those who will describe and administer the assessment. This information can be collected through:

1) A *Systematic Review* of existing literature to provide a critical evaluation of existing measures of the construct of interest, and help identify a suitable instrument for adaptation.

2) The use of a *Panel* of local professionals to develop a conceptual vocabulary. A culturally appropriate definition of the construct of interest can be developed by a panel of people with relevant expertise, in community mobilization, child development and/or behavior change. These professionals should also be familiar with the language of the proposed test takers. We have, for example, run workshops with groups of community nurses and field workers involved in health research. Through discussion and activities designed to develop lay assessments of the concept of interest, the panel then prepare *a glossary of relevant terms* in the language of the target population. We have found producing a glossary of terms prior to exposure to the original instrument important in training non-experts in how to undertake conceptual rather than literal translations. The production of conceptual translations is important to ensure that the sensitivity, content and face validity of an instrument is not compromised.

3) *Consultation with the Community*. While professional panels often consist of community members, we have observed that the process of training in modern techniques such as bio medical medicine will exert a significant influence upon the understanding of panel members, separating them from the experience of lay community members. To adequately infer the potential understanding of assessment tasks and questionnaire items by the target respondents it is also important to consult directly with members of the community whose exposure to education mirrors that of the majority of the population. Methods for eliciting community understanding include Focus Group Discussions (FGDs), individual in-depth interviews, and direct observation, outlined in some detail in Abubakar et al. (2008) Participant consultation contributes to developing an understanding of a relevant definition of the construct, provides face validity for the item/ tasks and helps evaluate the cultural

appropriateness of the items/tasks (Abubakar, van de Vijver, van Baar, Kitsao-Wekulo, and Holding, 2008).

STEP 2 - Item pool creation: The aim of this step is to prepare of a list of potentially acceptable items, in a clear and unambiguous language, and, where relevant, achieve item and semantic equivalence. Using the information on cultural practices and available vocabulary gathered in Step 1, original items from existing instruments are vetted for appropriateness, and additional items are added to the potential item pool. These additional items provide potential substitutes for discarded items. Where a test is being adapted, no item from the original test should be discarded until it has been rigorously evaluated, as this can lead to premature removal. An example is, in an evaluation of the home environment, our initial assumption that an item on exposure to television viewing would be irrelevant as a sample of daily activities. This item later proved to be sensitive to between household differences, as children had access to televisions within the neighborhood, even if few had one in their own houses.

To prepare materials in the appropriate language the WHO recommends the use of a single schedule translated by a bilingual panel comprising members with related competencies. In practice, we have found the translation procedure recommended by the WHO (2007) to have fundamental flaws, many of which are also outlined in Leplège and Verdier (1995). One limitation is the assumption that an adequate vocabulary exists in the target language, another that it is possible to identify a panel familiar with the relevant concepts. The reality is that a lack of relevant expertise and budget constraints mean that available personnel are often native speakers of the target language, but are neither professional translators, nor necessarily familiar with the topics to be investigated. For this reason, we have followed a translation system similar to the procedure described in Gandek et al. (1998).

To begin with an initial translation of the original test material is made using the glossary of relevant terms as a guide. The translation team aims at producing a conceptual rather than a literal translation. This version is then evaluated through comparison between different back translations to look for semantic and conceptual clarity. The steps of the evaluative process include:

- Production of at least two back-translations by two independent native speakers
- Evaluation of conceptual equivalence by a study panel through comparison of the similarities and differences between the multiple translations
- Production of a second draft that replaces problematic items or response choices and.
- Repetition of step 1) using a second set of translators for further evaluation.

The process continues until there are no more semantic differences between translations provided, showing that the essential meaning of the items has been understood. In the process some items that remain poorly understood are eliminated.

STEP 3 –*Developing the procedure*: The main aim is to produce a schedule of items of acceptable length, as well as clear guidelines for their administration, which will promote operational equivalence. An iterative process is used, with each version or sub-set of items trialled on 5-10 participants. Items are discarded and administrative procedures that produce little variation in response, or which elicit negative reactions from participants are modified. An example of the latter is an item evaluating the development of self-recognition through asking infants to look into a mirror. The item was removed when it was ascertained the activity was considered a taboo in several African societies. Administration techniques, such as one-to-one interaction between the assessor and the child, have also been modified following observation that this test set-up can reduce response variation (Harkness and Super, 2008). We have, in common with other researchers, also used testing rooms that allow a child to work independently, but in view of other children.

Table 1 outlines the principles that should govern the evaluation of the test development process, and summarizes the methods by which that evaluation can be carried out.

Table 1. Guidelines for Item Selection

Principle	Methods of Evaluation
• Relevance to the construct • Relevance to the community • Clarity of language being used. • Clarity of instructions • Acceptability of the chosen method of Administration • Suitability of the testing environment	• Item score variance • Participant feedback and community knowledge • Multiple translation process • Test session observations • Feedback from administrators • Error analysis • Correlation of responses with other assessments/measurements

STEP 4 – *Test Evaluation*: The aim of this step is to establish the basic psychometric properties of the instrument, and for adapted test, to evaluate measurement and functional equivalence. The performance of the test is evaluated following its administration to at least 35 participants, but preferably to more than 75. Standard psychometric evaluations, outlined in Table 2 are used to determine a final schedule of items, and the overall appropriateness of the instrument (Anastasia, 1988; Kline, 1993).

Issues of Interpretation of Results

It should be remembered that any changes, even minimal translations, bring into question the applicability of the initial standardization, and the possibility of the inappropriate use of normative tables, leading to the very real possibility of mis-interpretations and misdiagnoses (Losen, et al. 2002). An adequate control group will overcome many issues of interpretation and analysis within a new context. Between contexts, statistical techniques, such as effect sizes, enable us to make cross-cultural comparisons in the absence of directly comparable raw or standardized scores.

Table 2. A summary of the statistical consideration in test evaluation

Psychometric consideration	Description	Statistical Technique Recommended cut offs
Item Level Analysis		
Item variability	Distribution of test scores to look for floor, ceiling effects and overall distribution of responses	Descriptive Statistics; No one response selected in excess of 75% (or if interested in unusual behaviors 90%)
Internal reliability	Intercorrelation of items	Cronbach's alpha; split half reliabilities
Evaluation Whole Schedule/Summary Scores		
Test-retest reliability	Correlation of measures between two time-points	Intra class correlations, two-way consistency, the acceptable value is over .60. Also a Pearson-moment correlation that is significant.
Inter-tester reliability	Correlation of measures taken by 2 assessors	Intra class correlations, two-way absolute agreement, the acceptable value is over .60. Also a Pearson-moment correlation that is significant.
Inter-form reliability	Correlation between 2 parallel forms	A significant Pearson-moment correlation between the scores from 2 forms
Concurrent validity (including criterion validity)	Relationship between test under construction and alternative measures of same concept (e.g. current best practice) taken simultaneously	A significant Pearson-moment correlation between scores taken from 2 tests.
Convergent validity	Relationship between abilities theorized to be closely related	Correlation between subscales of new measure
Divergent validity	Lack of relationship between abilities theorized not to be closely related	Lack of correlation or lower correlation between other subscales

Specific Examples of Adaptation and Assembly

1. Maintaining Factor Equivalence

Table 3 illustrates the achievement of factor equivalence following both the Adaptation and Assembly of sub-tests based upon the K-ABC. The sub-tests of the K-ABC are designed to measure two constructs, Sequential and Simultaneous processing. Factor analysis suggested a close correspondence with the initial factor structure despite adaptations that involved the alteration of constituent items and modifications of procedures, as well as the inclusion replacement tests (in italics) assembled from novel procedures (Holding, 1998).

Table 3. Information Processing Dichotomy with factor loadings for the Kilifi Battery

Sub-tests	Factors	
	1	2
eigen values	4.40	1.23
% variance	48.40	13.60
Hand Movements	.15	.78
Number Recall	.27	.80
Word Order	.29	.72
Magic Window	.78	.26
Face Recognition	.69	.22
Gestalt Closure	.88	.15
Construction	.73	.39
Matrix Analogies	.58	.16
Visual Search	.21	.74

Reproduced from Holding 1998.

2. Maintaining Reliability

Lack of variability in response often characterizes the performance of imported items. After incorporating more common behaviors we have achieved high internal consistency, a measure of the relationship of each item on a test to a single underlying concept, in questionnaires and schedules. Examples include an adaptation of the Edinburgh Post Natal Depression Inventory (alpha .7- unpublished data) and the assembly of the Kilifi Developmental Inventory (.96) (Abubakar, Holding et al., 2008). Additional reliabilities i.e. test-re-test and interobserver, were all above the acceptable levels (Abubakar, van de Vijver, van Baar, Newton, and Holding, 2008). Sound underlying psychometrics allow for confidence in the identification and interpretation of disease effects.

Training and the Production of Manuals and Guidelines

We have already referred to the probable lack of experience and lack of prior training of assessment staff. Tasks and tests that are complex to administer and depend upon extensive previous training may not be suitable to the African context, and administrative manuals that are also piloted for clarity of language and procedures should be developed. We have further developed a curriculum to ensure adequate preparation of an assessment team (Holding 2005).

The training curriculum is divided into 2 parts. Part One introduces broad issues related to assessment in the context of developmental psychology; topics covered in this section include theories of child development, basic research methods, data collection techniques and ethical issues in research. This provides a conceptual background, found to be essential to understanding the rigors of standardized assessment procedures. In part two, the specific measures to be administered are then introduced, and test administration performance is evaluated according to a structured performance schedule that sets a minimum level of competence. We estimate that a minimum of 3-4 weeks of instruction, practice and evaluation for a team with no previous testing experience is required to achieve a minimum acceptable standard.

Conclusions

Our experience shows that, through following rigorous procedures, one can adequately adapt measures for use in Africa (Abubakar, Holding et al., 2008; Alcock, et al. 2008; Holding et al., 2004). However we strongly advocate the evaluation of assessments developed for SSA in multiple contexts. By combining the data across different sites normative samples can be built up that allow both researchers and clinicians to carry out their work without having to constantly undergo the time consuming process of test adaptation.

References

Abubakar, A., Holding, P., Van Baar, A., Newton, C. R., and van de Vijver, F. J. (2008). Monitoring psychomotor development of infant and toddlers in a resource poor setting: An evaluation of the Kilifi Developmental Inventory. *Annals of Tropical Pediatrics, 28*, 217-226

Abubakar, A., van de Vijver, F., van Baar, A., Kitsao-Wekulo, P. K., and Holding, P. A. (2009). Enhancing the Validity of Psychological Assessment in Sub-Saharan Africa through participant Consultation. In A. Gari & K. Mylonas (Eds.), Quod erat demonstrandum: From Herodotus' ethnographic journeys to cross-cultural research (pp. -). Athens: Atropos Editions.

Alcock, K. J., Holding, P. A., Mung'ala Odera, V., and Newton, C. R. J. C. (2008). Constructing tests of cognitive abilities for schooled and unschooled children *Journal of Cross-cultural Psychology 39*, 529-551.

Anastasi, A. (1988). *Psychological testing*. New York: MacMillian Publishing.

Ardila, A. (2005). Cultural values underlying psychometric cognitive testing. *Neuropsychology Review, 15*, 185-194.

Baddeley, A., Gardener, J. M., and Grantham-McGregor, S. (1995). Cross-cultural cognition: developing tests for developing countries. *Applied Cognitive Psychology, 9*, S173-S195.

Carter, J. A., Less, J. A., Murira, G., Gona, J., Neville, B. G. R., and Newton, C. R. J. C. (2005). Issues in the development of cross-cultural assessments of speech and language for children. *International Journal of Language and Communication Disorders*; 40:385-401.

Connolly, K., and Grantham-McGregor, S. (1993). Key issues in generating a psychological-testing protocol. *American Journal of Clinical Nutrition, 57*, 317-318.

de Vries, M. (1999). Babies, brains and culture: optimizing neurodevelopment on the savannah. *Acta Paediatrica, Suppl, 88*, 43-48.

Elliott, C. D., Smith, P., and McCullouch, K. (1990). *British Abilities Scales*. London: Harcourt Assessment

Foxcroft, C. D. (2002). *Ethical issues related to psychological testing in Africa; What I have learnt so far*. In W. J. Lonner, D. L. Dinnel, S. A. Hayes and D. N. Sattler (Eds.), *Online Readings in Psychology and Culture (Unit 5, Chapter 2 http://www.ac.wwu.edu/~culture*. Washington USA: Center for Cross-Cultural Research, Western Washington University, and Bellingham.

Gandek, B., Ware, J. E., Aaronson, N. K., Apolone, G., Bjorner, J. B., Brazier, J. E., et al. (1998). Cross-validation of item selection and scoring for the SF-12 Health Survey in nine countries: results from the IQOLA Project. International Quality of Life Assessment. *Journal of Clinical Epidemiology, 51*, 1171-1178.

Harkness, S., and Super, C. M. (2008) *Why African children are so hard to test?* Levine, R.A. and New R.S. (Eds.) Anthropology and child development: A cross-cultural reader. Malden, MA: Blackwell.

Herdman, M., Fox-Rushby, J, and Badia, X. (1998). A model of equivalence in the cultural adaptation of HRQoL instruments: the universalist approach. *Quality of Life Research, 7*, 323-335.

Holding, P. A. (1998). *Does cerebral malaria constitute a risk factor for special educational needs?*, University College, University of London, London.

Holding, P. A., Taylor, H. G., Kazungu, S. D., Mkala, T., Gona, J., Mwamuye, B., et al. (2004). Assessing cognitive outcomes in a rural African population: Development of a neuropsychological battery in Kilifi District, Kenya. *Journal of the International Neuropsychological Society, 10*, 246-260.

Kathuria, R., and Serpell, R. (1998). Standardization of the Panga Munthu Test- A nonverbal cognitive test developed in Zambia. *Journal of Negro Education, 67*, 228-241.

Kaufman, A. S., and Kaufman, N. L. (1983). Kaufman Assessment Battery for Children: Administration and scoring manual. *Circle Pines, MN: American Guidance Service*.

Kearins, J. (Ed.). (1976). *Skills of desert children* Canberra: Australian Institute of Aboriginal Study.

Kline, P. (1993). *The Handbook of Psychological Testing*. London: Routledge.

Leiderman, H. P., Babu, B., Kagia, J., Kraemer, H. C., and Leiderman, G. F. (1973). African infant precocity and some social influences during the first year. *Nature, 242*, 247-249.

Leplège, A., and Verdier, A. (1995). The adaptation of health status measures: methodological aspects of the translation procedure. In S. Shumaker and R. Berzon (Eds.), *The international assessment of health-related quality of life: Theory, translation, measurement and analysis*. Oxford: Rapid communications of Oxford.

Losen, D. J., Orfield, G., and Civil Rights Project. (2002). *Racial inequity in special education*. Cambridge, MA: Civil Rights Project at Harvard University, Harvard Education Press.

Lynn, R. (1998). New data on black infant precocity. *Personality and Individual Differences, 25*, 801-804.

Malda, M., and Van de Vijver, F. R. (2005). *Assessing cognition in nutrition intervention trials across cultures*. Paper presented at the 18th International congress on nutrition, Durban, SA.

Mpofu, E. (2001). Indigenization of the psychology of human intelligence in sub-Saharan Africa. In W. J. Lonner, D. L. Dinnel, S. A. Hayes and D. N. Sattler (Eds.), *Online Readings in Psychology and Culture (Unit 5, Chapter 2 http://www.ac.wwu.edu/*

~culture/index-cc.htm). Washington USA: Center for Cross-Cultural Research, Western Washington University, and Bellingham.

Neil, W. (1972). African infant precocity. *Psychological Bulletin, 78*, 353-367.

Nell, V. (2000). *Cross-cultural neuropsychological assessment: theory and practice.* Mahwah, NJ: Lawrence Erlbaum Associate.

Oluyomi, A. O., and Houser, R. F. (2002). Yoruba toddler's engagement in errands and cognitive performance on the Yoruba Mental Subscale. *International Journal of Behavioural Development, 26*, 145-153.

Pike, K. L. (1967). *Language in relation to a unified theory of structure of human behaviour.* The Hague Mouton.

Reynolds, C. R. (1983). Test bias: In God we trust: All others must have data. *Journal of Special Education, 17*, 242-260.

Sigman, M., Neumman, C., Carter, E., Cattel, D. J., D'Souza, S., and Bwibo, N. (1989). Home interactions and the development of Embu toddlers in Kenya. *Child Development, 59*, 1251-1261.

Sternberg, R. J., Nokes, C., Geissler, P. W., Prince, R., Okatcha, F., Bundy, D. A., et al. (2001). The relationship between academic and practical intelligence: A case study in Kenya. *Intelligence, 29*, 401-418.

Van de Vijver, F. J. R., and Tanzer, N. K. (2004). Bias and equivalence in cross-cultural assessment: An overview. *European Review of Applied Psychology, 54*, 119-135.

Weschsler, D. (1989). *Weschsler Preschool and Primary Scale of Intelligence- Revised.* New York: The Psychological Corporation.

WHO. (2007). Process of translation and adaptation of instruments from http://www.who.int/substance_abuse/research_tools/translation/en/index.

In: Encyclopedia of Neuroscience Research
Editors: Eileen J. Sampson and Donald R. Glevins

ISBN 978-1-61324-861-4
© 2012 Nova Science Publishers, Inc.

Chapter XXX

Paid Personal Assistance for Older Adults with Cognitive Impairment Living at Home: Current Concerns and Challenges for the Future

Claudio Bilotta*, Luigi Bergamaschini, Paola Nicolini, and Carlo Vergani

Department of Internal Medicine, Geriatric Medicine Unit,
Fondazione Ospedale Maggiore Policlinico,
Mangiagalli e Regina Elena IRCCS;
University of Milan, Milan, Italy

Abstract

Quality indicators recommend a comprehensive support for informal caregivers of elders with cognitive impairment in order to delay nursing home placement of care recipients, since care comes at high financial, psychological and physical costs for the caregivers. However, according to a recent European survey caregivers often have to pay for home support services, which can be afforded only by about one out of every three people with dementia. In the last two decades both the workforce providing non-institutional personal assistance and the funding for such services have dramatically increased. Also, over the next twenty years the population aged 50 to 75, which represents the main source of family caregiving, is expected to decrease compared to the population aged 85 and older, which includes the main care recipients. Private caregiving will thus gain even more importance. Improved satisfaction with personal assistance and fewer unmet needs were reported after receiving consumer-directed services than after receiving agency-directed services. There are several concerns related to the actual provision of paid personal assistance. Hard work, stress and high risk of job burnout, low wage levels and limited health care benefits for personal assistance workers can explain

---

* Corresponding author: Claudio Bilotta, Geriatric Medicine Unit, via Pace 9, 20122 Milan, Italy. Phone number: (+39) 02 5503 5412. Fax number: (+39) 02 5501 7492. E-mail: claudio.bilotta@gmail.com.

high job turnover rates. High turnover itself leads to the potential for substandard care, especially when dealing with vulnerable elders with cognitive impairment. Dementia-specific training can support the provision of high-quality care and also increase job satisfaction for workers. However, concerns have risen about inadequate education and training of care workers. Few studies have analyzed the quality of paid personal care for older adults with disabilities living at home, none of them specifically considering people with cognitive impairment. So far, providing personal aides with adequate training and acceptable job and general living conditions appears to be one of the strongest promoters of a better quality of care, especially for older adults with cognitive impairment.

Introduction

Dementia is a leading cause of disability and nursing home placement among older adults living in the community [Cumming 2003, Hendrie 1998]. Nursing home placement is associated with a poorer quality of life for the elderly [Kane 2001] and with emotional upheaval for caregiving families [Schulz 2004].

The most important provider of care for older people with cognitive impairment is the informal sector, such as families and community groups which "offer support without funding, without charging and often without recognition" [Knapp 2007].

About half of the total informal care for older adults with dementia consists of time spent on day and night surveillance [Wimo 2002]. The time spent caring increases with the severity of dementia and behavioral disturbances [Wimo 2002], amounting to more than 10 hours a day in 20% of the caregivers of elders with mild dementia and in 50% of those caring for elders with severe dementia [Georges 2008]. Actually, caring for vulnerable elders with cognitive impairment comes at high financial, psychological and physical costs for the caregivers [Cacabelos 1999, Dunkin 1998, Waite 2004, Jonsson 2005, Gruffydd 2006]. Moreover, the caregiver burden, if left untreated, comes at great financial costs for the society overall [Prigerson 2003].

Recent guidelines for the care of the elderly with dementia emphasize the need for a comprehensive support for caregivers in order to postpone the placement of patients in a nursing home [Freil 2007, Lyketsos 2006].

Formal Support for Older Adults with Cognitive Impairment Living at Home

The availability of informal care significantly influences the need for "formal" care, for which funding has to be raised in order to employ a workforce. It has been shown that "formal" care becomes particularly important as the severity of dementia and the needs of the elderly increase [Bonsang 2009, Knapp 2007]. In the United States the proportion of older individuals receiving both formal and informal assistance is greater for groups with the greatest predicted one-year risk of institutional placement: 1.9%, 11.2%, and 21.8% for low, moderate, and high one-year risk respectively [Davey 2005]. A study conducted on a sample of community-dwelling people aged 75 and over in France and Israel pointed out that the co-existence of formal and informal care is a common observation and that mixed provision occurs more frequently in situations of greater need [Litwin 2009], including the assistance to subjects with cognitive impairment. This study also highlighted that spouse caregivers had less formal home-care supports than either co-resident children or other family caregivers [Litwin 2009].

There are wider inter-country variations in the provision of home care services for people with dementia than in the provision of institutional services, and comparisons are limited by definitional differences (i.e. differences in the home services actually provided) [Knapp 2007]. Financial support for caregivers is becoming more common, being provided by different means, such as grants from social care budgets (e.g. in France, Australia and Sweden), tax credits (e.g. in Canada, Spain and the U.S.), pension credits (e.g. in Canada, Germany and the U.K.), consumer-directed payments (e.g. in Germany, the U.K. and in the Lombardy region of Italy) [Knapp 2007].

However, recent studies have shown that the availability of home support services for European informal caregivers is rather limited. One of them demonstrated that the policies in advanced European welfare states such as the U.K. and the Netherlands do not provide working caregivers with the support that would be necessary to avoid the risk of impoverishment [Arksey 2008]. A survey of a sample of working women in Belgium showed that the probability of ceasing work for those aged 50 or more was not increased by the simple fact of caring for frail older people, **but by the fact of providing a "heavy" kind of care** [Masuy 2009], such as the care needed by older adults suffering from cognitive impairment. Another survey pointed out that home support services are available only for 44% of the European citizens caring for people with dementia and are actually used by 30% of them, mainly as private home assistance (65% of cases) [Georges 2008]. In Scotland, an average 11% of people with dementia living in the community receive home care and 12% receive day care, against a working target of 28% [Alzheimer Scotland 2007].

Paid Personal Assistance for Elderly with Cognitive Impairment

Within the context of home care services, paid personal aides provide companionship, general supervision and also help elderly people perform both basic activities of daily living (i.e. transferring, feeding, bathing, dressing, toileting) and instrumental activities of daily living such as preparing food, housekeeping, taking medications and going out of the house. Moreover, they are valuable from a clinical point of view, especially for people suffering from cognitive impairment: in fact they often provide doctors with information on the physical, psychological, cognitive and behavioral status of the care recipients.

In the U.S. the demand for paid personal aides caring for community-dwelling older adults with any kind of disability has increased three-fold between 1989 and 2004, growing at a much faster rate than the population needing such services [Kaye 2006]. During the same period, Medicaid spending for non-institutional services showed an impressive increase, from about 3 billion U.S. dollars in 1989 to nearly 15 billion dollars in 2003, whereas spending for institutional services remained relatively stable at about 25-28 billion dollars [Kaye 2006]. A similar trend was observed in other developed countries, notwithstanding significant differences related to the availability and funding of home and social care services. In Italy, for example, the "economic dependency ratio" (i.e. the ratio of the population aged 65 and over to the population aged 15 to 64) was the highest among the developed countries in 2005 - 44.3% [OECD 2006] – and the prevalence of older people with dementia receiving paid personal care from formal assistants rose from 30% in 1999 to 41% in 2006 [Censis 2007]. An observational study conducted in Israel has shown that 56% of community-dwelling elderly with dementia as well as 79% of those with severe dementia were receiving home care services, and that 18% of all elderly with dementia had a 24-hour personal-care aide [Wertman 2005].

Future Increase in Demand for Paid Personal Aides

The assistance provided by paid personal aides not only has become very important for care-recipients and their families, but is also expected to gain further importance in the near future, since the supply of informal care is expected to decrease due to demographic trends.

Over the next twenty years in most developed countries the population aged 50 to 75, which constitutes the main source of family caregiving, is expected to decrease relatively to the population over 85, which includes the main care-recipients [Robine 2007]. In Italy, the "oldest old support ratio", which is the ratio between these two age groups [Robine 2007], has been predicted to decrease from 13 in 2006 to 8 in 2030 [Vergani 2007]. Also, the "economic dependency ratio" will increase by 2020 to 54.5% in Italy, 51.4% in France, 50.5% in Japan, 45.5% in Sweden and 45.2% in Germany [OECD 2006].

Private caregiving will thus gain increasing importance. According to the U.S. Department of Labor, the number of private aides will have risen by 51% by the year 2016, from 767,000 to 1,156,000 jobs [U.S. Department of Labor 2008].

Level of Satisfaction with Paid Personal Assistance

With reference to the experiences and the level of satisfaction of care recipients and their families who use paid personal assistance services, few studies are available and involve older adults with disabilities and not specifically people suffering from cognitive impairment.

A sample of informal carers from nine states of the U.S. reported general satisfaction with current paid personal assistance, although previous experiences with poor care as well as many unmet needs due to a shortage of hours provided by state programs were disclaimed [Grossman 2007]. In the U.S, consumer-directed care services are strongly preferred over agency-directed services, with more client satisfaction and fewer unmet needs [Clark 2008, Grossman 2007]. In a recent review on controlled trials concerning personal assistance for adults aged 65+ living in the community, with disabilities but without dementia, the personal assistance model was generally preferred over other services by older care-recipients and their carers, although it may cost government more than alternatives [Montgomery 2008].

In the U.K., where there has been a trend from the provision of home care services by "in-house" local government authorities towards models of services commissioned by independent providers, the older users of independent providers reported a lower level of satisfaction and a poorer perceived quality of care than the older users of in-house providers [Netten 2007].

Concerns about the Quality of Paid Personal Care

There are several concerns related to the current provision of paid personal assistance. First, little is still known about the quality of paid personal care and its correlates and outcomes. Recent observational studies from Italy investigated the correlates and the one-year outcomes of the perceived quality of care provided by paid personal aides for community dwelling older adults with disabilities [Bilotta 2008, Bilotta 2009]. Fifty-six per cent of the sample were suffering from dementia and sixty-nine per cent from cerebrovascular disease. At baseline, good language skills and non-distressing living conditions of the private aides appeared to be correlates of an optimal perceived quality of care; also, a better quality of life of the elderly and a lower distress of their informal caregivers appeared to be potential outcomes of the quality of personal care, without there being any significant difference between the quality of care groups in terms of social variables, functional and cognitive status, prevalence of depressive disorders and morbidity of the elderly [Bilotta 2008]. Interestingly, at a one-year follow-up a poor or fair quality of paid personal care turned out to be a predictor of placement in a long-term care facility [Bilotta 2009].

Second, the relevance of language skills and general living conditions of paid personal aides in determining the quality of personal care, as emerged from these studies, is not at all surprising since caring for elderly people is a very demanding task. In fact, according to the US Department of Labor, "**personal and home**-care aides should have a desire to help people and not mind hard work. They should be responsible, compassionate, emotionally stable and cheerful" [**U.S. Department of Labor** 2008]. Paid aides themselves recognize that their work requires a number of good qualities and especially patience, as demonstrated by a recent survey [Hokenstad 2006]. However, the actual situation of private aides in most developed countries is often far from easy. They are generally immigrant, middle-aged, married women with children who have left their families behind in their countries of origin, and have been employed for a job for which they receive low pay (according to the U.S. Department of Labor 2008 median hourly wages were 9.18 USD for services for the elderly and persons with disabilities and 7.19 USD for home healthcare services) and little or no training [Bilotta 2008, Fleming 2007, Jorgensen 2009, Montgomery 2005, Yamada 2002].

Third, the lack of training and the high turnover rates reported among these health care workers, as a result of hard work, high risk of job burnout, low wages and benefits and irregular work schedules [Montgomery 2005, US Department of Labour 2008, Yamada 2002], may also lead to disruptions in continuity of care for clients [Dawson 2000, Stone 2004] and a poor quality of care [Castle 2005]. A high staff turnover fosters the need for further investments in training new workers, and this could discourage providers from implementing training programs. However, training is a proven means to improve the quality of care as well as job satisfaction [Maas 1994], which was identified as a significant factor for retention of nursing home staff [Waxman 1984]. Therefore, attracting and retaining skilled workers could become increasingly difficult if job conditions of paid personal aides do not improve [Kaye 2006]: pay, retention and training of personal care workers seem to be closely linked.

Fourth, the three items previously discussed acquire greater importance when dealing with vulnerable older adults suffering from cognitive impairment, for whom continuity of appropriate care is an essential issue. In this perspective we must emphasize that abusive behaviors towards the elderly are unfortunately a well-documented reality that also involves family caregivers of people with dementia [Cooper 2009]. With reference to personal care workers, it has been shown that financial abuse is more likely in domiciliary settings, whereas physical abuse is more common in nursing homes [Hussein 2009]. Further studies will be necessary to investigate possible abusive behaviors of the private aides of elderly with dementia and to give us some insight into how they could be prevented.

Conclusion

Since there is no effective pharmacological treatment for dementia and other cognitive disorders, an optimal personal assistance represents a key component of interventions to improve the quality of life and reduce/delay nursing home placement for the elderly suffering from these diseases, as well as to support their informal caregivers. The literature concerning the quality of care provided by paid personal assistants is scant, especially with respect to the care for vulnerable older care recipients with cognitive impairment. Social policies should make paid personal assistance a more desirable occupation, by ensuring adequate training as well as by offering wages commensurate with the critical skills of personal aides for community-dwelling older adults. So far, providing personal aides with adequate training and acceptable job and general living conditions appears to be one of the strongest promoters of a better quality of care, especially for older adults with cognitive impairment.

References

Alzheimer Scotland. The dementia epidemic - Where Scotland is now and the challenge ahead. Edinburgh; 2007.

Arksey, H; Morèe, M. Supporting working carers: do policies in England and the Netherlands reflect 'doulia rights'? *Health Soc. Care Com*, 2008, 16, 649-657.

Bilotta, C; Vergani, C. Quality of private personal care for elderly people with a disability living at home: correlates and potential outcomes. *Health Soc. Care Com*, 2008, 16, 354-362.

Bilotta, C; Nicolini, P; Vergani, C. Quality of private personal care for elderly people in Italy living at home with disabilities: risk of nursing home placement at a 1-year follow-up. *Health Soc. Care Com*, 2009, early view, doi: 10.1111/j.1365-2524.2009.00853.x.

Bonsang, E. Does informal care from children to their elderly parents substitute for formal care in Europe? *J. Health Econ*, 2009, 28, 143-154.

Cacabelos, R; Takeda, M; Winblad, B. The glutamatergic system and neurodegeneration in dementia: preventive strategies in Alzheimer's disease. *Int. J. Geriatr. Psychiatry*, 1999, 14, 3-47.

Castle, NG; Engberg, J. Staff turnover and quality of care in nursing homes. *Med. Care*, 2005, 43, 616-626.

Censis. 41st Annual report on the social conditions of the country. Rome; 2007; 4-5. (in Italian).

Clark, MJ; Hagglund, HJ; Sherman, AK. A longitudinal comparison of consumer-directed and agency-directed personal assistance service programmes among persons with physical disabilities. *Disabil. Rehabil*, 2008, 30, 689-695.

Cooper, C; Selwood, A; Blanchard, M; Walker, Z; Blizard, R; Livingstone, G. Abuse of people with dementia by family carers: representative cross sectional survey. *BMJ*, 2009, 338, b155.

Cumming, JL; Cole, G. Alzheimer disease. *JAMA*, 2002, 287, 2335-2338.

Davey, A; Femia, EE; Zarit, SH; Shea, DG; Sundstrom, G; Berg, S; Smyer, MA; Savla, J. Life on the edge: patterns of formal and informal help to older adults in the United States and Sweden. *Journal of Gerontology Series B: Psychological Sciences and Social Sciences*, 2005, 60, S281-S288.

Dawson, SL; Surpin, R. The home health aide: scarce resource in a competitive marketplace. *Care Management Journals*, 2000, 2, 226-231.

Dunkin, JJ; Anderson-Hanley, C. Dementia caregiver burden: a review of the literature and guidelines for assessment and intervention. *Neurology*, 1998, 51, S53-S60.

Fleming, G; Taylor, BJ. Battle on the home care front: perceptions of home care workers of factors influencing staff retention in Northern Ireland. *Health Soc. Care Com*, 2007, 15, 67-76.

Freil, DG; Mac Lean, C; Sultzer, D. Quality indicators for the care of dementia in vulnerable elders. *J. Am. Geriatr. Soc*, 2007, 55, S293-S301.

Georges, J; Jansen, S; Jackson, J; Meyrieux, A; Sadowska, A; Selmes, M. Alzheimer's disease in real life – the dementia carer's survey. *Int. J. Geriatr. Psychiatry*, 2008, 23, 546-551.

Grossmann, BR; Kitchener, M; Mullan, JT; Harrington, C. Paid personal assistance services: an exploratory study of working age consumers' perspectives. *J. Aging Social Policy*, 2007, 19, 27-45.

Gruffydd, E; Randle, J. Alzheimer's disease and the psychosocial burden for caregivers. *Community Pract*, 2006, 79, 15-18.

Hendrie, HC. Epidemiology of dementia and Alzheimer's disease. *Am. J. Geriatr. Psychiatry*, 1998, 6, S3-S18.

Hokenstad, A; Hart, AY; Gould, DA; Halper, D; Levine, C. Closing the home care case: home health aides' perspectives on family caregiving. *Home Health Care Management and Practice*, 2006, 18, 306-314.

Hussein, S; Stevens, M; Manthorpe, J; Rapaport, J; Martineau, S; Harris, J. Banned from working in social care: a secondary analysis of staff characteristics and reasons for their referrals to the POVA list in England and Wales. *Health Soc. Care Com*, 2009, 17, 423-433.

Jonsson, L; Berr, C. Cost of dementia in Europe. *European Journal of Neurology*, 2005, 12, 50-53.

Jorgensen, D; Parsons, M; Gundersen, M; Weidenbohm, K; Parsons, J; Jacobs, S. The providers' profile of the disability support workforce in New Zealand. *Health Soc. Care Com*, 2009, 17, 396-405.

Kane, RA. Long-term care and good quality of life. Bringing them closer together. *Gerontologist*, 2001, 41, 293-304.

Kaye, S; Chapman, S; Newcomer, R; Harrington, C. The personal assistance workforce: trends in supply and demand. *Health Affairs*, 2006, 25, 1113-1120.

Knapp, M; Comas-Herrera, A; Somani, A; Banerjee, S. Dementia: international comparisons. Summary report for the National Audit Office. Personal Social Services Research Unit, London School of Economics and Political Science. London; May 2007. Available from: www.pssru.ac.uk (accessed on 5th April 2009).

Litwin, H; Attias-Donfut, C. The inter-relationship between formal and informal care: a study in France and Israel. *Ageing and Society*, 2009, 29, 71-91

Lyketsos, CG; Colenda, CC; Beck, C; Blank, K; Doraiswamy, MP; Kalunian, DA; Yaffe, K. Position Statement of the American Association for geriatric psychiatry regarding principles of care for patients with dementia resulting from Alzheimer disease. *Am. J. Geriatr. Psychiatry*, 2006, 14, 561-573.

Maas, M; Buckwalter, KC; Swanson, E; Mobily, PR. Training key to job satisfaction. *J. Long-Term Care Admin*, 1994, 22, 23-26.

Masuy, AJ. Effect of caring for an older person on women's lifetime participation in work. *Ageing and Society*, 2009, 29, 745-763.

Montgomery, P; Mayo-Wilson, E; Dennis, J. Personal assistance for older adults (65+) without dementia. *Cochrane Database Sys. Rev*, 2008, 1, CD006855.

Montgomery, RJ; Holley, L; Deichert, J; Kosloski, K. A profile of home care workers from the 2000 census. *Gerontologist*, 2005, 45, 593-600.

Netten, A; Jones, K; Sandhu, S. Provider and care workforce influences on quality of home-care services in England. *J. Aging Soc. Policy*, 2007, 19, 81-97.

OECD. OECD Factbook 2006: economic, enviromental and social statistics. Paris, 2006.

Prigerson, HG. Costs to society of family caregiving for patients with end-stage Alzheimer's disease. *N. Engl. J. Med*, 2003, 349, 1891-1892.

Robine, JM; Michel, JP; Herrmann, FR. Who will care for the oldest people in our ageing society? *BMJ*, 2007, 334, 570-571.

Schulz, R; Belle, SH; Czaja, SJ; McGinnis, KA; Stevens, A; Zhang, S. Long-term care placement of dementia patients and caregiver health and well-being. *JAMA*, 2004, 292, 961-967.

Stone, RI. The direct care worker: the third rail of home care policy. *Annu. Rev. Public Health*, 2004, 25, 521-537.

U.S. Department of Labor, Bureau of Labor Statistics. *Occupational Outlook Handbook, 2008-09 Edition*. Personal and Home Care Aides. Available from: http://www.bls.gov/oco/ocos173.htm (accessed on 2nd April 2009).

Vergani, C. La condizione anziana. In: Atti del Convegno "La condizione anziana: nuove politiche sociali per nuovi bisogni assistenziali". Regione Autonoma Valle d'Aosta, Italy; 15 novembre 2007.

Waite, A; Bebbington, P; Skelton-Robinson, M; Orrell, M. Social factors and depression in carers of people with dementia. *Int. J. Geriatr. Psychiatry*, 2004, 19, 582-587.

Waxman, HM; Carner, EA; Berkenstock, G. Job turnover and job satisfaction among nursing home aides. *Gerontologist*, 1984, 24, 503-509.

Wertman, E; Brodsky, J; King, Y; Bentur, N; Chekhmir, H. Elderly people with dementia: prevalence, identification of unmet needs and priorities in the development of services. Center for research on aging. Myers-JDC-Brookdale Institute 2005. Available from: http://brookdale.jdc.org.il/?catid={A114168B-9E1D-4CDE-A1E9-F23B3FFE968D} (accessed on 15th April 2009).

Wimo, A; von Strauss, E; Nordberg, G; Sassi, F; Johansson, L. Time spent on informal and formal caregiving for persons with dementia in Sweden. *Health Policy*, 2002, 61, 255-268.

Yamada, Y. Profile of home care aides, nursing home aides, and hospital aides: historical changes and data recommendations. *Gerontologist,* 2002, 42, 199-206.

Chapter XXXI

Neurotoxicity, Autism, and Cognitive Impairment

Rebecca Cicha, Brett Holfeld and F. R. Ferraro [*]
University of North Dakota, ND, USA

Abstract

Much research has been conducted in regard to identifying various etiological pathways of developmental disorders, particularly autism. The existing literature indicates potential roles of exposure to neurotoxic substances, including heavy metals and various synthetic chemicals, to the development and exacerbation of autistic symptoms. This paper will serve as a review of the relevant literature implicating environmental exposure to neurotoxic agents as a possible contributor to the development of autistic features.

Keywords: Autism, Neurotoxicity, Cognitive Impairment

Introduction

Within the past decade, a number of models have been proposed to account for the rising incidence (or at least the risen awareness) of developmental disorders, namely autism. The etiology of autism remains unknown but more research of late has focused on the hypothesis that autism is associated with early prenatal exposure to environmental toxins. Currently, though very few studies have been conducted examining possible influences of toxic agents on developmental disorders, neurotoxic exposures are seen to account for 3 to 25% of documented cases (Schmid and Rotenberg, 2005). Lathe (2006) argued that the large rise in

---

[*] Address all correspondence to: F. Richard Ferraro, Dept. Psychology - University of North Dakota, Corwin-Larimore Rm. 215, 319 Harvard Street Stop 8380, Grand Forks, ND 58202-8380, 701-777-2414 (O), 701-777-3454 (Fax), e-mail: f_ferraro@und.nodak.edu.

autism spectrum disorder (ASD) cases in recent years is a result of increased exposure to environmental toxins as well as a genetic predisposition that may increase biological vulnerability to said exposure. Fido and Al-Saad (2005) also implicated excessive neural trace elements as potentially affecting the development of autism as well as other psychological disorders (e.g., Down's syndrome, Parkinson's and Alzheimer's disease).

As such, prominent models implicate the exposure to exogenous sources of neurotoxicity including heavy metals, pesticides, flame retardants, plastics, and other chemicals, often within a diathesis-stress model of genetic vulnerability. The following sections will highlight factors potentially involved in the development of autistic symptoms, including heavy metal and synthetic chemical exposure.

Neurotoxic Exposure: Heavy Metals

A number of studies have demonstrated that exposure to heavy metals (e.g., lead and mercury), even at low doses, has often been associated with neuronal damage and subsequent developmental disorders. Fido and Al-Saad (2005) compared concentration levels of toxic metals (antimony, uranium, arsenic, beryllium, mercury, cadmium, lead and aluminum) in the hair of children and compared this with a control group. The researchers found that autistic children had higher levels of mercury, lead, and uranium evident within their hair follicles, but could not provide a direct causal link to developmental disorders such as autism (Fido and Al-Saad, 2005).

Lead

Additionally, lead is a prominent heavy metal implicated in a number of physical and mental complications; lead has been one of the only chemicals shown to play a significant role in the origin of some childhood disorders (Fido and Al-Saad, 2005). Upon exposure, it affects a number of bodily systems including hepatic, skeletal, and nervous (Schmid and Rotenberg, 2005). Within the nervous system, lead is implicated in substituting calcium, accelerating cell apoptosis, damaging cells, and interfering with neurotransmitter release (Schmid and Rotenberg, 2005). Physiologically, though adult individuals may effectively reverse damages purportedly caused by lead poisoning, the effects of exposed children may be more outstanding (Schmid and Rotenberg, 2005). In addition to having less developed means for detoxification upon exposure, children often are exposed more often and to heightened levels of lead, via increased hand-to-mouth activities as well as increased contact with dust (often containing lead particles) when crawling on the floor (Schmid and Rotenberg, 2005). Specifically, excessive levels of lead contained in the body can increase the risk of developing serious cognitive and behavioral deficits (Schroeder and Hawk, 1987); children exposed to lead often evidence prominent cognitive disruptions from significantly lower IQ points to confirmed cases of mental retardation (Schmid and Rotenberg, 2005).

Dietrich, Kraft, Shukla, Bornschein, and Succop (1987) also examined lead exposure using multiple assessments; however the authors concluded that further investigation was needed to make definitive conclusions. Overall, results from six cases of boys aged 3 to 5 years old with infantile autism show that lead poisoning or excessive lead consumption can be an influential factor that can lead to mental retardation, communication problems and other various autistic features (Accardo, Whitman, Caul, and Rolfe, 1988).

Mercury

Exposure to mercury has long been regarded as harmful within the scientific community. Viewed as a neurotoxin, various forms of mercury exist including elemental (e.g. thermometer fillers), inorganic (e.g. mercury salts), and organic (e.g. ethyl mercury) (Schmid and Rotenberg, 2005). Arguably, one of the most accessible (and therefore more hazardous) forms of mercury is of the organic variety; building up within the food chain, organisms can be exposed to organic mercury through ingesting other organisms that contain trace amounts of mercury within their tissues, particularly fish (Schmid and Rotenberg, 2005). Mercury acts upon physiologic systems by impairing cell migration, mitochondrial activity, plasma membrane lysis, and cell necrosis and apoptosis that are often evidenced in postmortem brain examinations of autistic humans (Schmid and Rotenberg, 2005; Hornig, 2002).

Historically, mercury exposure epidemics have significantly impacted the health of a number of populations (e.g. Minamata and Niigata, Japan; Iraq) and have resulted in documented cases of severe developmental disabilities characterized by mental retardation, seizures, and cerebral palsy (Schmid and Rotenberg, 2005). Some evidence points to abnormal amounts of mercury in the bodies of autistic children; however, little published research has shown a direct causal relationship between mercury exposure and incidence of autism (Aschner, 2002). Recently, Ip, Wong, Ho, Lee and Wong (2007) attempted to answer the question using a cross sectional study over a 5-month period in 2000 looking at hair and blood mercury levels of children with ASD. However, they found that mean mercury levels were the same in both experimental and control groups. As such, the authors concluded that there is no causal relationship between mercury as an environmental toxin and autism (Ip, Wong, Ho, Lee and Wong, 2007). Aschner (2002) examined the role of astrocytes in the central nervous system of autistic individuals in an attempt to find an explanation of the etiology of autism. Physiologically, astrocytes have high affinity uptake systems that transport glutamate and reactive astrogliosis that is found in the central nervous system of autistic individuals (Aschner, 2002). Ultimately, Aschner (2002) proposed that astrocytes do play some kind of role in the etiology of autism however further research is needed to address the issue.

In light of the growing body of literature indicating a potential role of mercury exposure in the development of developmental disorders, much debate has ensued to re-formulate a number of substances that children are commonly exposed to, namely vaccines. Specifically, the measles-mumps-rubella vaccine contained a significant amount of ethyl mercury, which may have exceeded recommended guidelines (Schmid, and Rotenberg, 2005). Conversely, the argument that increasing vaccinations increases incidence of autism was strongly opposed by Offit and Golden (2004). Components thimerosal and aluminum in vaccines are key elements in this theory. However, Offit and Golden (2004) concluded that avoiding vaccines will not reduce the chance of developing a developmental disorder; however, avoiding vaccines will subsequently increase the risk of contracting preventable diseases.

As a whole, though much of the literature has associated heavy metal exposure with physiological disruption, the mechanism responsible for the effect of toxic metals on the development of developmental disabilities, namely autism, remains unknown. However, other researchers have identified various synthetic chemicals, including dioxin-like compounds, pesticides, and plastic additives that may significantly impact biological processes related to the development of autistic features.

Neurotoxic Exposure: Synthetic Chemicals

Dioxin-like compounds (e.g., polychlorinated biphenyls and polybrominated diphenylethers) have also been implicated in the disruption of a number of organ systems upon exposure. Despite recent legislation proposed banning the usage of a number of Dioxin-like compounds, residual buildup in the environment (e.g. soil residue, buildup in fish tissue) may still serve as a source of exposure (Schmid and Rotenberg, 2005). Physiologically, dioxin-like compounds primarily affect endocrine systems through disruption of thyroid processes (e.g. disruption of circulating hormones, thyroid responsiveness, hormone-responsive genes, etc.), though they have been seen to significantly impact the integumentary organ system (e.g. skin, hair, nails), and affect cognitive processes, resulting in documented cases of lowered IQ, motor and cognitive delays, lethargy, and apathy (Schmid and Rotenberg, 2005).

Furthermore, documented cases of human neonatal exposure to dioxin compounds through maternal ingestion of contaminated foods has resulted in children with behavioral disruptions, up to a threefold likelihood of lower IQ scores, poor reading comprehension, impairments in memory and attention span, and other cognitive impairments (Gasiewicz, 1997). In mice, exposure to PBDEs, common flame retardants, has resulted in developmental impairments such as inhibited learning, memory, spontaneous behaviors and motor activity, habituation to new environments, and impaired reproductive functioning (Charboneau and Koger, 2008).

Pesticides

Pesticide (i.e. organophosphate pesticide, OP) exposure has also been implicated in a number of health-related concerns, particularly in regard to cognitive disturbances. Originally developed as neurotoxic agents for warfare, OPs act upon acetylcholinesterase (AChE) neurotransmitter systems by inhibiting the breakdown of AChE (Charboneau and Koger, 2008). Once released into the synaptic cleft, residual AChE continues to stimulate the receptor neuron, often resulting in nausea, vomiting, headache, anxiety, bronchioconstriction, paralysis, and death (Schmid and Rotenberg, 2005; Charboneau and Koger, 2008). Exposure to pesticides also disrupts psychophysiological systems through the disruption of the brainstem, forebrain, limbic system, hippocampus, basal ganglia, and cerebellum (Schmid and Rotenberg, 2005). Exposure in rats has resulted in documented cases of disrupted investigative and social behavior, working and reference memory, locomotor activity, and decreased conditioned learning time (Charboneau and Koger, 2008). In humans, building evidence suggests that neonatal exposure may lead to impairments including hyperactivity, cognitive deficits, and decreased attention span (Charboneau and Koger, 2008).

Chronic postpartum exposure to pesticides in humans has been associated with Parkinson's disease, decreased perceptual abilities (Charboneau and Koger, 2008). In a particular investigation conducted by Guillette et al. (1998), children living in a Mexican community in which pesticides were regularly sprayed on food crops evidenced deficient motor skills, memory, creative abilities, and social interactions when compared to unexposed peers. Overall, it is apparent that exposure to pesticides, even at low doses, can negatively impact individuals of all ages, particularly eliciting deficits commonly seen within the autistic spectrum.

Plastic Additives

Exposure to plastics, particularly compounds found within plastics that are used primarily to increase plastic pliability, has also been linked to negative health outcomes in mice and humans. Specifically, the additive Bisphenal A (BA) (i.e. phthalates), is viewed as an endocrine disruptor, and has been associated with a number of birth defects, reproductive abnormalities, and cancer (Charboneau and Koger, 2008). Though BA is seen to metabolize rather quickly (i.e. its half life is less than one day in humans), exposure has led to impaired hormonal, reproductive, enzyme, and brain functioning, and chromosomal abnormalities in mice and humans; mothers with a significant history of miscarriage were more likely to have higher BA blood levels than mothers who had never miscarried (Charboneau and Koger, 2008).

Neurotoxic Exposure in Critical Developmental Stages

In addition to the investigation of simple exposures environmentally-accessible heavy metals and synthetic chemicals, other researchers have identified crucial developmental stages in which the physiological effects of neurotoxic exposure may be maximized or minimized.

Hornig (2002) emphasized the notion of heightened sensitivity to potentially damaging exogenous factors during critical developmental stages. In a neonatal mouse model, mice exposed to the Borna disease virus (BDV) at 4 weeks gestation exhibited behavioral disturbances approximating that of autism (e.g. disturbed sensorimotor development, impaired social interaction, and emotional reactivity) (Hornig, 2002). It is speculated that neuronal loss resulting from the infection, particularly via accelerated apoptosis (i.e. programmed cell death), may account for the observed behavioral changes (Hornig, 2002). Furthermore, extending to a human model, postmortem examinations of brain tissue in autistic individuals indicate a possible biochemical explanation for neuronal losses exhibited in autism (Hornig, 2002). Specifically, autistic individuals evidenced an increase in factors that accelerate neuronal apoptosis (e.g., factor p53) as well as a decrease in anti-apoptic factors (e.g., factor bcl2); suggesting that accelerated apoptosis may be responsible for increased neuronal losses evidenced in autism (Hornig, 2002). As a whole, the author speculated that, considering heightened vulnerability of neonatal mice at 4 weeks gestation to apoptosis following BDV infection, cell apoptosis may significantly influence nervous system development, depending upon the current developmental stage of the individual when exposed to infection (Hornig, 2002).

Other researchers have proposed, in addition to mere exposure to toxic agents in certain critical developmental stages, individuals may be at risk for tissue damage, based upon anomalies in genetic structure. Specifically, D'Amelio et al. (2005) conducted a cross-cultural study that examined the expression of the paraoxonase gene variant (PON1), which is responsible for coding an enzyme (paraoxonase) that aids in the breakdown of OPs within the central nervous system, in Caucasian-Americans and Italians. Overall, it was demonstrated that Caucasian-Americans who tested positive for a less active PON1 gene variant were more likely to be diagnosed with autism (D'Amelio, Ricci, Sacco, Liu, D'Agruma, and Muscarella, 2005). Furthermore, Caucasian-Americans that exhibited autistic features also tested positive for other genetic variants involved with decreased paraoxonase enzymatic activity (e.g., the T-108 allele) (D'Amelio, Ricci, Sacco, Liu, D'Agruma, and Muscarella, 2005). As a whole, it is apparent that specific genetic functions may be responsible for increased vulnerability when exposed to particular neurotoxic agents, in this case, pesticides (D'Amelio, Ricci, Sacco, Liu, D'Agruma, and Muscarella, 2005).

Neurotoxic Exposure and Autism: Future Investigations

As is apparent, the literature examining the potential influence of neurotoxic exposure on the incidence of developmental disorders is somewhat limited, particularly in regard to the development of cognitive and behavioral autistic features. Considering the often accumulative and comprehensive nature of possible etiological factors to the development of autistic features external to neurotoxic exposure (e.g., nutritional deficiency, abnormality of social and educational environments, and other related genetic abnormalities), the definition of clear linear relationships between neurotoxic exposure and possible subsequent manifestation of autistic symptoms is inherently complex. As such, future investigations should continue to isolate possible contributing variables, including an extension of the current literature implicating exposure to the aforementioned heavy metals and synthetic chemicals (i.e., empirical replication and longitudinal examination) in addition to other naturally occurring and synthetic compounds which may similarly impact neurological systems involved in autism.

References

Accardo, P, Whitman, B., Caul, J., and Rolfe, U. Autism and plumbism: A possible association. *Clinical Pediatrics, 27*, 41-44.

Aschner, M. (2002). Toxicology of neurodevelopmental disorders: The neuropathogenesis of mercury toxicity. *Molecular Psychiatry, 7*, 40-41.

Charboneau, J. P. and Koger, S. M. (2008). Plastics, pesticides, and PBDEs: Endocrine disruption and developmental disabilities. *Journal of Developmental and Physical Disabilities, 20*, 115-128.

Dietrich, K. N., Kraft, K. M., Shukla, R., Bornschein, R. L., and Succop, P. A. (1987). The neurobehavioral effects of early lead exposure. In S. R. Schroeder (Ed.), *Toxic Substances and Mental Retardation: Neurobehavioral Toxicology and Teratology* (pp. 71-95). Washington, DC: American Association on Mental Retardation.

Fido, A. and Al-Saad, S. (2005). Toxic trace elements in the hair of children with autism. *Autism, 9*, 290-298.

Gasiewicz, T. A. (1997). Exposure to dioxin and dioxin-like compounds as a potential factor in developmental disabilities. *Mental Retardation and Developmental Disabilities, 3*, 230-238.

Hornig, M. (2002). Animal Models: Infectious and immune factors in neurodevelopmental damage. *Molecular Psychiatry, 7*, S34-S35.

Ip, P., Wong, V., Ho, M., Lee, J., and Wong, W. (2007). Mercury exposure in children with autistic spectrum disorder: A case-control study: Erratum. *Journal of Child Neurology, 22*, 1324.

Krasnegor, N. A. (1990). On the identification and measurement of chemical time bombs: A behavioral development perspective. In R. W. Russel, P. E. Flattau, and A. M. Pope (Eds.), *Behavioral Measures of Neurotoxicity: Report of a Symposium* (pp. 191-205). Washington, DC: National Academy Press.

Lawler, C. P., Croen, L. A., Grether, J. K., and Van de Water, J. (2004). Identifying environmental contributions to autism: Provocative clues and false leads. *Mental Retardation and Developmental Disabilities Research Reviews, 10*, 292-302.

Offit, P. (2004). Scientific Correspondence: Theimerosal and Autism. *Molecular Psychiatry, 9*, 644-645.

Schroeder, S. R., and Hawk, B. (1987) Psycho-social factors, lead exposure, and IQ. *Toxic Substances and Mental Retardation: Neurobehavioral Toxicology and Teratology.*